ELECTRIC-DIPOLE POLARIZABILITIES OF ATOMS, MOLECULES AND CLUSTERS

Keith D. Bonin
Department of Physics
Wake Forest University

Vitaly V. Kresin
Department of Physics and Astronomy
University of Southern California

Published by

World Scientific Publishing Co. Pte. Ltd.

P O Box 128, Farrer Road, Singapore 912805

USA office: Suite 1B, 1060 Main Street, River Edge, NJ 07661

UK office: 57 Shelton Street, Covent Garden, London WC2H 9HE

Library of Congress Cataloging-in-Publication Data

Bonin, Keith D., 1956–

Electric-dipole polarizabilities of atoms, molecules, and clusters / Keith D. Bonin, Vitaly V. Kresin.

p. cm.

Includes bibliographical references.

ISBN 9810224931

1. Atoms. 2. Molecules. 3. Microclusters. 4. Polarization (Electricity) I. Kresin, Vitaly V., 1962– . II. Title.

QC173.B547 1997

539--dc21 97-8243

CIP

British Library Cataloguing-in-Publication Data

A catalogue record for this book is available from the British Library.

Printed in Singapore by Uto-Print

PROFESSOR KEITH D. BONIN received his PhD degree in Physics from the University of Maryland in 1984 and was Assistant Professor of Physics at Princeton University from 1986 to 1992. Since 1992, he has been an Associate Professor at Wake Forest University. His current research interests include all aspects of light forces and polarizability measurements of atoms and clusters. A member of the American Physical Society and the Optical Society of America, he has been a consultant to the US Army and to private industry.

PROFESSOR VITALY V. KRESIN obtained his PhD degree in Physics from the University of California at Berkeley in 1991. He was a postdoctoral scientist at Berkeley and the Lawrence Livermore National Laboratory before his present appointment as Assistant Professor of Physics at the University of Southern California in 1994. His experimental research has focused on electron dynamics in metal and fullerene microclusters (molecular beam spectroscopy, optical resonances, electron scattering, and collision studies). He has also carried out theoretical work exploring the response properties and collective states in clusters, fullerenes, and nanostructures.

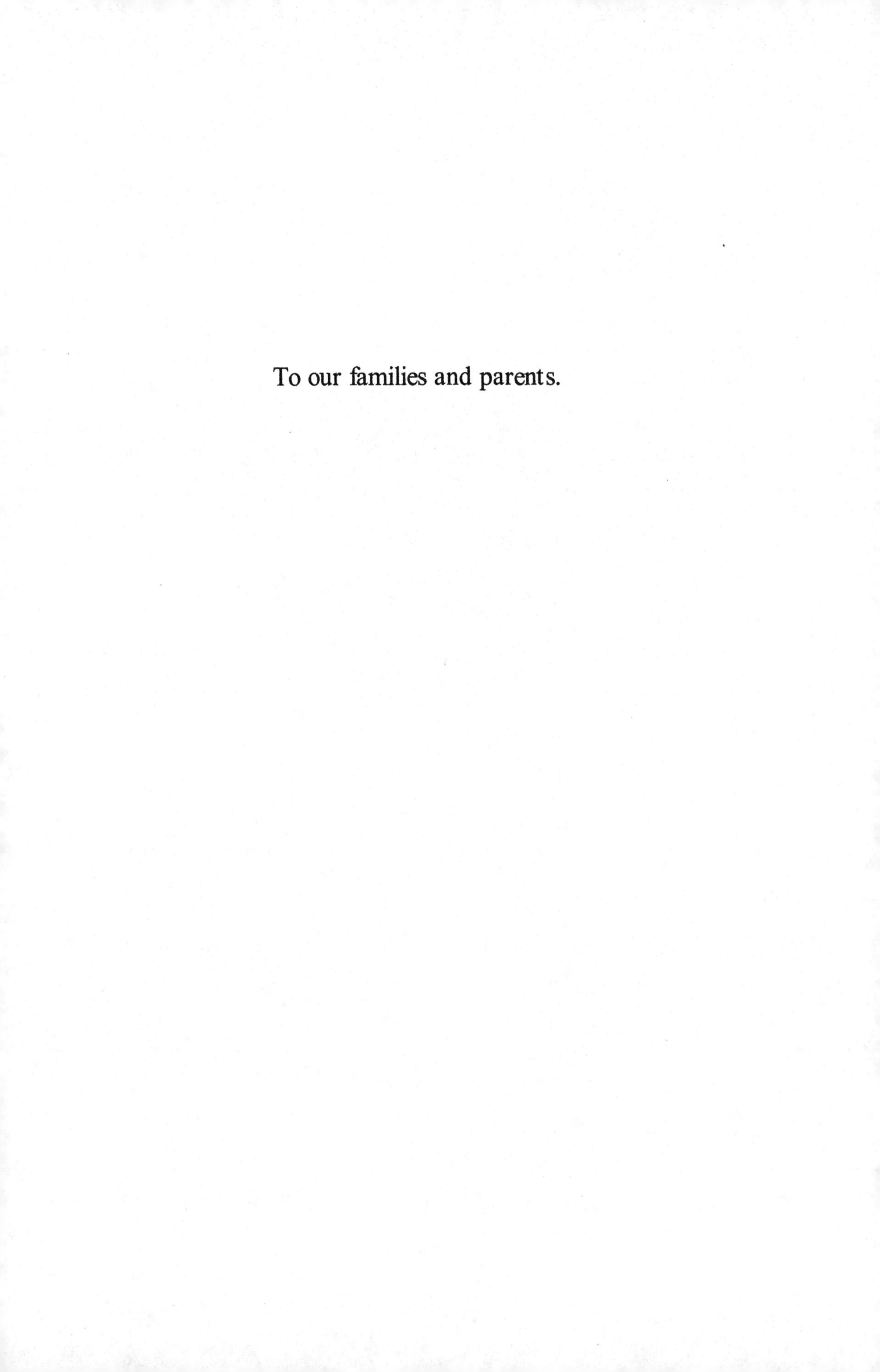

To our families and parents.

Preface

The seed for this book came from a review article published three years ago [1]. At the time a reviewer suggested that a book on polarizabilities was needed. Since the polarizability is a physical property of particles and materials that is so pervasive and important in describing and understanding the interaction of particles and electric fields, the need for such a book was clear. In this book we have aimed to clearly describe the polarizability as an electronic property, to discuss its importance in the physics and chemistry of atoms, molecules, and clusters, and to review theoretical and experimental techniques for deducing its value in a wide range of particle classes. A rigorous review of the frequency dependence and tensor properties of the polarizability is given. In addition, the last chapter attempts to provide some examples of the relation of the polarizability to important phenomena under recent study such as atom cooling and trapping, optical tweezers, and long-range interactions.

The ubiquitous occurrence of polarizabilities in physical phenomena has spawned a large literature that spans many fields: physics, chemistry, engineering, and biology. It is difficult to manage and maintain a complete command of this vast body of literature. However, we hope that one of the lasting effects of this book is to give an adequate summary of relevant work up to 1996 in the areas covered. We apologize in advance for not citing works that are relevant but that we missed. Also, as with any book, there are invariably mistakes that we inadvertently missed or that were made due to a lack of understanding on our part. In both of these cases, we ask our readers to provide us with a gentle reminder of our errors in exchange for the promise that we will try very hard to avoid the same mistakes in the future.

Many friends and colleagues have helped us in this endeavor, and we gratefully acknowledge their contributions. In particular, KDB thanks Will Happer for encouragement and moral support and especially for introducing him to this important subject in the first place. He is ever grateful to A.R. Marlow for long-term friendship and intellectual support. VVK learned about clusters, polarizabilities, and good science in general from Walter Knight, whose kind support and friendship are deeply appreciated. Joe Louderback was kind enough to read the entire manuscript twice and we thank him for his helpful feedback. We thank Thad Walker for many discussions on all aspects of atomic polarizabilities. We are also grateful to many others for their valuable comments on individual sections or whole chapters, viz., Jörg Becker (semiconductor clusters), Walt de Heer (position-sensitive time-of-flight polarizability technique), George Holzwarth (optical tweezers), Uwe Hohm and Dirk Goebel (refractive index measurements and dispersive Fourier-transform spectroscopy), Mark Pederson (theory chapter), Dave Pritchard (atom interferometry), Dany Shapiro (biomolecules and organic molecules), and Thad Walker (atom cooling and trapping). Others who gave important comments on the review article that were incorporated into the book include D. Bishop, A.D. Buckingham, and Tom Miller. We owe a debt of gratitude to Michael Kadar-Kallen for his contributions to the experimental techniques chapter (from the original article). His sections formed the core for roughly half of Chapter 5. Judy Swicegood is to be thanked for her help on the bibliography. Many colleagues graciously permitted us to reproduce figures from their papers and provided us with original copies. Work on this book was partially supported by the U.S. National Science Foundation Programs in Chemistry and Atomic, Molecular, Optical, and Plasma Physics (KDB) and Atomic, Molecular, Optical, and Plasma Physics (VVK). KDB thanks Wake Forest University for providing excellent computer facilities and personnel (in particular C.W. Yip).

Finally, a special appreciation goes to our families. KDB thanks his children, Alex, Claire, and Caroline, and his wife, Marian, for cheerfully accepting a father and husband whose frequent physical absences, blank looks, and empty stares all indicated a mind present elsewhere. VVK is infinitely grateful to his wife, Susan, for her patience and support.

Contents

List of Figures

List of Tables

ELECTRIC-DIPOLE POLARIZABILITIES OF ATOMS, MOLECULES AND CLUSTERS

Chapter 1

Introduction

The purpose of this book is to review experimental and theoretical techniques for determining electric-dipole polarizabilities of atoms, molecules, and clusters. Interest in polarizabilities of basic particles continues to grow, both to understand the electronic and optical response of new and old materials and to test computational methods for calculating electronic and optical properties. These methods and codes apply to other important material characteristics, such as magnetic response of particles and mechanical properties. Given the growing importance of light forces in atomic and molecular physics and chemistry that rely on the frequency-dependent polarizability, this book considers both static and frequency-dependent polarizabilities. However, for brevity and ease of understanding, in many cases we shall focus on the static polarizability.

The charge redistribution that occurs when a particle is exposed to an electric field is characterized by a set of constants called polarizabilities. The new charge distribution can be written in terms of electric multipole moments. The *lowest-order* moment of a *neutral* particle is a dipole moment $\mathbf{p}$. In a uniform electric field $\mathbf{E}$ the dipole moment of the particle is conveniently written as

$$\mathbf{p} = \mathbf{p}_0 + \boldsymbol{\alpha} \cdot \mathbf{E} + \frac{1}{2}\boldsymbol{\beta}{:}\mathbf{E}^2 + \frac{1}{6}\boldsymbol{\gamma}\vdots\mathbf{E}^3 + \cdots \tag{1.1}$$

The term $\mathbf{p}_0$ represents the permanent dipole moment. The polarizability $\boldsymbol{\alpha}$ is a second-rank Cartesian tensor that characterizes the lowest-order *induced* dipole moment in a species. It is this quantity which is the subject of this book. The hyperpolarizabilities $\boldsymbol{\beta}$ and $\boldsymbol{\gamma}$ represent third-and fourth-rank Cartesian tensors. For symmetric species, since $\mathbf{p}(\mathbf{E}) = -\mathbf{p}(-\mathbf{E})$, the permanent moment $\mathbf{p}_0 = \mathbf{0}$ and the hyperpolarizability $\boldsymbol{\beta} = \mathbf{0}$. Since the advent of lasers, nonlinear optics has blossomed and interest in values of the *frequency-dependent* hyperpolarizability $\boldsymbol{\gamma}$ for various nonlinear systems has increased.

For neutral species the polarizability is a basic property that characterizes the lowest-order response of the species to an applied electric field. The polarizability has units of volume and is of the same order of magnitude as the volume of the particle (e.g., $\sim 10^{-23}$ cm^3 for an atom). Polarizabilities are important in three broad areas of physics and chemistry:

- Electromagnetic field-matter interactions, where polarizabilities determine the response of neutral particles to applied fields, such as those produced by lasers.
- Collision phenomena, where polarizabilities determine the behavior of neutral particles as partners in interactions with other neutrals and/or charged particles.
- As an indication of physical size, structure, and shape.

Polarizabilities are also helpful in determining the electronic structure of atoms, molecules, and clusters. Some important physical properties of atoms and molecules that depend on the polarizability are given in Table 1.1. A more extensive list can be found in the Handbook of Chemistry and Physics [2]. A brief description of different notations and the units used to describe polarizabilities is given in Appendix A.

The earliest observation of an effect that could (subsequently) be directly related to a molecular polarizability was the first notice of anomalous dispersion by Fox Talbot (1840). It was Stokes who first made the crucial proposal that molecules could be treated as having their own natural vibrational frequencies that result in observable effects when interacting with light at a different frequency. This proposition, made in 1852 [4], was an attempt to explain fluorescence, a term coined by Stokes. In 1860 le Roux [5] reported in the literature his observations on anomalous dispersion of light propagating through iodine vapor. In 1871 Kundt [6–8] first made the general observation that anomalous dispersion occurred whenever an absorption band was present. Almost simultaneously, Maxwell in 1869 [9] and Sellmeier in 1872 [10] suggested a theoretical explanation based on the existence of harmonic oscillators in the material that would become forced when the wavelength (or frequency) of light approached the natural frequency of the harmonic oscillators. Hence, the famous Sellmeier formula for the refractive index of a material η as a function of wavelength λ

$$\eta^2 - 1 = \frac{e^2}{\pi m c^2} \sum_k N f_k \lambda_k^2 \left[1 + \frac{\lambda_k^2}{\lambda^2 - \lambda_k^2} \right], \tag{1.2}$$

where $N f_k$ is the number of electrons having a resonance at wavelength λ_k, c is the speed of light, e is the charge on an electron, and m is the mass of the electron. A modern, microscopic understanding of the response of atoms and molecules to electric fields began with the prediction of Voigt [11] in 1901 that an electric analog to the Zeeman effect should exist. In 1913, Lo Surdo [12] and Stark [13] independently reported the first observations of the splitting of atomic levels by application of electric fields. Although the Stark effect was first treated theoretically by [14–16], it was Kramers [17] who first treated an energy splitting that required an *induced*, field-dependent dipole moment (quadratic Stark effect).

The earliest measurements on free atoms that could be used (later) to deduce polarizabilities were the dielectric constant measurements of Hochheim [18] in 1908 and the refractive index measurements of Cuthbertson and Cuthbertson [19] in 1910. Both sets of measurements involved the inert gases. Koch [20] in 1908 measured the refractive indices of common molecules hydrogen and oxygen as well as air, determining the dispersion constants for these species that are useful in Sellmeier's dispersion

Table 1.1: Important physical quantities that depend on the scalar polarizability.

Electromagnetic Response Properties	
Quantity	Relation to α
1. Dielectric constant[a]	$\varepsilon = 1 + 4\pi\alpha$
2. Refractive index[a]	$\eta = 1 + 2\pi\alpha$
3. Energy shift	$U = -\alpha \mathrm{E}^2/2$
4. Phase shift of wavefunction[b]	$\Delta\phi = \int_{x_i}^{x_f} U dx/\hbar v$
Collision Properties	
Quantity	Relation to α
1. Long-range electron- or ion-atom interaction potential[c]	$U = -e^2\alpha/2r^4$
2. Ion mobility in gas[d]	$K = 13.876\sqrt{\alpha\mu}$ cm^2/volt sec
3. van der Waals constant between systems a, b (Slater-Kirkwood approximation)[e]	$C_6 = \frac{3}{2}\left(\frac{\alpha_a\alpha_b}{\sqrt{\alpha_a/n_a}+\sqrt{\alpha_b/n_b}}\right)$
Relation to Physical Structure[f]	
Particle	Expression for α
1. Classical metal sphere	R^3
2. Classical dielectric sphere	$R^3(\varepsilon - 1)/(\varepsilon + 2)$
3. Classical dielectric shell[g]	$\frac{(1-\rho)(\varepsilon-1)(2\varepsilon-1)}{(2\varepsilon+1)(\varepsilon+2)-2\rho(\varepsilon-1)^2}R^3$
4. Small metal cluster[h]	$R^3 f(R)$

[a]This relation holds for a dilute, nonpolar gas ($\varepsilon \approx 1$ limit of the Clausius-Mossotti equation).

[b]The quantity $x_f - x_i$ is the length of the electric field region (in quantity 3, previous line).

[c]The quantity r is the separation between the colliding pair of particles; e is the charge on an electron and we have assumed singly-ionized ions.

[d]Assumes α in units of Å^3; μ is the reduced mass of the ion-atom pair in atomic mass units.

[e]Here $n_{a,b}$ refers to the number of outer shell electrons in species a and b respectively.

[f]Here R is the physical radius of the particle.

[g]The quantity $\rho = \delta R/R$, where the thickness of the shell wall is δR. This expression is derived in Ref. [3].

[h]For a discussion of the function $f(R)$ see Sec. 4.10.4.

formula Eq. (1.2). A straightforward technique for measuring polarizabilities is to deflect the atoms in an inhomogeneous electric field and measure the resulting deflection. This technique was first tried by Scheffers and Stark [21] in 1934. They measured the scalar polarizability (α_0) of alkali atoms to accuracies of 20-50%. To this day the alkali and alkaline earth atoms and those elements which are a gas at room temperature account for most of the accurately measured atomic polarizabilities, with the exceptions being Hg, In, Tl, Al and most recently, U [22], Ga and As [23], and Zn [24]. For most elements, theoretical values are the only ones available.

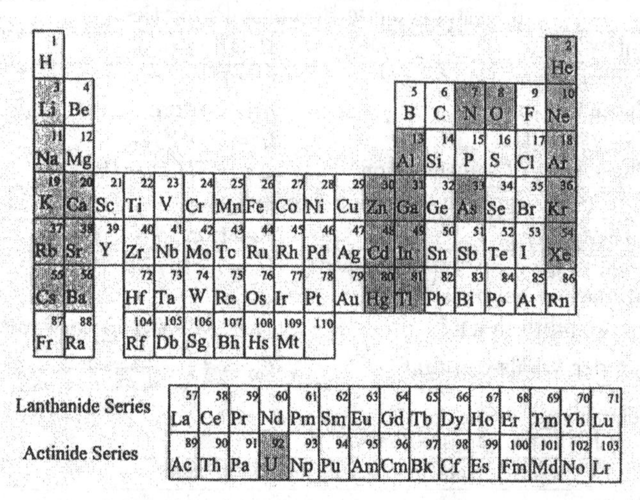

Figure 1.1: Periodic table of elements indicating which elements have measured polarizabilities.

A number of reviews of polarizability data, including experimental details and calculations have been published [1, 25–30]. The CRC Handbook of Chemistry and Physics contains an annually-updated list of polarizability values for the elements and many molecules. The book by Atkins [31] contains a clear elementary introduction into the properties and manifestations of molecular polarizabilities. The review paper of Miller and Bederson [25] is particularly useful and includes a discussion of techniques and results on excited states which are not discussed here. They also give an extensive review of theoretical calculations up to 1976; here, we will mention some recent techniques used to calculate polarizabilities. This book reviews the frequency dependence of the polarizability, discusses and compares experimental techniques and calculations, and provides a review of work on clusters (both metal and dielectric). The recent experimental techniques that are discussed in detail include

- Dispersive Fourier transform spectroscopy: used to measure the frequency-dependent polarizability over a broad frequency spectrum.
- M-lines method: used to measure the refractive index and polarizability of thin films and molecules in a thin-film matrix.
- Time-of-flight: used to accurately measure the polarizability of aluminum clusters.

- Light force: used to measure the polarizability of a highly refractory element — uranium.
- Atom interferometry: used to make the most accurate measurements of α for an alkali atom (sodium).

A summary of some of the most important experimental techniques is given in Table 1.2.

Table 1.2: Summary of experimental techniques for measuring ground-level polarizabilities.

Technique	Pol.[a]	Error[b] α_0 (%)	Advantages	Disadvantages
Bulk				
Dielectric const.	α_0	0.05	Accurate	Limited to inert gases/simple molecules
Refractive index	α_0	0.02	Accurate	Limited mainly to inert/molecular gases
Rayleigh scatt.	κ^2	—	Accurate for anisotropic pol.	measurement of α_0 difficult
Kerr effect	$\alpha_0\kappa$	—	Can measure hyperpols.	Cannot measure α_0
Beam				
Deflection	α_0	4.4	Can be applied to clusters	Somewhat sensitive to velocity distribution; large field uncertainties
E-H balance	α_0, α_2	2	Independent of velocity distrib.	Limited to species with magnetic moments
Beam-resonance	α_2	—	Sensitive to $\alpha_2 < 10^{-6}$ Å^3	Can only measure α_2
Light-force	$\alpha_0, \alpha_1, \alpha_2$	6	Good for ions, refractories	Must extrapolate to obtain α(dc)
Atom interferometry	α_0	0.3	Accurate	

[a]The quantities in column two are the scalar polarizability (α_0), the vector polarizability (α_1), the tensor polarizability (α_2), and the anisotropy of the polarizability tensor (κ). See text for details.
[b]This column represents the lowest errors for these techniques, reflecting the accuracy of the best measurements.

The polarizability of bulk matter is a macroscopic quantity whose relationship to the microscopic polarizabilities of the constituent atoms and molecules is quite complex. Bulk matter polarizabilities are only briefly mentioned when needed to discuss an experimental result. A detailed discussion of the dielectric properties of bulk materials can be found elsewhere [32, 33].

Practical use of the polarization properties of particles and bulk materials is quite extensive. Most readers are probably aware of the importance of the polarization properties of dielectrics used in capacitors and other electronic devices. Fewer are aware of the use of these properties in geophysics. A well-established technique, called Induced Polarization (IP), was discovered by the French scientist, Conrad Schlumberger, in the 1910's. Although he never exploited or developed his discovery, such work began in the mid-1930's and became commercial in the 1950's. Induced Polarization is one of several techniques used in mining and geothermal exploration and in ground-water exploration. Due to its applied nature in a field outside the perview of most scientists, we shall briefly describe the technique of IP. An exhaustive review of IP is given in the two-volume set by Bertin and Loeb [34]. The discussion here is based on a more recent review by Ward [35].

One basic physical setup consists of four electrodes placed in the Earth's surface as shown in Fig. 1.2. The two electrodes on the left are supplied signals which result in constant currents for fixed periods of time [see Fig. 1.3(a)]. The electrodes on the right measure the potential response of the surrounding subsurface region of the Earth [see Fig. 1.3(b)]. The main physical property of the soil that these signals measure is the resistivity. This can be seen by noting that the resistivity ρ is given by the relationship [35]

$$\rho = K\frac{V}{I}, \tag{1.3}$$

where K is a factor dependent on the electrode geometry, V is the potential difference measured between the two right electrodes when a current I flows between the two left electrodes. Information about the soil conductivity is also present in the decay part of the potential curve [Fig. 1.3(b)].

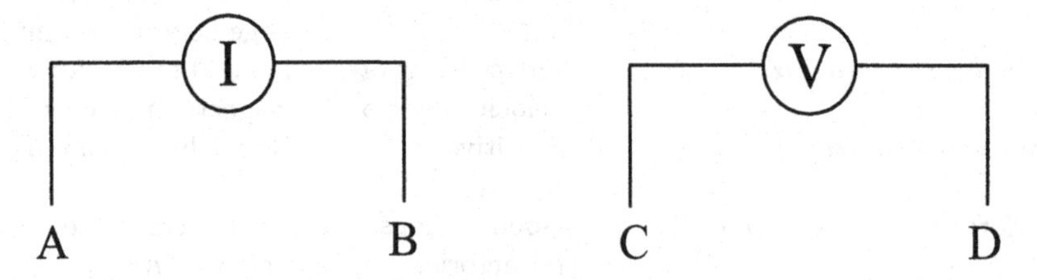

Figure 1.2: A typical dipole-dipole array of electrodes for an Induced Polarization measurement of a region of the Earth's surface to determine subsurface conductivity and other properties. A current is sent through the subsurface of the Earth with a current source connected at points A and B. The resulting electric potential between points C and D is measured.

Two types of polarization play a role in the signals produced with the Induced Polarization method. The first, called electrode polarization, occurs at the boundary of a metal electrode in soil. In the soil layer adjacent to the electrode charge separation occurs that results in a polarization. Here, current conduction occurs in two ways: through the direct flow of ions and through charging and discharging of the charge layers surrounding the electrode. The second type of polarization important in measuring soil resistivity via the IP method is membrane polarization of clay particles in soil. Clay particles are normally negatively charged in soils and so they accumulate a layer of cations near their surface that results in a separation of

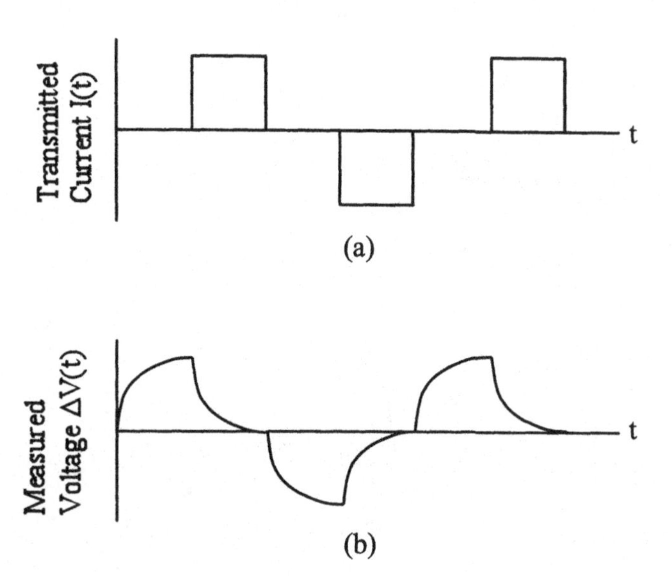

Figure 1.3: The transmitted and received signals for a typical dipole-dipole array of electrodes for an Induced Polarization measurement of a region of the Earth's surface.

charges — the so-called membrane polarization. Clay particles dramatically affect soil conductivity due to their low resistivity. When an electric potential is applied to the soil, i.e. a current flows, the clay particles act as an ion-selective membrane that only allows cationic flow. This produces a region of high concentration of cations and anions at one end of the membrane zone and a deficit at the other end of the zone. This concentration gradient reduces ion mobility and hence results in an increase in resistivity. This fascinating phenomenon is lucidly explained on a microscopic scale, with helpful diagrams, by Ward [35]. Hence, one can view the Induced Polarization technique as a method for measuring soil resistivity which depends on the dielectric constant or 'polarizability' of subsurface soil.

Chapter 2

General Properties of the Linear Polarizability

2.1 Basic Polarizability Relations

This chapter and the next give general, detailed descriptions of the polarizability in scattering theory and for many different systems (atoms, molecules, and clusters). Before starting such a detailed description, we thought it would be useful to summarize, in one section, many of the basic definitions and relations involving the polarizability. These relations will only be cursorily described, and references to more detailed discussions will be given in each case.

The polarizability α is the fundamental quantity that relates dipole moment $\mathbf{p}$ that is produced when a neutral particle is placed in a weak, external electric field $\mathbf{E}$

$$\mathbf{p} = \alpha \mathbf{E}. \tag{2.1}$$

Here we have introduced the polarizability as a constant of proportionality between the dipole moment induced in a particle and the field that induced the moment. However, in general the polarizability is a tensor (see Section 2.2) and it depends on the frequency ω of the inducing field (see Section 2.3). The dynamical polarizability for a particle in the ground state, denoted by $|0>$, can be derived from perturbation theory [see Section 2.4 and Eq. (2.67)]

$$\alpha(\omega) = \frac{2e^2}{3\hbar} \sum_l \frac{\omega_{l0} | < l|\mathbf{r}|0 > |^2}{(\omega_{l0} - \omega - i\Gamma_l/2)(\omega_{l0} + \omega + i\Gamma_l/2)}. \tag{2.2}$$

Here ω_{l0} corresponds to the Bohr frequency between the two energy levels $|l>$ and $|0>$. The full width at half-maximum linewidth of the excited level $|l>$ is denoted by Γ_l, e is the electronic charge, and

$$\mathbf{r} = \sum_{\substack{\text{all} \\ \text{electrons}}} \mathbf{r}_i \tag{2.3}$$

where $\mathbf{r}_i$ is the position of the ith electron in the particle. The above expression corresponds to the polarizability for the most common case, that of an interaction with the electric field in which the *scalar* polarizability dominates. A very convenient way to express the polarizability is to use oscillator strengths (see Section 3.2.1), where the absorption oscillator strength for the transition $|0> \rightarrow |l>$, f_{0l}, is defined

$$f_{0l} = \frac{2m_e\omega_{l0}}{3\hbar}| < 0|\hat{\mathbf{r}}|l > |^2. \tag{2.4}$$

Here $\hat{\mathbf{r}}$ represents the quantum-mechanical position operator. The resulting expression for the polarizability is

$$\alpha(\omega) = \frac{e^2}{m_e}\sum_l \frac{f_{l0}}{(\omega_{l0} - \omega - i\Gamma_l/2)(\omega_{l0} + \omega + i\Gamma_l/2)}. \tag{2.5}$$

Since for many systems oscillator strengths have been measured or calculated, Eq. (2.5) is a good starting point for an estimate of the polarizability of such systems.

It is apparent from the expressions for the polarizability given above that α is a complex quantity. The real part corresponds to forward light scattering by the particle while the imaginary part corresponds to off-axis scattering or absorption by the particle. The photoabsorption cross section at frequency ω, $\sigma(\omega)$, is directly proportional to the imaginary part of the polarizability (see Section 2.5)

$$\sigma(\omega) = \frac{4\pi\omega}{c}\mathrm{Im}[\alpha(\omega)], \tag{2.6}$$

where c is the speed of light.

As already mentioned, the polarizability is a tensor, whose Cartesian components are denoted by α_{ij}. Two very useful relations are the definition of the scalar polarizability α_0 from the tensor and the definition of the polarizability anisotropy, $\Delta\alpha$, from the polarizability tensor. The scalar polarizability is defined to be

$$\alpha_0 = \frac{1}{3}\mathrm{Tr}(\boldsymbol{\alpha}) = \frac{1}{3}\sum_i \alpha_{ii}, \tag{2.7}$$

and the polarizability anisotropy is given by

$$(\Delta\alpha)^2 = \frac{1}{2}\left[3\mathrm{Tr}(\boldsymbol{\alpha}^2) - (\mathrm{Tr}\boldsymbol{\alpha})^2\right]. \tag{2.8}$$

The polarizability anisotropy is important in cases where a particle produces a significant difference in the scattered intensity of light in different directions.

The polarizability is related to many important bulk properties of a collection of particles including the dielectric constant ε, the refractive index η, the extinction coefficient κ, and the electric susceptibility χ [32]. The polarizability is related to the dielectric constant by the Debye equation (see Section 5.1.1)

$$\alpha = \frac{3}{4\pi n}\left(\frac{\varepsilon - 1}{\varepsilon + 2}\right) - \frac{\mu_0^2}{3kT}, \tag{2.9}$$

where n is the number density, μ_0 is the permanent electric dipole moment of the molecule, k is Boltzmann's constant and T is the absolute temperature. The special case when $\mu_0 = 0$ is called the Clausius-Mossotti relation.

The index of refraction is related to the average polarizability α_0 by the Lorentz-Lorenz relation (see Section 5.1.2)

$$\alpha_0 = \frac{3}{4\pi n}\left(\frac{\eta^2 - 1}{\eta^2 + 2}\right) \tag{2.10}$$

where n is the number density of the gas or liquid. This equation is valid for non-polar molecules or at frequencies high enough that the permanent dipole moments cannot follow the electric field. A dipole term must be added if the frequency of the radiation is less than or comparable to rotational frequencies of the molecule. The Lorentz-Lorenz equation and Clausius-Mossotti equations are related by the Maxwell relation $\varepsilon = \eta^2$. For a dilute gas, where the refractive index is nearly one, we have

$$\eta = 1 + 2\pi n\alpha_0 \qquad \text{dilute gas.} \tag{2.11}$$

The relations between the polarizability and the extinction coefficient and electric susceptibility are given in Section 6.1.

2.2 Scalar and Tensor Polarizabilities

A neutral particle placed in a *static* external electric field $\mathbf{E}$ will have its positively charged constituents (nuclei) displaced from the center of the electron distribution and the particle will acquire a dipole moment $\mathbf{p}$. If the field is not too large, and the response of the particle is isotropic, the induced dipole moment will be proportional to the electric field, and one can write

$$\mathbf{p} = \alpha\mathbf{E}. \tag{2.12}$$

The constant of proportionality α is called the polarizability of the particle. Figure 2.1 illustrates the simplest case, that of an atom.

For many particles with spherical symmetry (nearly all atoms, some molecules, and some clusters), the polarizability is well-approximated by a single constant — a scalar quantity. In these cases, the polarizability is the same regardless of the direction of the applied field — the polarizability is isotropic. However, for a few atoms, for many molecules, and for a significant number of clusters the redistribution of charge under application of an electric field cannot be characterized by a single constant. This can be most clearly understood by considering a simple diatomic molecule like H_2. It has no permanent dipole moment, but the charge distribution along the internuclear axis is very long compared to the distribution perpendicular to this axis. Hence, we expect the charge separation induced by an external field to be greater along the internuclear axis than along a perpendicular axis. Indeed, for H_2, the polarizability along the symmetry axis is 30% larger than the polarizability perpendicular to the

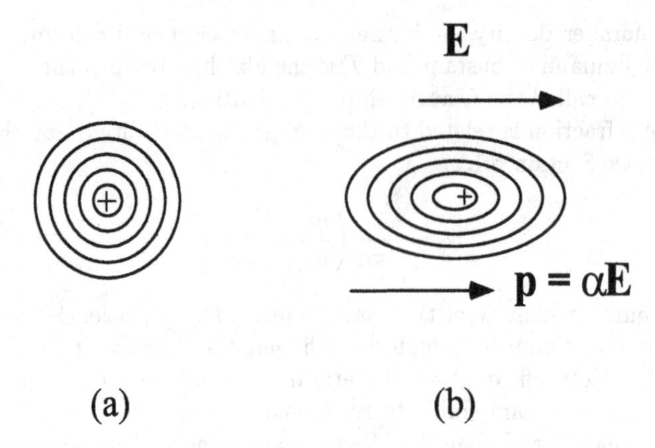

Figure 2.1: The polarizability reflects the size of the charge separation that occurs when an external electric field is applied to a particle. In (a) the spherical charge distribution in a hydrogen atom is sketched. In (b) the center of electron distribution is displaced from the nuclear core, giving rise to an induced dipole moment.

axis of symmetry. In N_2 and O_2, the symmetry axis polarizability is 100% larger than the polarizability in the plane perpendicular to the symmetry axis. Thus, in general, a single number is not sufficient to describe the polarizability.

The polarizability relates the external electric field **E** to the induced dipole moment **p**. The most general way to transform one vector (**E**), into another (**p**) is by a second-rank Cartesian tensor or matrix

$$\mathbf{p} = \boldsymbol{\alpha} \cdot \mathbf{E}. \tag{2.13}$$

In matrix notation

$$\begin{pmatrix} \mathrm{p}_x \\ \mathrm{p}_y \\ \mathrm{p}_z \end{pmatrix} = \begin{pmatrix} \alpha_{xx} & \alpha_{xy} & \alpha_{xz} \\ \alpha_{yx} & \alpha_{yy} & \alpha_{yz} \\ \alpha_{zx} & \alpha_{zy} & \alpha_{zz} \end{pmatrix} \begin{pmatrix} \mathrm{E}_x \\ \mathrm{E}_y \\ \mathrm{E}_z \end{pmatrix} \tag{2.14}$$

In the more compact component notation, where repeated indices are summed over, we can write

$$\mathrm{p}_i = \alpha_{ij}\mathrm{E}_j, \tag{2.15}$$

The second-rank Cartesian tensor α_{ij} can be written as a sum of a symmetric tensor α^s_{ij} and an antisymmetric tensor α^a_{ij}, i.e.

$$\alpha_{ij} = \alpha^s_{ij} + \alpha^a_{ij}, \tag{2.16}$$

where $\alpha^s_{ij} \equiv \alpha_{ij} + \alpha_{ji} = \alpha^s_{ji}$ and $\alpha^a_{ij} \equiv \alpha_{ij} - \alpha_{ji} = -\alpha^a_{ji}$. For a static field, we will show that the polarizability tensor is real and symmetric ($\alpha_{ij} = \alpha_{ji}$). First, it is helpful to consider the energy of an induced dipole in an electric field.

Inducing a dipole moment in a particle will affect its energy and this in turn determines the behavior of the particle in the external field. Thus the polarizability determines the size of the energy shift of a particle in an external field. To express the change in energy of the particle in terms of its polarizability, consider the work done on the particle by the external field. The infinitesimal work done on the induced dipole in changing the electric field from $\mathbf{E}$ to $\mathbf{E} + d\mathbf{E}$ is

$$dW = -\mathbf{p} \cdot d\mathbf{E} = -\mathbf{E} \cdot \boldsymbol{\alpha} \cdot d\mathbf{E}. \tag{2.17}$$

To find the total work done on the dipole in ramping the field from 0 to its final value $\mathbf{E}$, we integrate Eq. (2.17) to obtain

$$W = -\frac{1}{2}\mathbf{E} \cdot \boldsymbol{\alpha} \cdot \mathbf{E}. \tag{2.18}$$

Since the average polarizability of all species in their ground states is positive for static fields, the energy of the particle is lowered by the presence of the external field.

To show that the polarizability tensor is symmetric, consider two different ways of ramping up the electric fields that result in the same final fields and hence the same change in energy. In the first instance, let $\mathbf{E}$ first build up to its full value along x and then afterward it builds up to its final value along y. In the second case, the order is reversed (see Fig. 2.2).

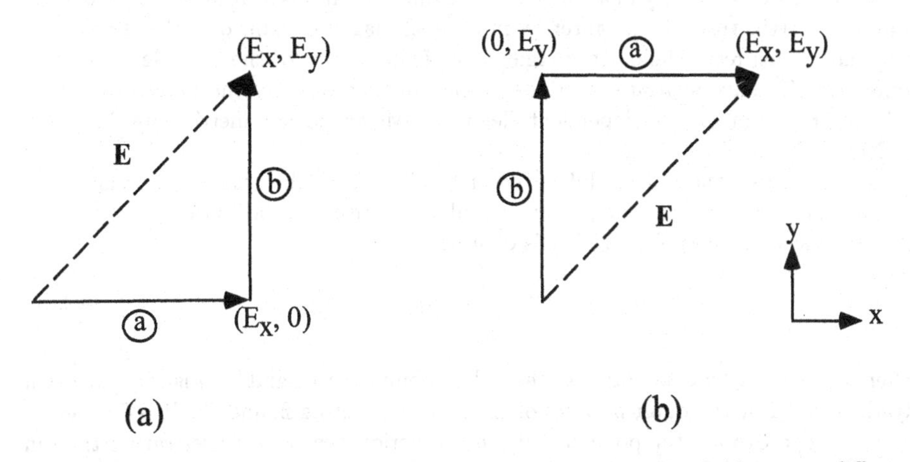

Figure 2.2: The energy is computed by applying the electric field in two different sequences, (a) first increase the field along $\hat{x}$ then along $\hat{y}$. (b) Increase the field along $\hat{y}$ first, then along $\hat{x}$. Since the energy is conservative it must be path independent.

In the first case the total work done by the external field is

$$\begin{aligned} W_1 &= -\int_0^{\mathrm{E}_x} \mathbf{p} \cdot d\mathbf{E} \, |_{path\ a} - \int_0^{\mathrm{E}_y} \mathbf{p} \cdot d\mathbf{E} \, |_{path\ b} \\ &= -\int_0^{\mathrm{E}_x} \alpha_{xx} \mathrm{E}_x d\mathrm{E}_x - \int_0^{\mathrm{E}_y} \alpha_{yx} \mathrm{E}_x d\mathrm{E}_y - \int_0^{\mathrm{E}_y} \alpha_{yy} \mathrm{E}_y d\mathrm{E}_y \\ &= -\frac{1}{2}\alpha_{xx}\mathrm{E}_x^2 - \alpha_{yx}\mathrm{E}_x\mathrm{E}_y - \frac{1}{2}\alpha_{yy}\mathrm{E}_y^2. \end{aligned} \tag{2.19}$$

The second case yields

$$\begin{aligned} W_2 &= -\int_0^{\mathrm{E}_y} \mathbf{p}\cdot d\mathbf{E}\,|_{path\ b} - \int_0^{\mathrm{E}_x} \mathbf{p}\cdot d\mathbf{E}\,|_{path\ a} \\ &= -\int_0^{\mathrm{E}_y} \alpha_{yy}\mathrm{E}_y d\mathrm{E}_y - \int_0^{\mathrm{E}_x} \alpha_{xy}\mathrm{E}_y d\mathrm{E}_x - \int_0^{\mathrm{E}_x} \alpha_{xx}\mathrm{E}_x d\mathrm{E}_x \\ &= -\frac{1}{2}\alpha_{yy}\mathrm{E}_y^2 - \alpha_{xy}\mathrm{E}_y\mathrm{E}_x - \frac{1}{2}\alpha_{xx}\mathrm{E}_x^2. \end{aligned} \tag{2.20}$$

Since the interaction is a conservative one, the energy change must be path independent. That is, the final change in energy must be the same if the starting and ending points (fields) are the same. Thus $W_1 = W_2$ and this implies $\alpha_{xy} = \alpha_{yx}$. Similar relations can be derived for the other off-diagonal components so the polarizability tensor is symmetric. To see that it is real, the diagonal terms are obviously real since the energy must be real and the electric field is as well. To prove it for off-diagonal terms, consider once again the energy shift for a particle placed in an external field of the form $\mathbf{E} = \mathrm{E}_x\hat{x} + \mathrm{E}_y\hat{y}$. The expressions for W_1 or W_2 imply that α_{xy} and α_{yx} are both real. Similar arguments can be applied to the other off-diagonal terms.

Since the polarizability in a static field is symmetric it has at most six independent components (the three diagonal terms and the off-diagonal terms on either side of the diagonal). This result holds true regardless of the nature of the particle. However, some particles have spatial symmetries which further simplify the polarizability by reducing the number of independent elements. Molecular symmetries are discussed in Section 3.3.1.

In this section the polarizability tensor has been described using Cartesian basis vectors, i.e. we have assumed a rectangular, Cartesian coordinate system. In a Cartesian system the polarizability is expressed by

$$\boldsymbol{\alpha} = \sum_{ij} \alpha_{ij}\hat{x}_i\hat{x}_j, \tag{2.21}$$

where α_{ij} is the ijth component of the polarizability tensor and $\hat{x}_i\hat{x}_j$ is the Cartesian dyadic formed by the outer product of the two unit vectors $\hat{x}_i$ and $\hat{x}_j$. It is frequently convenient to express the polarizability in a spherical tensor basis set rather than in a Cartesian basis set. This depends on the particles under investigation and on the physical setup of the experiment. In a spherical basis set, the polarizability tensor can be written

$$\boldsymbol{\alpha} = \sum_{LM} (-1)^M \alpha_M^L \mathbf{Q}_{-M}^L, \tag{2.22}$$

where α_M^L is the Mth component of the spherical tensor of rank L and $\mathbf{Q}_{-M}^L$ is the $-M$th component of the Lth-rank *spatial* tensor. Here the spherical tensor ranks are in the range $0 \leq L \leq 2$ and for a given tensor of rank L the components are in the range $-L \leq M \leq L$. The two descriptions are equivalent and relationships between the two can be derived (for example see the text by Weissbluth [36] and the paper by Happer [37]). The relation between the polarizability tensor components in the two coordinate systems is

$$\alpha_0^0 = -\frac{1}{3}[\alpha_{xx} + \alpha_{yy} + \alpha_{zz}], \tag{2.23}$$

$$\alpha_{\pm1}^1 = \frac{1}{2}[\alpha_{zx} - \alpha_{xz} \pm i(\alpha_{zy} - \alpha_{yz})], \tag{2.24}$$

$$\alpha_0^1 = \frac{i}{2}[\alpha_{xy} - \alpha_{yx}], \tag{2.25}$$

$$\alpha_{\pm2}^2 = \frac{1}{2}[\alpha_{xx} - \alpha_{yy} \pm i(\alpha_{xy} + \alpha_{yx})], \tag{2.26}$$

$$\alpha_{\pm1}^2 = \pm\frac{1}{2}[\alpha_{zx} - \alpha_{xz} \pm i(\alpha_{yz} + \alpha_{zy})], \tag{2.27}$$

$$\alpha_0^2 = \frac{1}{\sqrt{6}}[3\alpha_{zz} - (\alpha_{xx} + \alpha_{yy} + \alpha_{zz})]. \tag{2.28}$$

The relation between the spatial tensors and the dyadic is found by using the relation between unit vectors in the two coordinate systems, viz.

$$\hat{x}_1 = -\frac{1}{\sqrt{2}}(\hat{x} + i\hat{y}), \tag{2.29}$$

$$\hat{x}_0 = \hat{z}, \tag{2.30}$$

$$\hat{x}_{-1} = \frac{1}{\sqrt{2}}(\hat{x} - i\hat{y}). \tag{2.31}$$

Finally, the spatial tensor components or the spherical-basis dyadics can be written in terms of outer products of the unit vectors in spherical coordinates

$$\mathbf{Q}_M^L = \sum_{\mu=\pm1,0} \hat{x}_\mu(\hat{x}_{\mu-M})^*(-1)^{(\mu-1)}\sqrt{2L+1}\begin{pmatrix} 1 & 1 & L \\ \mu & M-\mu & -M \end{pmatrix}, \tag{2.32}$$

where

$$\begin{pmatrix} 1 & 1 & L \\ \mu & M-\mu & -M \end{pmatrix} \tag{2.33}$$

is a 3-j symbol. A more extensive discussion of the relation between the tensors in the two coordinate systems can be found in the Refs. [36] and [37]. We will return to a discussion of representation in coordinate systems in the section on diatomic molecules (Sec. 3.3.2).

2.3 Static and Dynamic Polarizabilities

The electronic response of particles to time-varying electric fields is more complex than the static behavior. Hence, a more sophisticated description of the polarizability is needed here than in the static case. In particular:

- the *antisymmetric* part of the tensor polarizability can be nonzero,

- the particle can absorb and emit photons so the polarizability tensor consists of two parts: a *dispersive* part and an *absorptive* part,
- the polarizability depends on the *frequency* of the light inducing the dipole moment.

The dispersive part accounts for the absorption and instantaneous reemission (stimulated emission) of photons that corresponds to the scattering of light by molecules. The absorptive part accounts for absorption of light by the particle, with subsequent deexcitation due to spontaneous emission or collisional quenching. That the polarizability depends on the frequency of the driving field is physically reasonable. However, as the frequency increases to extremely high values, eventually the charges are unable to follow the changing field, and the polarizability then drops to zero. In this section and the next, we review fundamental aspects of the polarizability that were first treated by Placzek [38]. A more accessible treatment, based on the work of Placzek, is given in the book by Berestetski, et al. [39]. We have drawn on both of these sources for the material in these sections.

Before presenting the fully general result for $\alpha(\omega)$, we give a short derivation and discussion of the expression for the simpler case of the static polarizability. To simplify as much as possible, we will assume a scalar polarizability only. To find an expression for $\alpha(\omega = 0) \equiv \alpha$, consider the energy shift of a particle placed in a static electric field $\mathbf{E}$. From time-independent perturbation theory [40], the energy shift of the kth level of the system experiencing a perturbing potential V is

$$W_k = W_k^0 + V_{kk} + \sum_{l(\neq k)} \frac{|V_{kl}|^2}{W_k^0 - W_l^0} + \ldots, \tag{2.34}$$

where only terms up to second-order in V have been retained. Here W_k^0 corresponds to the unperturbed energy of the kth level and V_{kl} is the klth matrix element of the perturbing potential. In quantitative form

$$V_{kl} = \int \psi_k^{0*} \, \hat{V} \, \psi_l^0 d^3r, \tag{2.35}$$

where ψ_l^0 is the unperturbed eigenstate for the lth level. In the present case, the perturbing Hamiltonian is the electric-dipole interaction

$$\hat{V} = -\hat{\mathbf{p}} \cdot \mathbf{E}, \tag{2.36}$$

where the dipole moment $\hat{\mathbf{p}}$ is a quantum-mechanical operator (so indicated by the hat over the symbol for the operator and not to be confused with the momentum operator) that is given by

$$\hat{\mathbf{p}} = \sum_n q_n \hat{\boldsymbol{r}}_n, \tag{2.37}$$

where q_n is the charge on the nth particle (electron or ion) in the molecule and $\hat{\boldsymbol{r}}_n$ is the position operator of the nth particle. Ignoring cases where two or more levels are nearly degenerate (e.g. H atom), the first-order energy correction is zero, i.e. V_{kk}=0.

The matrix element V_{kk} is zero since $\hat{V}$ has odd parity, i.e it changes sign under an inversion of coordinates, $\hat{\mathbf{r}} \to -\hat{\mathbf{r}}$. The next term in the energy shift Eq. (2.34) can be rewritten using our specific perturbing potential

$$W_k^{(2)} = \sum_{l(\neq k)} \frac{(\hat{\mathrm{p}}_z^*)_{kl}(\hat{\mathrm{p}}_z)_{lk}}{\hbar\omega_{kl}}\mathbf{E}^2, \tag{2.38}$$

where we have taken the electric field to be along the z-axis and the superscript (2) indicates the second-order term. This can be rewritten in the traditional format as

$$W_k^{(2)} = -\frac{1}{2}\alpha\mathbf{E}^2. \tag{2.39}$$

From Eqs. (2.38) and (2.39), a simple expression for the static polarizability of a particle in the kth state is

$$\alpha = 2\sum_{l(\neq k)} \frac{(\hat{\mathrm{p}}_z^*)_{kl}(\hat{\mathrm{p}}_z)_{lk}}{\hbar\omega_{lk}}. \tag{2.40}$$

We could also write this as $\alpha = (\alpha_0)_{kk}$ to explicitly retain the dependence of the scalar polarizability on the state considered.

To find a quantitative expression for the frequency-dependent polarizability, consider a molecule exposed to an external electric field $\mathbf{E}(t)$ oscillating with frequency ω with the form

$$\mathbf{E}(t) = \frac{1}{2}\left(\mathbf{E}_0 e^{-i\omega t} + c.c.\right), \tag{2.41}$$

where c.c. represents the complex conjugate of the preceding terms and the polarization is contained in the vector amplitude $\mathbf{E}_0$. The polarizability is found by considering the lowest order correction to the electric-dipole moment $\hat{\mathbf{p}}$ of the molecule — see Eq. (2.37). Rigorously speaking, the polarizability is defined by only considering the first-order correction to the *diagonal matrix element* of the dipole operator. Since we are considering an external electric field that is harmonic, we shall also consider the oscillating dipole moment operator that this induces. The dipole operator matrix elements oscillating at frequency ω can be written

$$\hat{\mathbf{p}}_{kk}^{(1)}(t) = \frac{1}{2}\left\{(\hat{\mathbf{p}})_{kk}e^{-i\omega t} + c.c.\right\}. \tag{2.42}$$

Here the diagonal matrix element of the ith component of the electric dipole moment operator at frequency ω is related to the polarizability tensor by

$$(\hat{\mathrm{p}}_i)_{kk} = (\hat{\alpha}_{ij})_{kk}(\mathrm{E}_0)_j. \tag{2.43}$$

The quantity $(\hat{\alpha}_{ij})_{kk}$ is the ijth component of the polarizability tensor at frequency ω for a molecule in the kth state. To simplify the language used to describe matrix elements of operators, we shall refer to the quantum-mechanical matrix element of the dipole moment operator $(\hat{\mathbf{p}})_{kk}$ as the transition moment; likewise a quantum-mechanical matrix element of the polarizability tensor operator $(\hat{\alpha}_{ij})_{kk}$ is called the transition polarizability. An expression for the transition polarizability can be found by evaluating the correction to the dipole-moment operator using perturbation theory. In Appendix B.2, we have derived an expression for the first-order correction to an operator $\hat{G}$ [see Eq. (B.20)]. In the present case the operator $\hat{G} = \hat{\mathrm{p}}_i$ and we get from substitution into Eq. (B.20) with $m = k$

$$\begin{aligned}(\hat{\mathrm{p}}_i)_{kk} &= \sum_{l(\neq k)} \left\{ \left[\frac{(\hat{\mathrm{p}}_i)_{kl}\hat{V}_{lk}}{\hbar(\omega_{lk} - \omega)} + \frac{(\hat{\mathrm{p}}_i)_{lk}\hat{V}_{kl}}{\hbar(\omega_{lk} + \omega)} \right] e^{-i\omega t} \right. \\ &+ \left. \left[\frac{(\hat{\mathrm{p}}_i)_{kl}\hat{V}^*_{kl}}{\hbar(\omega_{lk} + \omega)} + \frac{(\hat{\mathrm{p}}_i)_{lk}\hat{V}^*_{kl}}{\hbar(\omega_{lk} - \omega)} \right] e^{i\omega t} \right\}, \end{aligned} \tag{2.44}$$

where $\hat{V}$ is the perturbation due to the external field. Substituting $\hat{V} = -\frac{1}{2}\hat{\mathbf{p}} \cdot \mathbf{E}_0$ into Eq. (2.44) and comparing the result to Eq. (2.42) and Eq. (2.43) we find

$$(\hat{\alpha}_{ij})_{kk} = \sum\!\!\!\!\!\!\int_{l(\neq k)} \left[\frac{(\hat{\mathrm{p}}_i)_{kl}\,(\hat{\mathrm{p}}_j)_{lk}}{\hbar(\omega_{lk} - \omega - i\Gamma_l/2)} + \frac{(\hat{\mathrm{p}}_j)_{kl}\,(\hat{\mathrm{p}}_i)_{lk}}{\hbar(\omega_{lk} + \omega + i\Gamma_l/2)} \right]. \tag{2.45}$$

The symbol $\sum\!\!\!\!\!\!\int$ denotes a sum over discrete states and an integral over continuum levels. For completeness, we have included the natural linewidth in the energy denominators for this expression. For a brief discussion of this point see Appendix B.2. Note that in the dc limit, where the frequency is zero, we obtain the static result Eq. (2.40). This expression includes the case where the frequency of the applied field may be close to resonances of the molecule. In this book, we will generally consider the nonresonant case, which is the dispersive part of the polarizability. However, expressions for the absorptive part of the polarizability can be easily derived using the technique discussed in Appendix B.2. For instance, in the section on atoms we give explicit expressions for both parts of the polarizability when close to a resonance.

Equation (2.45) is the polarizability at frequency ω for a particle in state ψ_k. If the state is degenerate, for example if there are several magnetic sublevels associated with the state, then the polarizability is an average obtained by taking the trace of the polarizability over all magnetic sublevels. For example, denoting the magnetic quantum numbers associated with state ψ_k by m_k we have

$$(\hat{\alpha}_{ij})_{kk} = \frac{1}{2J_k + 1} \sum_{m_k} (\hat{\alpha}_{ij})_{m_k m_k}, \tag{2.46}$$

where $2J_k + 1$ is the degeneracy of the sublevels and $(\hat{\alpha}_{ij})_{m_k m_k}$ is the polarizability associated with sublevel m_k.

With the help of Eq. (2.45) some important observations about the properties of the transition polarizability can now be made. Consider the case of large detuning, where the natural linewidth can be ignored, i.e. $\Gamma_l \sim 0$. Since the electric dipole operator is real, then $(\hat{\mathrm{p}})^*_{kl} = (\hat{\mathrm{p}})_{lk}$ and therefore the transition polarizability tensor is Hermitian

$$(\hat{\alpha}_{ij})_{kk} = (\hat{\alpha}^*_{ji})_{kk}. \tag{2.47}$$

From the discussion of symmetric and antisymmetric components of a Cartesian tensor in Section 2.2 and noting that the transition polarizability tensor is Hermitian, we can conclude that the symmetric part of the transition polarizability Eq. (2.45) must be real and the antisymmetric part must be imaginary. Finally, if the state ψ_k is nondegenerate, then it can be chosen real and this means the transition polarizability must be real. In this case the antisymmetric part is zero.

To summarize, the polarizability is a physical property of the particle describing its induced dipole moment in an external field. An expression for the polarizability as a function of frequency can be derived using perturbation theory. Quantum mechanically, the polarizability is associated with a specific state of the particle. Classically, as well as quantum-mechanically, an oscillating dipole moment will radiate electromagnetic energy. Thus the polarizability, which characterizes the oscillating dipole moment, is important in the scattering of electromagnetic radiation. It is in the context of scattering that the concept of polarizability can be generalized. We shall make this generalization in the next section.

2.4 Generalization of the Polarizability: the Scattering Tensor

The scattering of electromagnetic radiation can be characterized by the polarizability, or more generally, by the scattering tensor. In this section, we briefly describe the electromagnetic scattering process and derive an expression for the scattering tensor. In the end we will relate the scattering tensor to the polarizability.

To begin, consider the most general case of light scattering where light of frequency ω is incident on a collection of particles. Each particle can absorb light (a photon of energy $\hbar\omega$) and then emit light (a photon of energy $\hbar\omega'$). This process is shown diagrammatically in Fig. 2.3. Two different energy-level diagrams are shown: Fig. 2.3(a), where the scattered photon has less energy than the absorbed photon ($\omega' < \omega$ — Stokes scattering) and Fig. 2.3(b), where the scattered photon has more energy than the absorbed photon ($\omega' > \omega$ — antiStokes scattering)

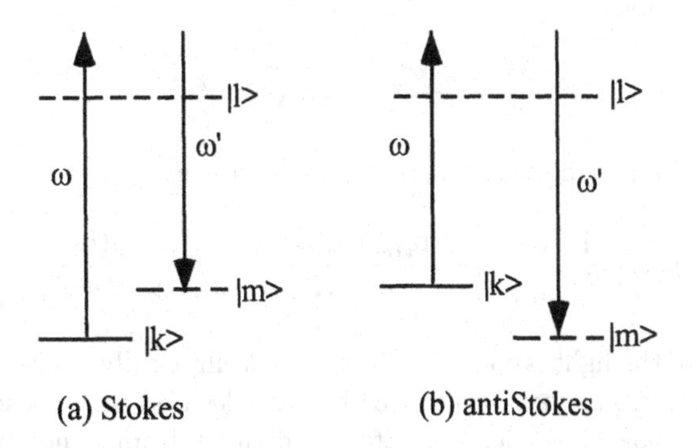

Figure 2.3: Energy level diagram showing the Stokes (a) and antiStokes (b) scattering processes.

Two cases can be distinguished:

(a) Rayleigh scattering, where the absorbed and emitted photons have the same frequency ($\omega = \omega'$) and the particle does not change in energy.

(b) Raman scattering, where the absorbed and emitted photons have different frequencies ($\omega \neq \omega'$) and the particle ends in a higher energy state ($\omega > \omega'$) or in a lower energy state ($\omega' > \omega$), as in Fig. 2.3.

The formulation here will cover both cases. The power P' of the light emitted at frequency ω' into a unit solid angle is given by

$$P' = I\left(\frac{\omega'}{\omega}\right)\frac{d\sigma}{d\Omega}, \tag{2.48}$$

where I is the intensity (power/area) of the incident light at ω and $d\sigma/d\Omega$ is the differential scattering cross section. In destroying a photon of frequency ω and creating a photon of frequency ω', the particle, initially in state ψ_k makes a transition to the final state ψ_m via intermediate states ψ_l. Using second-order perturbation theory, an expression for the differential scattering cross section can be obtained (see Berestetskii [39] or Sakurai [41])

$$\frac{d\sigma}{d\Omega} = \left|\sum_{l(\neq k,m)}\left\{\frac{(\mathbf{p}_{ml}\cdot\boldsymbol{\varepsilon}'^{*})\,(\mathbf{p}_{lk}\cdot\boldsymbol{\varepsilon})}{\omega_{lk}-\omega-i\Gamma_l/2}+\frac{(\mathbf{p}_{ml}\cdot\boldsymbol{\varepsilon})\,(\mathbf{p}_{lk}\cdot\boldsymbol{\varepsilon}'^{*})}{\omega_{lm}+\omega'-i\Gamma_l/2}\right\}\right|^2\frac{\omega\omega'^3}{\hbar^2c^4}, \tag{2.49}$$

where $\mathbf{p}_{ml}$ is the ml-th matrix element of the dipole moment operator and c is the speed of light in vacuum. Here $\boldsymbol{\varepsilon}$ and $\boldsymbol{\varepsilon}'$ correspond to the polarization vectors of the photons at ω, ω' respectively and the symbol $*$ denotes the complex conjugate. The full-width at half-maximum linewidth of the lth energy level in the particle is denoted by Γ_l. The minimum linewidth corresponds to $\Gamma_l(\text{min}) = 1/\tau$, where τ is the natural lifetime of a particle in the lth level. The linewidth Γ_l can be broadened by collisions or by power-broadening that is fundamentally an ac Stark shift. This formula can be rewritten more succinctly

$$\frac{d\sigma}{d\Omega} = \frac{\omega\omega'^3}{c^4}\,|(c_{ij})_{mk}\varepsilon_i'^{*}\varepsilon_j|^2, \tag{2.50}$$

where $(c_{ij})_{mk}$ represents the scattering tensor which is given by

$$(c_{ij})_{mk} = \frac{1}{\hbar}\sum_{l(\neq k,m)}\left[\frac{(\hat{\mathbf{p}}_i)_{ml}(\hat{\mathbf{p}}_j)_{lk}}{\omega_{lk}-\omega-i\Gamma_l/2}+\frac{(\hat{\mathbf{p}}_j)_{ml}(\hat{\mathbf{p}}_i)_{lk}}{\omega_{lm}+\omega'-i\Gamma_l/2}\right] \tag{2.51}$$

The polarization of the light is complex if the light is elliptically or circularly polarized. A description of electromagnetic scattering by particles that uses the scattering tensor is exact. In the case of Rayleigh scattering of light from a nondegenerate level, the transition polarizability and the scattering tensor are identical. In other cases, such as an ensemble of free molecules having degenerate states, a direct relationship between the scattered radiation and the polarizability does not exist (see the extensive discussion of this case in the book by Berestetskii, *et al.* [39]). Also, Raman scattering, where the initial and final states are different, is important for molecules. Hence, a rigid definition of the polarizability that requires identical initial and final states does not apply. We can generalize the definition of the polarizability to include Raman cases by considering the first-order corrections to the off-diagonal matrix elements of the electric-dipole operator $\hat{\mathbf{p}}$ when the particle is placed in an electric field oscillating at frequency ω. This is worked out using time-dependent perturbation theory in Appendix B.2 [see Eq. (B.20)]. From this result, with the substitutions $\hat{G} \rightarrow \hat{\mathrm{p}}_i$ and $\hat{V} \rightarrow -\frac{1}{2}\,\hat{\mathrm{p}}_j(\mathrm{E}_0)_j$, we can write the first-order correction to the off-diagonal elements of $\hat{\mathrm{p}}_i$ as

$$\begin{aligned}\int \psi_m^* \hat{\mathrm{p}}_i \psi_k \, d^3\boldsymbol{x}\,\Big|_{1st-order} = \\ \frac{1}{2\hbar} \sum_{l(\neq k,m)} \Bigg\{ \left[\frac{(\hat{\mathrm{p}}_i)_{ml}(\hat{\mathrm{p}}_j)_{lk}\,(\mathrm{E}_0)_j}{(\omega_{lk}-\omega)} + \frac{(\hat{\mathrm{p}}_i)_{lk}(\hat{\mathrm{p}}_j)_{ml}\,(\mathrm{E}_0)_j}{(\omega_{lm}+\omega)} \right] e^{-i(\omega_{km}+\omega)t} \\ + \left[\frac{(\hat{\mathrm{p}}_i)_{ml}(\hat{\mathrm{p}}_j)^*_{kl}(\mathrm{E}_0)^*_j}{(\omega_{lk}+\omega)} + \frac{(\hat{\mathrm{p}}_i)_{lk}(\hat{\mathrm{p}}_j)^*_{ml}(\mathrm{E}_0)^*_j}{(\omega_{lm}-\omega)} \right] e^{-i(\omega_{km}-\omega)t} \Bigg\}. \end{aligned} \tag{2.52}$$

Note that Eq. (2.52) ignores the term that arises from a permanent dipole moment. Here ψ_m represents the perturbed wavefunction for the mth eigenstate and the matrix elements on the right hand side of Eq. (2.52) are matrix elements evaluated using the unperturbed eigenstates ϕ of the system, i.e.

$$(\hat{\mathrm{p}}_i)_{mk} = \int \phi_m^* \hat{\mathrm{p}}_i \phi_k \, d^3\boldsymbol{x}. \tag{2.53}$$

This expression for the matrix element of the dipole operator Eq. (2.52) will agree with the result for classical radiation of a real dipole only if we consider *real* combinations of such matrix elements [38, 42], that is

$$Re\left(\int \psi_m^* \hat{\mathrm{p}}_i \psi_k \, d^3\boldsymbol{x} \Big|_{1st} \right) = \int \psi_m^* \hat{\mathrm{p}}_i \psi_k \, d^3\boldsymbol{x} \Big|_{1st-order} + \text{c.c.} \tag{2.54}$$

$$\equiv (\hat{\mathrm{p}}_i')_{mk} \tag{2.55}$$

The terms in Eq. (2.52) oscillate at different frequencies: either $(\omega_{km}+\omega)$ or $(\omega_{km}-\omega)$. According to Placzek [38], a proper physical interpretation in terms of scattered radiation for these two terms requires positive frequencies, i.e. $(\omega_{km}+\omega) > 0$ and $(\omega_{km}-\omega) > 0$. The two terms correspond to dipole moments oscillating at the two different frequencies. These oscillating dipoles will produce scattered light at the corresponding frequencies. We remind the reader that ω_k and ω_m are the Bohr frequencies associated with the initial and final states respectively. The two cases in Eq. (2.52) can be described separately:

Case 1: Raman scattering $(\omega_{km}+\omega) > 0$. Here two cases can be distinguished depending on whether the final state is higher in energy (Stokes scattering: $\omega_{km} < 0$) or lower in energy (anti-Stokes scattering: $\omega_{km} > 0$) than the initial state. An energy level diagram of these cases is given in Fig. 2.3.

Case 2: Two-photon emission $(\omega_{km} - \omega) > 0$. The second term in Eq. (2.52) represents the induced emission of two photons, one of frequency ω and the other of frequency $(\omega_{km} - \omega)$. Obviously, this term only occurs when the initial state is higher in energy than the final state, i.e. $\omega_k - \omega_m > 0$. To understand this refer to Fig. 2.4. The scattered light from this term has frequency $\omega_{km} - \omega$ and this can only be produced by starting in a level $|k>$ and emitting a pair of photons, one at the driving frequency ω and one at the scattering frequency $\omega_{km} - \omega$. This case will be ignored since it is rarely important.

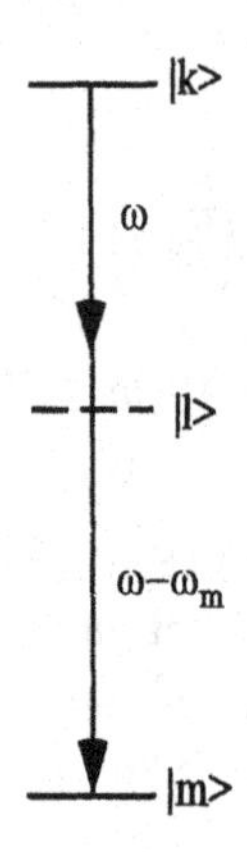

Figure 2.4: Energy level diagram showing the two-photon emission process characterized by the second term of Eq. (2.52).

The Raman part of the transition moment is obtained by combining the first term of Eq. (2.52) with Eq. (2.55)

$$(\hat{p}'_i)_{mk} = \frac{1}{2\hbar} \sum_{l(\neq k,m)} \left[\frac{(\hat{p}_i)_{ml}(\hat{p}_j)_{lk}}{(\omega_{lk} - \omega)} + \frac{(\hat{p}_i)_{lk}(\hat{p}_j)_{ml}}{(\omega_{lm} + \omega)} \right] (E_0)_j \, e^{-i(\omega_{km}+\omega)t} + c.c. \quad (2.56)$$

$$= \frac{1}{2}(\alpha_{ij})_{mk}(E_0)_j \, e^{-i(\omega_{km}+\omega)t} + c.c. \quad (2.57)$$

Thus the generalized (complex) transition polarizability is

$$(\alpha_{ij})_{mk} = \frac{1}{\hbar} \sum_{l(\neq k,m)} \left[\frac{(\hat{p}_i)_{ml}(\hat{p}_j)_{lk}}{(\omega_{lk} - \omega - i\Gamma_{lk}/2)} + \frac{(\hat{p}_i)_{lk}(\hat{p}_j)_{ml}}{(\omega_{lm} + \omega + i\Gamma_{lm}/2)} \right], \quad (2.58)$$

where

$$\Gamma_{lk} = \Gamma_{kl} = \frac{(\Gamma_{kk} + \Gamma_{ll})}{2} + \Gamma'_{lk} \quad (2.59)$$

is the full width at half-maximum linewidth of the $|l> \rightarrow |k>$ transition. The term Γ'_{lk} is the dephasing rate constant, associated with interactions that will interrupt or change the relative phase between the wavefunctions associated with levels $|l>$

and $|k>$. This expression for the transition polarizability $(\alpha_{ij})_{mk}$ is the same as that of the scattering tensor Eq. (2.51) for the case $\omega' = \omega$ (and where the detuning from resonance is large, i.e. $|\omega_{lk} - \omega| \gg \Gamma_l$). Physically, this generalized transition polarizability can account for Raman scattering (both Stokes and anti-Stokes) as well as Rayleigh scattering ($\omega_m = \omega_k$). In the case of Rayleigh scattering in a system with degenerate states, the scattering process can be coherent ($m = k$), which is the classical case, or incoherent ($m \neq k$), which is a purely quantum-mechanical result. In the limit of a static field ($\omega = 0$), the transition polarizability when $\omega_m \neq \omega_k$ corresponds to the correction to the spontaneous emission probability of an atom in the presence of a static external field [43].

As mentioned in Section(2.2), any Cartesian tensor can be decomposed into symmetric and antisymmetric parts. By convention, the symmetric part is further decomposed into a diagonal, *scalar* part and a traceless, *tensor* part. Such a decomposition is convenient because the three resulting parts (one antisymmetric and two symmetric) each represent irreducible tensors in a spherical basis set. Symbolically, this decomposition is written

$$\alpha_{ij} = \alpha^{sc}\delta_{ij} + \alpha^{v}_{ij} + \alpha^{t}_{ij}, \tag{2.60}$$

where

$$\alpha^{sc} = \frac{1}{3}\alpha_{ii} \tag{2.61}$$

is the scalar polarizability (symmetric),

$$\alpha^{t}_{ij} = \frac{1}{2}(\alpha_{ij} + \alpha_{ji}) - \alpha^{sc}\delta_{ij} \tag{2.62}$$

is the tensor polarizability (symmetric and traceless), and

$$\alpha^{v}_{ij} = \frac{1}{2}(\alpha_{ij} - \alpha_{ji}) \tag{2.63}$$

is the vector polarizability (antisymmetric). Using Eq. (2.58) and the definitions Eqs. (2.61-2.63), the corresponding transition polarizabilities are

$$(\alpha^{sc})_{mk} = \frac{1}{3\hbar}\sum_{l(\neq k,m)} \frac{(\omega_{lk} + \omega_{lm})}{(\omega_{lk} - \omega)(\omega_{lm} + \omega)} (\hat{p}_i)_{ml}(\hat{p}_i)_{lk}, \tag{2.64}$$

$$(\alpha^{t}_{ij})_{mk} = \frac{1}{2\hbar}\sum_{l(\neq k,m)} \frac{(\omega_{lk} + \omega_{lm})[(\hat{p}_i)_{ml}(\hat{p}_j)_{lk} + (\hat{p}_j)_{ml}(\hat{p}_i)_{lk}]}{(\omega_{lk} - \omega)(\omega_{lm} + \omega)} - \alpha^{sc}\delta_{ij}, \tag{2.65}$$

$$(\alpha^{v}_{ij})_{mk} = \frac{1}{2\hbar}\sum_{l(\neq k,m)} \frac{(2\omega + \omega_{km})[(\hat{p}_i)_{ml}(\hat{p}_j)_{lk} - (\hat{p}_j)_{ml}(\hat{p}_i)_{lk}]}{(\omega_{lk} - \omega)(\omega_{lm} + \omega)}, \tag{2.66}$$

We note that the spherical tensor description of the polarizability tensor given in Sec. 2.2 [see Eqns. (2.22)-(2.28)] is identical to that here if we make the connection that the scalar part is the $L = 0$ rank spherical tensor, the vector part is the $L = 1$ rank spherical tensor, and the tensor part is the $L = 2$ rank spherical tensor.

We can make one further generalization: to include absorption or the non-Hermitian parts of the polarizability. The procedure is to add a linewidth parameter Γ_l, as discussed in Appendix (B.2). The transition polarizabilities then have the form

$$(\alpha^{sc})_{mk} = \frac{1}{3\hbar}\sum_{l(\neq k,m)} \frac{(\omega_{lk}+\omega_{lm})(\hat{p}_i)_{ml}(\hat{p}_i)_{lk}}{(\omega_{lk}-\omega-i\Gamma_l/2)(\omega_{lm}+\omega+i\Gamma_l/2)}, \tag{2.67}$$

$$(\alpha^{t}_{ij})_{mk} = \frac{1}{2\hbar}\sum_{l(\neq k,m)} \frac{(\omega_{lk}+\omega_{lm})[(\hat{p}_i)_{ml}(\hat{p}_j)_{lk}+(\hat{p}_j)_{ml}(\hat{p}_i)_{lk}]}{(\omega_{lk}-\omega-i\Gamma_l/2)(\omega_{lm}+\omega+i\Gamma_l/2)} - \alpha^{sc}\delta_{ij}, \tag{2.68}$$

$$(\alpha^{v}_{ij})_{mk} = \frac{1}{2\hbar}\sum_{l(\neq k,m)} \frac{(2\omega+\omega_{km})[(\hat{p}_i)_{ml}(\hat{p}_j)_{lk}-(\hat{p}_j)_{ml}(\hat{p}_i)_{lk}]}{(\omega_{lk}-\omega-i\Gamma_l/2)(\omega_{lm}+\omega+i\Gamma_l/2)}, \tag{2.69}$$

These expressions will be used in the next section, where we discuss the relation between the real and imaginary parts of the polarizability.

2.5 Kramers-Krönig Relations and the Optical Theorem

The dispersive and absorptive parts of the polarizability tensor are related to each other through a set of relations first worked out by Krönig and Kramers in 1926-27 [44–46]. To be precise, the dispersive part of the polarizability corresponds to the Hermitian part of the polarizability tensor and the absorptive part of the polarizability corresponds to the anti-Hermitian part of the polarizability tensor. That these associations are the proper ones is clear from quantum mechanics, where dissipative phenomena, such as absorption, are represented by anti-Hermitian parts of the Hamiltonian. These relations will be made explicit when we discuss light scattering in detail in Section 6.1.1.

The Kramers-Krönig relations have their physical origin in the *linear* response of the material to an applied, time-varying electric field and to the requirements of *causality.* Causality requires that the induced dipole moment cannot begin to exist until the inducing field has been applied. A clear derivation of the Kramers-Krönig relations using linearity and causality is given in Appendix C. As discussed in this appendix, the Kramers-Krönig relations only apply to the polarizability tensor that describes Rayleigh transitions, not the generalized polarizability tensor useful in a description of Raman transitions. Hence we consider the transition polarizabilities given at the end of the last section for Rayleigh transitions ($\omega_{km} = 0, \omega_{lk} = \omega_{lm}$). Again we include a natural linewidth term $i\Gamma_l/2$ in the denominators to explicitly account for the case of absorption, writing

$$(\alpha^{sc})_{mk} = \frac{2}{3\hbar}\sum_{l(\neq k)} \frac{\omega_{lk}\,(\hat{p}_i)_{ml}(\hat{p}_i)_{lk}}{(\omega_{lk}-\omega-i\Gamma_l/2)(\omega_{lk}+\omega+i\Gamma_l/2)}, \tag{2.70}$$

$$(\alpha^{t}_{ij})_{mk} = \frac{1}{\hbar}\sum_{l(\neq k)} \frac{\omega_{lk}\,[(\hat{p}_i)_{ml}(\hat{p}_j)_{lk}+(\hat{p}_j)_{ml}(\hat{p}_i)_{lk}]}{(\omega_{lk}-\omega-i\Gamma_l/2)(\omega_{lk}+\omega+i\Gamma_l/2)} - \alpha^{sc}\delta_{ij}, \tag{2.71}$$

$$(\alpha^{v}_{ij})_{mk} = \frac{1}{\hbar}\sum_{l(\neq k)} \frac{\omega\,[(\hat{p}_i)_{ml}(\hat{p}_j)_{lk} - (\hat{p}_j)_{ml}(\hat{p}_i)_{lk}]}{(\omega_{lk} - \omega - i\Gamma_l/2)(\omega_{lk} + \omega + i\Gamma_l/2)}. \tag{2.72}$$

Each component of the polarizability can be written as $\alpha = \alpha' + i\alpha''$, where α' is the real part and α'' is the imaginary part of the polarizability. For the two transition polarizabilities above that are symmetric (α^{sc} and α^{t}), the following Kramers-Krönig relations hold

$$\alpha'_{ij}(\omega_0) = \frac{1}{\pi}\mathcal{P}\int_{-\infty}^{+\infty} \frac{\alpha''_{ij}(\omega)}{\omega - \omega_0} d\omega, \tag{2.73}$$

$$\alpha''_{ij}(\omega_0) = -\frac{1}{\pi}\mathcal{P}\int_{-\infty}^{+\infty} \frac{\alpha'_{ij}(\omega)}{\omega - \omega_0} d\omega, \tag{2.74}$$

$$\tag{2.75}$$

where $\mathcal{P}$ denotes the principal value of the integral. Since in the laboratory we only deal with positive, real frequencies, it is useful to rewrite these relations so the range of integration is $(0, \infty)$. Noting that α'' is an odd function of frequency ω, we can write

<u>Symmetric Case</u>

$$\alpha'_{ij}(\omega_0) = \frac{1}{\pi}\mathcal{P}\int_{0}^{+\infty} \frac{\alpha''_{ij}(\omega)}{\omega + \omega_0} d\omega + \frac{1}{\pi}\mathcal{P}\int_{0}^{+\infty} \frac{\alpha''_{ij}(\omega)}{\omega - \omega_0} d\omega, \tag{2.76}$$

$$= \frac{2}{\pi}\mathcal{P}\int_{0}^{+\infty} \frac{\omega\alpha''_{ij}(\omega)}{\omega^2 - \omega_0^2} d\omega. \tag{2.77}$$

Similarly, as α' is an even function of frequency ω, we can write

<u>Symmetric Case</u>

$$\alpha''_{ij}(\omega_0) = -\frac{2\omega_0}{\pi}\mathcal{P}\int_{0}^{+\infty} \frac{\alpha'_{ij}(\omega)}{\omega^2 - \omega_0^2} d\omega. \tag{2.78}$$

Here the real part of the polarizability, denoted by α' is the Hermitian or dispersive part of the polarizability. Likewise, the imaginary part of the symmetric polarizability, denoted by α'', is the anti-Hermitian or absorptive part of the polarizability. These relations allow us to calculate the value of the dispersive or absorptive part of the polarizability from a knowledge of the other part over a wide range of frequencies. For example, these relations are frequently used to estimate the dispersive part of the polarizability at a specific frequency (related to the refractive index of the material at that frequency) from a knowledge of the absorption over a wide band of frequencies.

The behavior of the real and imaginary parts of the polarizability is interesting when the frequency is close to a resonance frequency of the molecule (ω_{lk}). Using the expressions Eq. (2.70)-(2.72), expressions for the frequency dependence of α' and α'' near a resonance can be found

$$\alpha'(\omega) \propto \frac{(\omega_{lk}^2 - \omega^2 + \Gamma^2/4)}{[(\omega_{lk}^2 - \omega^2 + \Gamma^2/4)^2 + \Gamma^2\omega^2]}, \tag{2.79}$$

$$\alpha''(\omega) \propto \frac{\Gamma\omega}{[(\omega_{lk}^2 - \omega^2 + \Gamma^2/4)^2 + \Gamma^2\omega^2]}. \tag{2.80}$$

A plot of these two lineshapes near a resonance is given in Fig. 2.5.

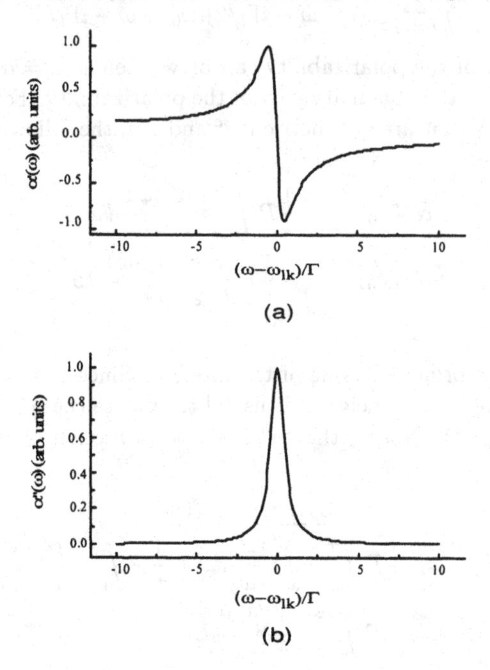

Figure 2.5: A plot showing the lineshapes characteristic of the real and imaginary parts of the polarizability in the region near a molecular resonance $\omega \simeq \omega_{lk}$. The real part corresponds to the refractive index while the imaginary part corresponds to absorption. Note that in the region near a resonance, the real part α' is antisymmetric *about the resonance* while the imaginary part α'' is symmetric about the resonance.

In general, the Kramers-Kronig relations imply $\alpha^*_{ij}(\omega) = \alpha_{ij}(-\omega^*)$. This relation also holds for the polarizability tensor operator, i.e. $\hat{\boldsymbol{\alpha}}^*(\omega) = \hat{\boldsymbol{\alpha}}(-\omega^*)$. For a purely scalar polarizability, these last relations imply that for real frequencies ω, the real part α'_0 and the imaginary part α''_0 of the polarizability are even and odd functions of ω respectively, that is

$$\alpha'_0(\omega) = \alpha'_0(0) + \frac{\omega^2}{2!}\frac{d^2}{d\omega^2}\alpha'_0(0) + ... \tag{2.81}$$

and

$$\alpha''_0(\omega) = \omega\frac{d}{d\omega}\alpha''_0(0) + \frac{\omega^3}{3!}\frac{d^3}{d\omega^3}\alpha''_0(0) + ... \tag{2.82}$$

Since the Kramers-Krönig relations involve an integral of one part of the polarizability over frequency space to determine the other part, we need to carefully consider the frequency-dependence of the antisymmetric part of the polarizability to write down the correct Kramers-Krönig relations for it. By comparing the frequency dependence of the antisymmetric part of the polarizability $\alpha^a = \alpha^v$ [see Eq. (2.72)] with either of the parts that are symmetric [see Eqs. (2.70)-(2.71)], we see that α^a/ω has the same frequency dependence as the symmetric parts of α. Hence, we simply substitute α^a/ω in each place that α^s occurs. The resulting Kramers-Krönig relations for the antisymmetric parts, equivalent to Eq. (2.77) and Eq. (2.78) for the symmetric parts are

Antisymmetric Case

$$\alpha'_{ij}(\omega_0) = \frac{2\omega_0}{\pi}\mathcal{P}\int_0^{+\infty} \frac{\alpha''_{ij}(\omega)}{(\omega^2-\omega_0^2)}d\omega. \tag{2.83}$$

$$\alpha''_{ij}(\omega_0) = -\frac{2\omega_0^2}{\pi}\mathcal{P}\int_0^{+\infty} \frac{\alpha'_{ij}(\omega)}{\omega(\omega^2-\omega_0^2)}d\omega. \tag{2.84}$$

Finally, the Kramers-Krönig relations are also useful in the regime where intensities are large enough to modify the polarizability. For example, near atomic resonances, even modest intensities of $\sim 100\mu\text{W/cm}^2$ will change the absorption coefficient (corresponding to α'') due to depletion of the ground state. A description of such depletion and its use in saturated-absorption spectroscopy can be found in the book by Demtröder [47]. Similarly, changes in carrier density in bulk semiconductors due to intensity modify the absorption coefficient and refractive index of these materials [48]. As discussed by Butcher and Cotter [48], in these cases Kramers-Krönig relations can frequently be applied to good effect. Following their treatment with some modification, we can write the intensity-dependent polarizability as

$$\alpha'(\omega,0) + \Delta\alpha'(\omega,I) + i[\alpha''(\omega,0) + \Delta\alpha''(\omega,I)], \tag{2.85}$$

where I is the intensity of the light field and $\Delta\alpha'(\omega,I)$ and $\Delta\alpha''(\omega,I)$ are the modifications to the linear values of the real and imaginary parts of the polarizability respectively. By substitution into the Kramers-Krönig relation Eq. (2.77), we can derive the relation

$$\Delta\alpha'(\omega_0,I) = \frac{2}{\pi}\mathcal{P}\int_0^{+\infty} \frac{\omega}{(\omega^2-\omega_0^2)}\Delta\alpha''(\omega,I)d\omega. \tag{2.86}$$

This relation and its counterpart, equivalent to Eq. (2.78), give good results if the frequency range is restricted to the finite range ($\omega_1 < \omega < \omega_2$) over which $\Delta\alpha''(\omega,I)$ is significant. The limits in the integral Eq. (2.86) have to be restricted to this range. For further discussion, see the book by Butcher and Cotter and references therein [48].

The Kramers-Krönig relations belong to a general class called *dispersion relations* that relate dispersive and absorptive parts of scattering processes involving particles (photons, muons, nuclei, etc.). For further details on the general nature of dispersion relations see the detailed discussions in [49], the advanced treatment in [50], or consult one of the books on dispersion theory [51]. For a discussion of dispersion relations for hyperpolarizabilities in nonlinear scattering see [48] or [52].

The frequency-dependent polarizability tensor operator

$$\hat{\boldsymbol{\alpha}}(\omega) = \hat{\boldsymbol{\alpha}}'(\omega) + i\hat{\boldsymbol{\alpha}}''(\omega) \tag{2.87}$$

is the sum of a Hermitian part $\hat{\boldsymbol{\alpha}}'(\omega)$ and an anti-Hermitian part $i\hat{\boldsymbol{\alpha}}''(\omega)$. The anti-Hermitian part of the polarizability tensor operator accounts for the dissipation of the incident field by the absorption of light. It reaches maximum values at atomic or molecular resonances (where $\boldsymbol{\alpha}'$ is zero). The optical theorem relates the total cross section for scattering to the anti-Hermitian part of the polarizability. The optical theorem states [41] that the total (inelastic and elastic) photon scattering cross section $\sigma_T(\omega)$ is proportional to the imaginary part of the quantum-mechanical elastic scattering amplitude f, i.e.

$$\sigma_T(\omega) = \frac{4\pi}{k}\mathrm{Im}[f(\omega)], \tag{2.88}$$

where $k = c/\omega$ is the wavevector magnitude at frequency ω and c is the speed of light. The quantum-mechanical elastic scattering amplitude at frequency ω, $f(\omega)$, is directly related to the polarizability by

$$f(\omega) = \alpha_{ij}(\omega)\varepsilon_i^*\varepsilon_j, \tag{2.89}$$

where ε_i is the ith component of the polarization of the absorbed photon and ε_j is the jth component of the polarization of the emitted photon. There is an implied summation over repeated indices. Thus the optical theorem gives the result that the total elastic and inelastic scattering cross section is determined by the anti-Hermitian part of the polarizability, i.e.

$$\sigma_T(\omega) = \frac{4\pi\omega}{c}\alpha_{ij}''(\omega)\varepsilon_i^*\varepsilon_j, \tag{2.90}$$

We note that at zero frequency ($\omega = 0$) the anti-Hermitian part of the polarizability operator is negligibly small and so the scattering cross section is zero.

2.6 The Stark Shift

The fundamental result of placing a particle in an applied electric field $\mathbf{E}_a$ is a change in the *energy* of the particle. In Sec. 2.2, a classical expression for the energy shift W was given in Eq. (2.18). If we include the shift that occurs when a particle has a permanent dipole moment $\mathbf{p}_0$, then the shift is

$$W = -\mathbf{p}_0 \cdot \mathbf{E}_a - \frac{1}{2}\alpha \mathrm{E}_a^2, \tag{2.91}$$

where, for simplicity, we assume the polarizability is isotropic. The energy shift due to an induced dipole (corresponding to the second term) has been of practical importance a least since the first half of the eighteenth century. During this period Leyden jars became popular as a means of storing charge (or energy) on a pair of conducting plates separated by an insulating, bulk material used as a dielectric. The first quantitative measurements of the increased capacity due to the dielectric were published by Faraday in 1837 [53–55]. However, this energy shift was first observed on a microscopic scale in 1913 by Stark and LoSurdo (independently — see Sec. 1). The first microscopic observations involved line splittings in atomic spectra.

The polarizability was defined in terms of the *response* of the *particles* to an *external field.* A complete treatment of the polarizability must include the *response* of the *field* to the redistribution of charge in the *particles.* The whole particle-field interaction must be considered. For example, in the static case the energy of the particle is typically lowered ($\alpha < 0$). Since the particle *lowered* its energy, the field must have increased in energy. To see this more clearly and quantitatively, Fig. (2.6) shows the dipoles induced by the field in a dilute collection of gas particles.

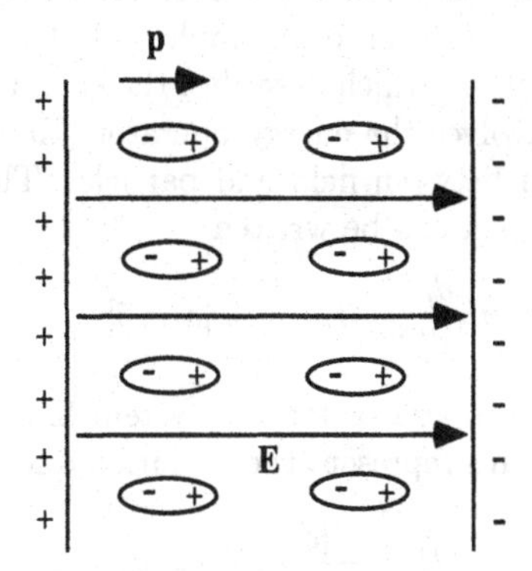

Figure 2.6: Dilute gas particles between a pair of charged condenser plates. The induced dipoles of the particles line up with the applied field to lower the energy of the particles in the field.

From classical electromagnetism (see [56])

$$\mathbf{E}_{new} = \mathbf{E}_a + 4\pi\mathbf{P}, \tag{2.92}$$

where $\mathbf{E}_a$ is the applied field and the macroscopic polarization $\mathbf{P}$ is

$$\mathbf{P} = n\mathbf{p}. \tag{2.93}$$

Here n is the number density of gas particles. The energy/volume in the field is given by

$$\frac{\mathrm{E}^2}{8\pi} = \frac{\mathrm{E}_a^2}{8\pi} + \frac{4\pi\mathbf{P}\cdot\mathbf{E}_a}{8\pi} + 16\pi^2\mathrm{P}^2, \tag{2.94}$$

$$\simeq \frac{\mathrm{E}_a^2}{8\pi} + \frac{1}{2}n\mathbf{p}\cdot\mathbf{E}_a. \tag{2.95}$$

Here, we have only retained the lowest-order term in the induced dipole moment $\mathbf{p}$. Thus the increase in the field energy is

$$\delta W = \left[\frac{\mathrm{E}^2}{8\pi} - \frac{\mathrm{E}_a^2}{8\pi}\right]\mathrm{V} = N\left(\frac{1}{2}\alpha\mathrm{E}_a^2\right), \tag{2.96}$$

where V is the volume and $N = n\text{V}$ is the total number of gas particles. Thus the increased energy in the field is equal in magnitude to the decrease in energy of the particles.

Physically, the polarizability is the *fundamental* property that characterizes the particle-field interaction. As such, the polarizability is measurable from an energy shift or a momentum shift in the particles *or* as a change in the energy or momentum of the field (as in the scattering and absorption of light). We shall return to these topics at the end the book in Sections 6.1.1 and 6.1.2.

A rigorous, quantum-mechanical description of the interaction of the applied field and the particles (system) must be described using the density-matrix formalism. The evolution of particles under the action of an applied electric field can be understood by solving the Liouville equation, which describes the evolution of the density matrix ρ in time. This equation involves the energy operator (Hamiltonian) that describes the fundamental interaction between field and particle. The Liouville equation for the cases described in this book can be written

$$i\hbar\frac{d\hat{\rho}}{dt} = [\hat{H}_0, \hat{\rho}] + [\delta\hat{H}, \hat{\rho}], \tag{2.97}$$

where $\hat{H}_0$ is the Hamiltonian operator for the system in the absence of an applied field and the second set of terms represent the evolution due to the applied field. Here

$$\delta\hat{H} = -\frac{|\mathcal{E}|^2}{4}\boldsymbol{\varepsilon}^* \cdot \hat{\boldsymbol{\alpha}} \cdot \boldsymbol{\varepsilon}, \tag{2.98}$$

where we have assumed an electric field of the form

$$\boldsymbol{E}(\boldsymbol{r}, t) = \frac{\mathcal{E}}{2}[\boldsymbol{\varepsilon} e^{i(\boldsymbol{k}\cdot\boldsymbol{r} - \omega t)} + c.c.]. \tag{2.99}$$

The most general formulation retains the non-Hermitian parts of $\delta\hat{H}$ as well as the Hermitian parts. The Hermitian part of the perturbing Hamiltonian will cause an energy shift with a corresponding operator

$$\begin{aligned} \delta\hat{H}_{shift} &= \frac{1}{2}(\hat{H} + \hat{H}^\dagger), & (2.100)\\ &= -\frac{|\mathcal{E}|^2}{8}(\boldsymbol{\varepsilon}^* \cdot \hat{\boldsymbol{\alpha}} \cdot \boldsymbol{\varepsilon} + \boldsymbol{\varepsilon} \cdot \hat{\boldsymbol{\alpha}}^\dagger \cdot \boldsymbol{\varepsilon}^*). & (2.101) \end{aligned}$$

The non-Hermitian part of the energy operator, corresponding to absorption of light by the molecules, can be written

$$\begin{aligned} \delta\hat{H}_{abs} &= \frac{1}{2}(\hat{H} - \hat{H}^\dagger), & (2.102)\\ &= -i\frac{|\mathcal{E}|^2}{8}(\boldsymbol{\varepsilon}^* \cdot \hat{\boldsymbol{\alpha}} \cdot \boldsymbol{\varepsilon} - \boldsymbol{\varepsilon} \cdot \hat{\boldsymbol{\alpha}}^\dagger \cdot \boldsymbol{\varepsilon}^*). & (2.103) \end{aligned}$$

The total perturbing Hamiltonian is given by

$$\delta\hat{H} = \delta\hat{H}_{shift} - i\delta\hat{H}_{abs}, \tag{2.104}$$

and so the Liouville equation can be rewritten

$$i\hbar\frac{d\hat{\rho}}{dt} = [(\hat{H}_0 + \delta\hat{H}_{shift}), \hat{\rho}] - i[\delta\hat{H}_{abs}, \hat{\rho}]. \tag{2.105}$$

The density matrix $\hat{\rho}$ is used later to evaluate the polarizability — see, for example, Section 3.3.2.

Chapter 3

Polarizable Systems

3.1 Elementary Particles

The most elementary system, but not always the simplest, is the vacuum of space. Here the polarizability has its lowest possible magnitude ($\alpha = 0$), with a refractive index of 1 and no absorption. This simplest picture of the response of the vacuum breaks down when the field becomes intense enough to polarize the vacuum. Here the light is so intense that you must treat the vacuum by including the creation of virtual positron-electron pairs that produce a polarizability. Phenomena such as light-by-light scattering can be explained in this way. However, this is the subject of nonlinear optics and high-energy physics (see [41]).

The next most elementary system is a free 'point' particle that has no internal structure. The electron is one of the most elementary particles. A free electron in space has a frequency-dependent polarizability that can be easily derived classically. Consider a charged particle of mass m and charge $-e$ in an oscillating electric field (a scalar field treatment will suffice)

$$\mathrm{E}(t) = \mathrm{E}_0 \cos \omega t. \tag{3.1}$$

The resulting motion of the particle is described by the expression

$$m\ddot{x} = -e\mathrm{E}_0 \cos \omega t, \tag{3.2}$$

where $\ddot{x}$ means the second time-derivative of position x with respect to time t. Physically, the charge follows the electric field, oscillating back and forth in space about some average position. This oscillating charge produces a time-varying dipole moment (see Fig. 3.1).

The induced electric dipole can be written

$$\mathrm{p}(t) = -ex(t) = -\frac{e^2}{m\omega^2}\mathrm{E}_0 \cos \omega t. \tag{3.3}$$

Thus the frequency-dependent polarizability is given by

$$\alpha(\omega) = -\frac{e^2}{m\omega^2}. \tag{3.4}$$

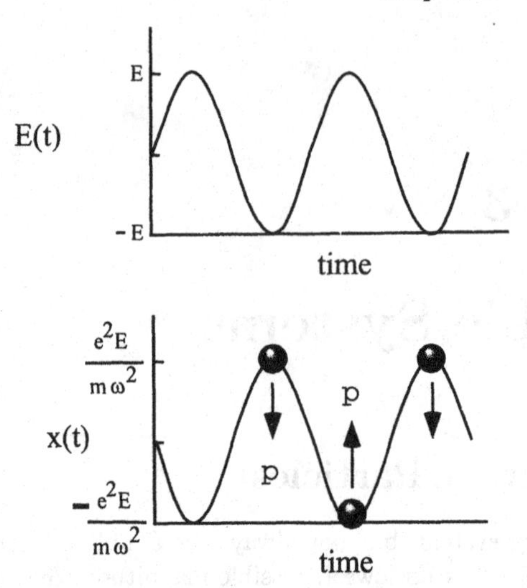

Figure 3.1: Plots showing the time-varying electric field that induces a time-varying electric dipole due to the motion of the charged particle in the field. The electric field is shown in (a). The motion of the particle is shown in (b), where the arrows indicate the direction of the dipole moment at the peaks and troughs of the particle motion.

This is the classical polarizability of a charged particle responding to a time-varying electric field. It is possible that the electron could have an extremely tiny contribution to its polarizability due to charge redistribution upon application of an electric field. To understand just how small such a contribution must be, assume we can assign a value that corresponds to the 'volume' of the electron as determined using the classical electron radius ($r_e = 2.8 \times 10^{-13}$ cm). The corresponding polarizability is $\alpha = r_e^3 = 5 \times 10^{-15}$ Å^3. Compare this to the value of the classical polarizability taken at the Compton wavelength of the electron ($\lambda_C = 0.024$ Å), $\alpha = e^2/m\omega^2 = 4.3 \times 10^{-10}$ Å^3. The classical polarizability still dominates by 5 orders of magnitude!

The contribution of the classical polarizability to observed phenomena shows up in plasma physics where the pondermotive force (or Miller force) is sometimes important. In regions of a plasma where inhomogeneous electric fields occur, pondermotive forces must be taken into account to understand the motion of charges in the plasma. For a derivation of the pondermotive force in plasma physics see the books by Cap [57].

The neutron is a fundamental, non-point particle whose polarizability has only recently been accurately measured [58]. In a series of neutron beam transmission experiments in ^{208}Pb, Schmiedmayer et al. [58] measured the neutron polarizability to be $\alpha_n = (1.20 \pm 0.15 \pm 0.20) \times 10^{-3}$ fm^3, where the error is quoted as $\pm$ statistical $\pm$ systematic and 1 fm = 10^{-15}m. Slightly later, the proton polarizability was accurately measured by a different group by measuring the Compton scattering cross section on hydrogen of photons in the energy range 32-72 MeV [59]. The quantity actually measured by Compton scattering is the total electronic polarizability, labelled $\bar{\alpha}$.

This total polarizability is the sum of the proper polarizability α and a "retardation" term of the form $e^2 < r >^2 /3Mc^2$, that is

$$\bar{\alpha} = \alpha + \frac{e^2 < r >^2}{3Mc^2}, \tag{3.5}$$

where $< r >^2$ is the mean-square charge radius of the proton and M is the proton mass. For details on the source of the retardation term, see the nonrelativistic treatment given in Ref. [60, 61]. The result for the proton, $\alpha_p = (7.0 \pm 2.2 \pm 1.3) \times 10^{-4}$ fm^3, is about half that of the neutron and the error is somewhat larger. This unexpected result, that there is a large difference in the proton-neutron polarizabilities, has sparked theoretical interest. Recent theoretical papers have attempted to explain the polarizability values for these hadrons [62, 63], with some of the work focused mainly on the large difference between the two values [64].

3.2 Atoms

3.2.1 Introduction

An atom in a ground state of zero total angular momentum, i.e. $J = 0$, has an isotropic polarizability and α is therefore a scalar that can be represented by a single number, regardless of the direction of the applied electric field. If the atomic electrons have a net angular momentum $J \geq 1/2$ in the ground state of the atom, the induced dipole moment will depend on the spin polarization of the atoms and will be anisotropic. That is, the polarization of the atom will be proportional to the external field, but the direction of the induced dipole moment may not be the same as the direction of the applied field. We must then describe the dipole moment induced in the atom using a polarizability tensor, α_{ij}, where the indices label the axes of a Cartesian coordinate system.

It is frequently convenient to deal with the effective polarizability *operator* rather than the transition polarizability. For atoms in particular, the effective polarizability operator is an excellent approximation to the rigorous result. For molecules, use of a polarizability operator is not as good an approximation and usually the transition polarizability is preferred. A very detailed discussion of the effective operator formalism as applied to atomic polarizabilities and Stark shifts is given in the work of Happer [37]. The ith component of the amplitude of the oscillating electric dipole moment operator at frequency ω is [see Eq. (2.43)]

$$\hat{\mathrm{p}}_i = \hat{\alpha}_{ij}(\mathrm{E}_0)_j, \tag{3.6}$$

where $\hat{\alpha}_{ij}$ is the effective polarizability operator. As discussed at the end of Section 2.3, both the dipole moment and the polarizability should be thought of as expectation values, which depend on the spin polarization of the atoms in their ground state[37]. Mathematically, we write this as $\alpha_{ij} = \mathrm{Tr}\,(\rho\,\hat{\alpha}_{ij})$ where ρ is the density matrix of the polarized atoms, $\hat{\alpha}_{ij}$ is the polarizability tensor operator, and the trace operation extends over all spin sublevels of the atom.

We can easily write the expression for the dynamic polarizability operator at frequency ω by adapting the result for the transition polarizability obtained from time-dependent perturbation theory [see Eq. (2.45)]

$$\hat{\alpha}_{ij}(\omega) = \frac{1}{\hbar} \sum_{l \neq k} \frac{\hat{\mathrm{p}}_i|l><l|\hat{\mathrm{p}}_j}{\omega_{lk} - \omega - i\Gamma_l/2} + \frac{\hat{\mathrm{p}}_j|l><l|\hat{\mathrm{p}}_i}{\omega_{lk} + \omega + i\Gamma_l/2}, \tag{3.7}$$

where we still assume atoms at rest and we have included the linewidths, Γ_l, of resonance lines. The full-width at half-maximum (FWHM) of the natural linewidth of the lth line is determined by the natural lifetime, τ_l, of the lth excited state, that is $\Gamma_l = 1/\tau_l$.

The Cartesian polarizability operator can be most conveniently expressed as a sum of three terms that represent the three irreducible components in a spherical-tensor basis set. The change in representation from a Cartesian basis to a spherical basis was discussed in detail in Section 2.4. In a spherical basis set, the frequency-dependent polarizability can be described in terms of three complex functions of frequency: the scalar polarizability α_0, which describes the index of refraction of a vapor of atoms with randomly oriented spin directions; the vector polarizability α_1, which can be nonzero if $J > 0$, and which describes physical phenomena like the paramagnetic Faraday rotation of plane polarized light by spin polarized atomic vapors; and the tensor polarizability α_2, which can be nonzero for $J > 1/2$ and which describes the birefringence of a vapor of spin-aligned atoms. In the spherical-basis set terminology of Section 2.4, the traceless, symmetric part of the polarizability tensor α^s_{ij} corresponds to the tensor polarizability α_2 and the antisymmetric part of the polarizability tensor α^a_{ij} corresponds to the vector polarizability α_1. In the literature[65], the complex coefficients α_0, α_1, and α_2 are sometimes labeled α_s, α_v, and α_t, where s, v, t correspond to scalar, vector, and tensor. The frequency-dependent polarizability operator can be written in terms of the electronic angular momentum operators $\hat{J}_i$ of the atom as

$$\hat{\alpha}_{ij} = \alpha_0 \hat{\delta}_{ij} - \alpha_1(\hat{J}_i\hat{J}_j - \hat{J}_j\hat{J}_i) + \alpha_2 \frac{\frac{3}{2}(\hat{J}_i\hat{J}_j + \hat{J}_j\hat{J}_i) - J(J+1)\hat{\delta}_{ij}}{J(2J-1)}. \tag{3.8}$$

Here $\hat{\delta}_{ij} = \delta_{ij}\hat{I}$, where δ_{ij} is the Kronecker-delta function and $\hat{I}$ is the identity operator. The complex coefficients $\alpha_0, \alpha_1, \alpha_2$ in Eq. (3.8) do not depend on the particular magnetic sublevel, m, of the ground state, since the dependence of $\hat{\alpha}_{ij}$ on geometry is given by the angular momentum operators explicit in Eq. (3.8). Expressions for these coefficients can be found using Eq. (3.7) and first-order time-dependent perturbation theory:

$$\alpha_0 = \frac{1}{3\hbar g_J} \sum_{l \neq k} |<\gamma_k J||\hat{\mathrm{p}}||\gamma_l J_l>|^2 \; G_+(l,k,\omega) \; \phi_0(J, J_l) \tag{3.9}$$

$$\alpha_1 = \frac{1}{2\hbar g_J} \sum_{l \neq k} |<\gamma_k J||\hat{\mathrm{p}}||\gamma_l J_l>|^2 \; G_-(l,k,\omega) \; \phi_1(J, J_l) \tag{3.10}$$

$$\alpha_2 = \frac{1}{3\hbar g_J} \sum_{l \neq k} | < \gamma_k J || \hat{\mathrm{p}} || \gamma_l J_l > |^2 \; G_+(l,k,\omega) \; \phi_2(J, J_l) \tag{3.11}$$

where $g_J = 2J + 1$ is the degeneracy of the ground state and $\hat{p}$ is the magnitude of the electric-dipole operator. In the ground-state wavefunctions, $|\gamma_k J\rangle$, the quantity γ_k represents a set of ground state quantum numbers (including, for example, the energy) and J is the ground state angular momentum. Excited-state wavefunctions are similarly written $|\gamma_l J_l\rangle$, where γ_l represents a set of excited state quantum numbers and J_l is the total angular momentum of the lth excited state. The matrix elements $| < \gamma_k J || \hat{\mathrm{p}} || \gamma_l J_l > |^2$ are the usual reduced matrix elements associated with the Wigner-Eckart theorem. Accordingly, the matrix element of a tensor operator, e.g., $\hat{\mathrm{p}}^1_q$ which is the qth component of the vector operator $\hat{\mathbf{p}}$ (a tensor of rank 1, as indicated by the superscript), can be written as the product of a reduced matrix element and a geometric factor (Clebsch-Gordan coefficient or 3-j symbol)

$$< \gamma_k J m_k | \hat{\mathrm{p}}^1_q | \gamma_l J_l m_l > = \frac{(J_l m_l, 1q | J m_k)}{(2J+1)^{1/2}} \times < \gamma_k J || \hat{\mathrm{p}} || \gamma_l J_l > . \tag{3.12}$$

Clebsch-Gordan coefficients, denoted by $(J_l m_l, 1q|Jm_k)$, and reduced matrix elements, denoted by $< \gamma_k J || \hat{p} || \gamma_l J_l >$, are discussed elsewhere [66]. The quantities m_k, m_l are the magnetic quantum numbers of the ground and excited state respectively. They correspond to the projections of total angular momentum along the quantization axis, which is taken to be the direction of the electric field in the case of a dc field and in the case of linearly polarized light. For circularly-polarized light the propagation axis of the light is usually taken to be the quantization axis.

The functions $\phi_i(J, J_l)$ in Eqs. (3.9)-(3.11) are defined as follows

$$\begin{aligned}
\phi_0(J, J_l) &= \delta_{J_l,J-1} + \delta_{J_l,J} + \delta_{J_l,J+1} \\
\phi_1(J, J_l) &= \frac{-1}{J}\delta_{J_l,J-1} - \frac{1}{J(J+1)}\delta_{J_l,J} + \frac{1}{J+1}\delta_{J_l,J+1} \\
\phi_2(J, J_l) &= -\delta_{J_l,J-1} + \frac{2J-1}{J+1}\delta_{J_l,J} - \frac{J(2J-1)}{(J+1)(2J+3)}\delta_{J_l,J+1}
\end{aligned} \tag{3.13}$$

Here, the δ's represent Kronecker delta functions. The frequency-dependent parts of Eqs. (3.9)-(3.11) are given by

$$G_\pm(l,k,\omega) = \frac{1}{(\omega_{lk} - \omega - i\Gamma_l/2)} \pm \frac{1}{(\omega_{lk} + \omega + i\Gamma_l/2)}. \tag{3.14}$$

In the limit of large detuning from resonance ($\omega \ll \omega_{lk}, \Gamma_l \to 0$), these frequency coefficients $G_\pm$ simplify to

$$G_+(\omega) = \frac{2\omega_{lk}}{(\omega_{lk}^2 - \omega^2)} \tag{3.15}$$

$$G_-(\omega) = \frac{2\omega}{(\omega_{lk}^2 - \omega^2)}. \tag{3.16}$$

When the frequency of the field is close to an atomic ($\omega \simeq \omega_{lk}$), these frequency coefficients $G_\pm$ simplify to

$$G_\pm(\omega) \simeq \frac{i2}{\Gamma_l}. \tag{3.17}$$

In the limit of large detuning the polarizability components $(\alpha_0, \alpha_1, \alpha_2)$ are real and the diagonal terms of the polarizability operator Eq. (3.8), which correspond to the diagonal elements of the transition polarizability, are given by

$$< Jm|\hat{\alpha}_{ij}|Jm >= \alpha_0\delta_{ij} - m\alpha_1\varepsilon_{ijk}\delta_{kz} + \alpha_2\frac{[3m^2 - J(J+1)]}{J(2J-1)}\delta_{ij}\left(-\frac{1}{2}\right)^{(1-\delta_{iz})}, \tag{3.18}$$

where ε_{ijk} is the Levi-Civita antisymmetric tensor defined by

$$\varepsilon_{ijk} = \begin{cases} 1 & \text{if } ijk \text{ is cyclic in xyz} \\ -1 & \text{if } ijk \text{ is anticyclic in xyz} \\ 0 & \text{otherwise.} \end{cases}$$

In the dc limit, the coefficients α_i from Eqs. (3.9)-(3.11) substituted into the expression Eq. (3.18) reproduce the results given by Miller and Bederson [25]. The diagonal elements of the zz-component of the polarizability operator are the only nonzero components for two important cases: for linearly-polarized light and for a static field (where the electric field is along the z-axis). The resulting expression for this component can be written from Eq. (3.18) as

$$< Jm|\hat{\alpha}_{zz}|Jm >= \alpha_{zz}(m) = \alpha_0 + \alpha_2\frac{[3m^2 - J(J+1)]}{J(2J-1)}. \tag{3.19}$$

Hence the polarizability becomes a real, symmetric tensor.

A relation for the frequency-dependent polarizability in terms of oscillator strengths can be found by substituting the expression for ϕ_0 from Eq. (3.13) and the expression for $G_+(\omega)$ from Eq. (3.15) into the expression for α_0 in Eqs. (3.9)-(3.11). The familiar result is

$$\alpha_0(\omega) = \frac{e^2}{m_e}\sum\!\!\!\!\!\!\int_{l\neq k} \frac{f_{lk}}{(\omega_{lk}^2 - \omega^2)}, \tag{3.20}$$

where the oscillator strength of the $k \to l$ transition f_{kl} is given by

$$f_{kl} = \frac{2m_e\omega_{lk}}{3\hbar g_J e^2}| < \gamma_k J||\hat{\mathrm{p}}||\gamma_l J_l > |^2. \tag{3.21}$$

Of course, all polarizability components can be conveniently expressed using oscillator strengths, and we have

$$\alpha_i(\omega) = \frac{e^2}{m_e}\sum\!\!\!\!\!\!\int_{l\neq k} \frac{f_{kl}}{(\omega_{lk}^2 - \omega^2)}\phi_i(J_k, J_l), \quad i = 0, 2; \tag{3.22}$$

$$\alpha_1(\omega) = \frac{3}{2}\frac{e^2}{m_e}\sum\!\!\!\!\!\!\int_{l\neq k} \frac{\omega f_{kl}}{\omega_{lk}(\omega_{lk}^2 - \omega^2)}\phi_1(J_k, J_l). \tag{3.23}$$

3.2.2 Fine and hyperfine structure

Thus far atomic polarizabilities have been described assuming that the eigenstates of the atoms are described by quantum numbers associated with the total *electronic* angular momentum J and by the projection of this angular momentum along a quantization axis m_J. The quantum operator $\hat{\boldsymbol{J}}$ in Eq.(3.8) representing the total electronic angular momentum can be written, in the L-S coupling representation, as the sum of the electronic orbital angular momentum $\hat{\boldsymbol{L}}$ and the electronic spin angular momentum $\hat{\boldsymbol{S}}$: $\hat{\boldsymbol{J}} = \hat{\boldsymbol{L}} + \hat{\boldsymbol{S}}$. L-S coupling is an approximation that is good for atoms of small atomic number Z, but for large Z the increasing size of the spin-orbit interaction causes the orbital angular momentum of each individual electron (denoted by quantum number l) to couple to its spin (s) to produce a total electron angular momentum j. The electronic eigenstates are then j-j coupled to give the final eigenstates that depend on the total final angular momentum J (and m_J). At intermediate values of Z an intermediate picture between L-S and j-j coupling occurs. For convenience, most atomic levels are referred to using the *term* notation of $^{2S+1}L_J$, which presumes L-S coupling. In this section we will discuss the change in the expressions for the atomic polarizabilities when the fine structure interaction is ignored. In addition, we will discuss the changes that are needed to describe the polarizabilities when we include the interaction of the electronic total angular momentum $\hat{\boldsymbol{J}}$ with the nuclear spin $\hat{\boldsymbol{I}}$. Here the proper eigenstates of the atom are described by the quantum numbers associated with the total *atomic* angular momentum $\hat{\boldsymbol{F}} = \hat{\boldsymbol{J}} + \hat{\boldsymbol{I}}$ and its projection along the quantization axis m_F.

The fine-structure (or spin-orbit) interaction

$$\hat{H}_{SO} = \frac{1}{2m^2c^2}\frac{1}{r}\frac{dV}{dr}\hat{\boldsymbol{S}} \cdot \hat{\boldsymbol{L}} \tag{3.24}$$

is a relativistic energy correction that arises from the interaction between the spin magnetic moment of the electron and the effective magnetic field seen in the rest frame of the electron (due to the motion through the electric field produced by the nuclear charges). This relativistic correction has a magnitude of order $\sim 10^{-4}$ eV. This interaction forces us to use eigenstates represented by the good quantum numbers J, m_J. If the spin-orbit interaction were ignored and the orbital motion of the electrons were uncoupled from their spin angular momenta, then we could represent the polarizability tensor operator as

$$\hat{\alpha}_{ij} = \alpha_0\hat{\delta}_{ij} - \alpha_1(\hat{L}_i\hat{L}_j - \hat{L}_j\hat{L}_i) + \alpha_2\frac{\frac{3}{2}(\hat{L}_i\hat{L}_j + \hat{L}_j\hat{L}_i) - L(L+1)\hat{\delta}_{ij}}{L(2L-1)}. \tag{3.25}$$

Expressions for the complex coefficients $\alpha_k(L)$, where $i = 0, 1, 2$, are identical to those previously given in Eqs. (3.9)-(3.11) for the $\alpha_k(J)$'s, i.e.

$$\alpha_0 = \frac{1}{3\hbar g_L} \sum_{l \neq k}\!\!\!\!\!\!\int \; |< \gamma_k L||\hat{\mathrm{p}}||\gamma_l L_l >|^2 \; G_+(l,k,\omega) \; \phi_0(L, L_l)$$

$$\alpha_1 = \frac{i}{2\hbar g_L} \sum_{l \neq k}\!\!\!\!\!\!\int \; |< \gamma_k L||\hat{\mathrm{p}}||\gamma_l L_l >|^2 \; G_-(l,k,\omega) \; \phi_1(L, L_l)$$

$$\alpha_2 = \frac{1}{3\hbar g_L} \sum_{l \neq k}\!\!\!\!\!\!\!\int \; |<\gamma_k L||\hat{\mathrm{p}}||\gamma_l L_l>|^2 \, G_+(l,k,\omega)\, \phi_2(L,L_l) \tag{3.26}$$

where $g_L = 2L+1$ is the degeneracy of the state whose polarizability operator is represented here. These relations ignore the fine-structure splittings in energy that occur in the denominators of the expressions for $G_\pm$ [see Eqs.(3.14) and (3.15)].

Similarly, if we include the hyperfine interaction that occurs due to the interaction of the electronic magnetic moment with the magnetic field produced by the spin associated with the nucleus, then the polarizability operator can be further refined. In this case, we have the same relations as given for the $L-S$ coupled case where J, m_j are good quantum numbers [see Eq.(3.8)], except the total electronic angular momentum J is everywhere replaced by the total atomic angular momentum $F = I + J$, where I is the nuclear spin angular momentum. We can then write

$$\hat{\alpha}_{ij} = \alpha_0\hat{\delta}_{ij} - \alpha_1(\hat{F}_i\hat{F}_j - \hat{F}_j\hat{F}_i) + \alpha_2\frac{\frac{3}{2}(\hat{F}_i\hat{F}_j + \hat{F}_j\hat{F}_i) - F(F+1)\hat{\delta}_{ij}}{F(2F-1)}. \tag{3.27}$$

Expressions for the complex coefficients $\alpha_k(F)$, where $i = 0, 1, 2$, are identical to those given in Eq.(3.26) for the $\alpha_k(L)$'s, except F replaces L.

Relations among the three different sets of coefficients $\alpha_k(L)$, $\alpha_k(J)$, and $\alpha_k(F)$ can be derived using angular momentum algebra (for a complete treatment see the book by Weissbluth [36] or Brink and Satchler [67]). These relations for the scalar and tensor polarizabilities have been derived in the literature by Angel and Sandars [68]. We reproduce those relations here and, for completeness, we provide the same relations for the vector polarizability. The relations for the scalar coefficients are

$$\alpha_0(L) = \alpha_0(J) = \alpha_0(F). \tag{3.28}$$

The relations for the vector coefficients are

$$\alpha_1(F) = (-1)^{F+J+L+1}\sqrt{\frac{(2F+1)(2J+1)(J+1)J}{(F+1)F}}\begin{Bmatrix} J & J & 1 \\ F & F & I \end{Bmatrix}\alpha_1(J), \tag{3.29}$$

and

$$\alpha_1(J) = (-1)^{J+L+S+1}\sqrt{\frac{(2J+1)(2L+1)(L+1)L}{(J+1)J}}\begin{Bmatrix} L & L & 1 \\ J & J & S \end{Bmatrix}\alpha_1(L). \tag{3.30}$$

Here the symbol

$$\begin{Bmatrix} J & J & 1 \\ F & F & I \end{Bmatrix} \tag{3.31}$$

represents a $6-j$ symbol [36] that arises in situations where three angular momenta are coupled, such as I and $J = L + S$ to produce $F = I + J$. The relations for the tensor coefficients are

$$\alpha_2(F) = (-1)^{F+J+L}\sqrt{Q(F,J)}\begin{Bmatrix} J & J & 2 \\ F & F & I \end{Bmatrix}\alpha_2(J), \tag{3.32}$$

and

$$\alpha_2(J) = (-1)^{J+L+S}\sqrt{Q(J,L)}\left\{\begin{array}{ccc} L & L & 2 \\ J & J & S \end{array}\right\}\alpha_2(L). \tag{3.33}$$

Here the symbol $Q(a,b)$ for angular momenta a and b is defined

$$Q(a,b) \equiv \frac{(2a+1)a(2a-1)(2b+3)(2b+1)(b+1)}{(2a+3)(a+1)b(2b-1)}. \tag{3.34}$$

For the vector and tensor polarizabilities, these relations can be further simplified. Expanding the $6-j$ symbols and doing some algebra will give

$$\alpha_1(F) = -(-1)^{2(F+J+L)}\frac{X}{2F(F+1)}\alpha_1(J), \tag{3.35}$$

and

$$\alpha_2(F) = (-1)^{2(F+J+L)}\frac{3X(X+1) - 4(F+1)F(J+1)J}{2(2F+3)(F+1)J(2J-1)}\alpha_2(J), \tag{3.36}$$

where

$$X = I(I+1) - J(J+1) - F(F+1). \tag{3.37}$$

Similar relations can be derived for $\alpha_i(J)$ in terms of $\alpha_i(L)$.

3.3 Small Molecules

Describing the frequency-dependent polarizability of a molecule is more complicated than that of an atom. The additional degrees of freedom for molecular motion, viz. vibration and rotation, cause electronic distributions to vary with time. Another obvious complication is that some molecules have many possible structures. Some of these molecules cannot be treated as rigid and the molecules may transform between their different structures at room temperatures. Large alkali and semiconductor clusters at room temperature or higher, whose isomers are not known, are excellent examples of such molecules. For molecules that can be treated as rigid we must consider their symmetries under point group operations (which will be discussed shortly). Also, in contrast to atoms, where the dynamic polarizability mainly concerns Rayleigh scattering, the dynamic polarizability for molecules includes the possibility of Raman transitions. Since the study of Raman transitions is itself a broad field, we shall limit ourselves to Rayleigh transitions. We have already given fundamental expressions for the frequency-dependent polarizability that account for Raman transitions (see Section 2.4). Also, we will limit the discussion to the *dispersive* part of the polarizability. This means we assume frequencies are well detuned from molecular resonances and linewidths can be ignored ($\Gamma_k \to 0$). The dispersive and absorptive parts of the polarizability are related by Kramers-Krönig relations as discussed in Section 2.5.

This section will follow a natural progression that discusses the molecular transition polarizabilities that survive for rigid molecules in different symmetry classes. Next, the simplest real molecules (diatomics) are discussed in detail, including vibration and rotation. Finally, polyatomic molecules are discussed and we end with a brief section on large molecules of biological interest.

3.3.1 Molecular symmetries

The number of independent elements of the tensor polarizability for a molecule depends on the molecular spatial symmetries. Here we only consider atoms and molecules in *nondegenerate* states and with no applied static magnetic fields. Molecules that are not symmetric under any of the point group operations belong to the C_1 group (Schönflies notation) and the frequency-dependent polarizability α_{ij} has nine independent components (the static polarizability has six independnet components). An example is a carbon atom with four other atoms attached via tetrahedral bonds (see Fig. 3.2).

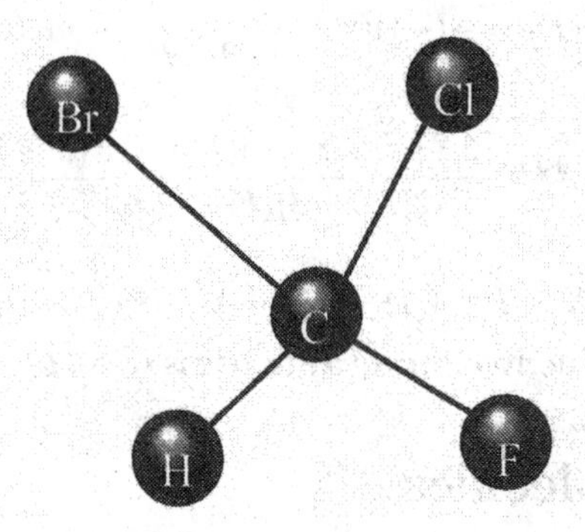

Figure 3.2: Shown is an example of a totally nonsymmetric molecule: bromochlorofluoromethane. It is a methane molecule with three hydrogens replaced by halogen atoms. To describe its frequency-dependent polarizability requires nine independent components (the static polarizability requires six).

In contrast, atoms belong to the group K_h and hence have one independent component (the scalar polarizability α_0). We reiterate that this only rigorously applies to nondegenerate atoms — for atoms with degenerate levels, such as magnetic sublevels, the polarizability requires α_1 and α_2 for a complete description. The number of independent components needed to describe the polarizability tensor for the 32 point groups associated with allowed crystal classes has been analyzed using group theoretical methods [69, 70]. However, the recently discovered fullerene molecules belong to other classes. For example, the fullerenes C_{60} and C_{70} belong to the point groups I_h and D_{5h} respectively. Other fullerene symmetries have been discussed by Diederich and Whetten [71]. Table 3.1 provides a list of the allowed polarizability tensor components based on point group symmetry considerations alone; it includes the standard 32 crystal groups and a few additional molecular groups.

Other physical considerations will further reduce the number of nonzero polarizability components. For example, we have not discussed symmetries under time-reversal, which will affect the number of transition polarizabilities that survive. It can be shown that a molecule with an even number of electrons (integral J) satisfies $(\alpha_{ij})_{mn} = (\alpha_{ij})^*_{mn}$ and so the polarizability is real , i.e. $(\alpha_{ij})_{mn} = (\alpha_{ij})_{mn}$. In addition, if the initial and final states are the same ($m = n$) then the polarizability is symmet-

Table 3.1: Polarizability components for molecular symmetry groups. The symbol $\cdot$ in the table indicates an independent, nonzero element. When the symbol xx occurs in another column, e.g. the yy column, it means that $\alpha_{yy} = \alpha_{xx}$.

System	Schönflies symbol	ind. comp.	xx	yy	zz	xy	yx
Triclinic[a]	C_1, C_i	9	$\cdot$	$\cdot$	$\cdot$	$\cdot$	$\cdot$
Monoclinic	C_2, C_s, C_{2h}	5	$\cdot$	$\cdot$	$\cdot$	$\cdot$	$\cdot$
Orthorhombic	$D_2, C_{2\nu}, D_{2h}$	3	$\cdot$	$\cdot$	$\cdot$	0	0
Tetragonal	C_4, S_4, C_{4h}	3	$\cdot$	xx	$\cdot$	$\cdot$	-xy
	$D_4, C_{4\nu}, D_{2d}, D_{4h}$	2	$\cdot$	xx	$\cdot$	0	0
Trigonal	C_3, S_6	3	$\cdot$	xx	$\cdot$	$\cdot$	-xy
	$D_3, C_{3\nu}, D_{3d}$	2	$\cdot$	xx	$\cdot$	0	0
Hexagonal	C_6, C_{3h}, C_{6h}	3	$\cdot$	xx	$\cdot$	$\cdot$	-xy
	$D_6, C_{6\nu}, D_{3h}, D_{6h}$	2	$\cdot$	xx	$\cdot$	0	0
Cubic	T, T_h, O, T_d, O_h	1	$\cdot$	xx	xx	0	0
Icosahedral	I_h	1	$\cdot$	xx	xx	0	0
Linear	$C_{\infty\nu}, D_{\infty\nu}$	2	$\cdot$	xx	$\cdot$	0	0
Spherical	K_h	1	$\cdot$	xx	xx	0	0
Other	D_{5h}	2	$\cdot$	xx	$\cdot$	0	0

[a]This class has 9 independent components; for all other classes the missing components in the table are zero, i.e. $\alpha_{xz} = \alpha_{zx} = \alpha_{yz} = \alpha_{zy} = 0$.

ric, i.e. $(\alpha_{ij})_{mn} = (\alpha_{ij})^s_{mn}$, where the superscript s indicates the symmetric part of the polarizability. For details about other specialized cases the literature should be consulted (see [72] and references therein). For static fields, the polarizability is real and symmetric. Thus, even the most nonsymmetric molecule (see Fig. 3.2) reduces to only 3 independent components for the polarizability tensor. These values are the ones usually quoted in the literature, e.g. the extensive tabulation in [73]. This case is identical to that of the moment of inertia tensor in classical physics, where any rigid body can be described by the 3 eigenvalues of the moment of inertia tensor, expressed in terms of the principal axes of the rigid body.

3.3.2 Diatomic molecules

The polarizability of diatomic molecules has been treated rigorously by Breiger [74] and in an excellent review article by Bishop [75]. Other treatments that complement those in Breiger and Bishop can be found in the review of the Stark effect by Buckingham [27] and in the review of atomic and molecular polarizabilities by Bogaard and Orr [29]. The present section relies heavily on these references. The presence of vibrational and rotational degrees of freedom greatly complicates a rigorous description of the polarizability of molecules. Vibrational motion typically changes the polarizability by less than a few percent; rotational motion affects it even less. However, if the frequency of the applied field is close to a particular vibrational or rotational frequency, then the contributions to α can be substantial. Also, there are particular cases where the contribution of vibrational motion to the polarizability actually exceeds the electronic part. For example, self-consistent field calculations of the dipole moments of particular hydrogen-bonded complexes as a function of electric field strength predict large polarizability contributions from the movement of the protons in the hydrogen bond [76, 77]. In weak fields, this "proton polarizability" dominates the polarizability contributions due to electronic distortions by almost two orders of magnitude. There are other cases in which the vibrational contribution to the polarizability can become quite large — on the order of 10-20% (e.g., see [78] and [79]). This can also be true of large polyatomic molecules. For example, recent calculations by Whitehouse and Buckingham [80] on the polarizability of $[C_{60}M]^{n+}$ endohedral complexes provide a quantitative basis for the physically reasonable expectation that vibrational contributions to α should dominate the electronic contributions in these complexes.

To properly discuss polarizabilities of diatomic molecules, the issue of coordinate frames must be addressed.

The Hamiltonian that describes the field-particle interaction depends on both field properties [polarization vectors $\hat{\varepsilon}$ — see Eq. (2.101)] and molecular properties [the polarizability, which depends on molecular dipole moments — see Eq. (2.101) and Eq. (2.45)]. Hence, two different, "natural" coordinate systems are convenient. Experimentally, the "natural" choice is to define the quantization axis (z-axis) to be along the external electric field (or along the propagation axis if it is a light field). This is called the space-fixed coordinate system or laboratory frame. Theoretically, the "natural" choice is to define the quantization axis (z-axis) to be along a symmetry axis of the molecule (for diatomic molecules, the internuclear axis). This is called the molecule-fixed coordinate system or molecule frame. These two natural choices are distinguished in Fig. 3.3. In experimental measurements on free molecules, the molecular symmetry axis is assumed to be randomly oriented with respect to the applied field. It is therefore most appropriate to describe the angular dependence of α on the ground-state magnetic quantum number M by a choice of quantization axis parallel to the electric field polarization axis (a space-fixed axis) — see Eq. (3.18) in Section 3.2.1. The relationship between the molecule-fixed axes and the space-fixed axes is independent of the particular model used to calculate the polarizability.

For diatomic molecules, a simple way to account for the dependence of the polarizability on the the internuclear coordinate R is to expand α in a Taylor series in R. For α defined in molecule-fixed axes we have [81]

$$\begin{aligned} \alpha_{||} &= \alpha_{||eq} + \alpha'_{||eq}\xi + \frac{1}{2}\alpha''_{||eq}\xi^2 + \ldots \\ \alpha_{\perp} &= \alpha_{\perp eq} + \alpha'_{\perp eq}\xi + \frac{1}{2}\alpha''_{\perp eq}\xi^2 + \ldots, \end{aligned} \tag{3.38}$$

where $\xi = (R - R_{eq})/R_{eq}$ is the relative displacement of the two nuclei from their equilibrium separation R_{eq}. Note that here a prime indicates a single derivative with respect to the relative displacement coordinate ξ and a double prime indicates two

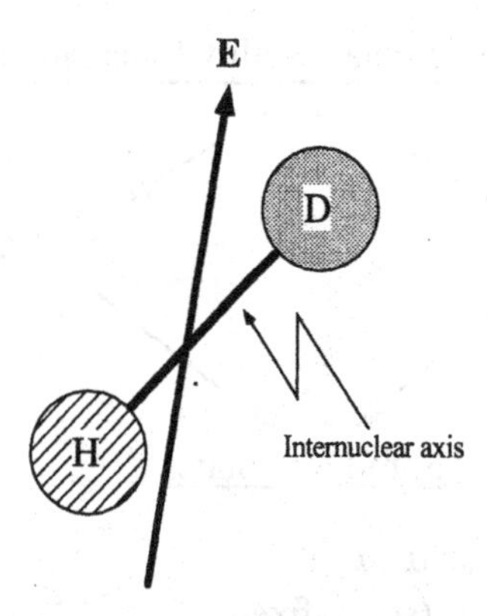

Figure 3.3: Here is a diagram showing a dc electric field applied to a heteronuclear diatomic molecule (HD). From the field's viewpoint a quantization axis aligned with it is the most "natural" selection. In this case, the polarization vector of the field is simply $\hat{\varepsilon} = \hat{z}$, whereas the polarizability is a nondiagonal matrix. From the viewpoint of the molecule, the most "natural" quantization axis is its internuclear axis. In this case, the polarizability is simply a diagonal matrix with two independent elements, whereas the polarization vector is a linear combination of $\hat{x}, \hat{y}$, and $\hat{z}$ with direction cosines as coefficients.

derivatives. We can use this Taylor's series to calculate the expectation value of the polarizability in the rovibrational state corresponding to vibrational quantum number ν and rotational quantum number J. For an anharmonic oscillator potential of the form $U(\xi) = f\xi^2 - g\xi^3$ we have [82]

$$\begin{aligned} \alpha_{\nu J} &= < \nu J|\alpha|\nu J > \\ &= \alpha_{eq} + (\nu + \frac{1}{2})(B_{eq}/\omega_{eq})(\alpha''_{eq} - 3a\alpha'_{eq}) + 4J(J+1)(B_{eq}/\omega_{eq})^2\alpha''_{eq}, \end{aligned} \quad (3.39)$$

where $a = g/f$ is a measure of the relative size of the cubic to the quadratic term in the potential $U(\xi)$. Here B_{eq} is the rotational energy constant in $W_{rot} = hcB_{eq}J(J+1)$ and ω_{eq} is the vibrational energyvibrational constant in $W_{vib} = hc\omega_{eq}(\nu + \frac{1}{2})$. To understand the size of these corrections, consider the results for hydrogen (H_2). Then [82] $B_{eq} = 60.8$ cm^{-1}, ω_{eq}= 4395 cm^{-1}, $a = -1.60$, $\alpha''_{eq}/\alpha_{eq} = -0.1$ gives the following results

$$\alpha_{eq} = 0.77\ \text{Å}^3, \alpha'_{eq} = 1.0\ \text{Å}^3, \alpha''_{eq} = -0.077\ \text{Å}^3. \quad (3.40)$$

The corrections for heavier diatomic molecules are probably smaller than for H_2 since $B_{eq} \propto 1/m$ and $\omega_{eq} \propto 1/\sqrt{m}$ so $B_{eq}/\omega_{eq} \propto 1/\sqrt{m}$.

Before discussing the relationship between polarizabilities described using molecule-fixed axes and those using space-fixed axes, we summarize the techniques for calculating α for diatomic molecules. A summary chart of the techniques is given on the next page. In all cases, the polarizability can be written as three separate contributions: electronic (α^e), vibrational (α^v), and rotational (α^r) [see also Section 4.9]. A brief description of the *static* techniques is given in a second chart that follows the first. It is important to note that the sum-over-states method (space-fixed axes) and

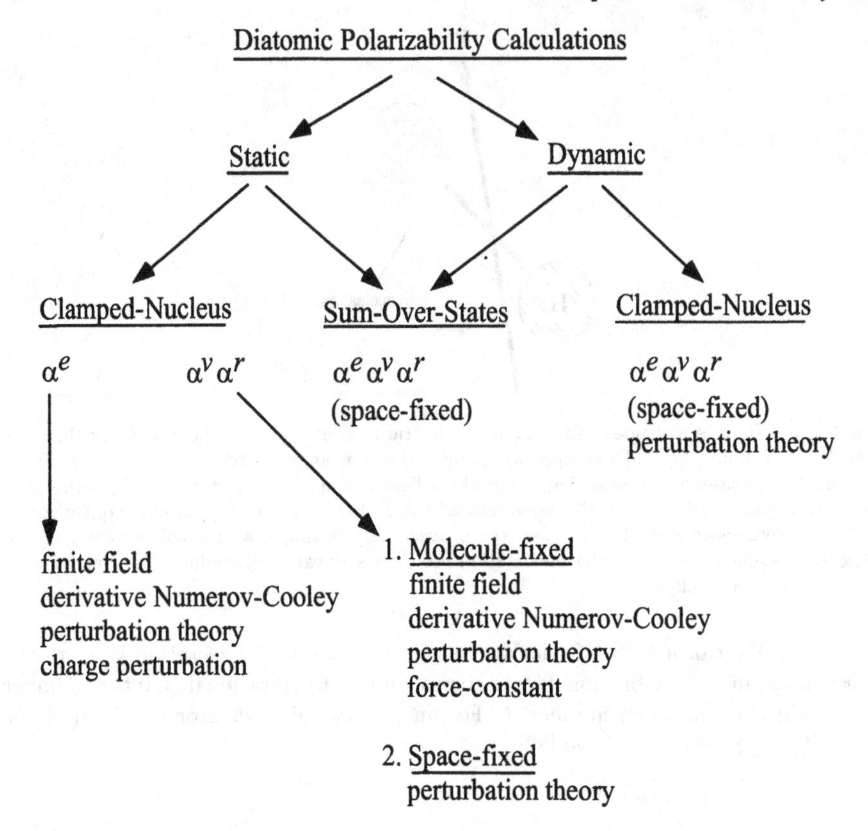

the clamped-nucleus method that utilizes space-fixed axes give identical results for the vibrational and rotational polarizabilities. They differ in their expressions for the *electronic* contribution to α. In the sum-over-states method, the denominators are differences in energies of rovibronic *states*, while in the clamped-nucleus perturbation theory method the electronic polarizability has denominators that are energy differences between electronic states at fixed internuclear separations R. Despite these differences, the two techniques yield results that are quantitatively very similar [75].

In the rest of this section we first give the classical relationship between α in space-fixed and molecule-fixed axes. Then we derive expressions for the dynamic polarizability components $\alpha^e, \alpha^v, \alpha^r$ using the sum-over-states technique.

Consider a molecule in the νth vibrational state of the ground electronic state of a molecule. Classically, for a molecule in thermal equilibrium at temperature T, the relation between the polarizabilities in the two frames is given by [83]

$$\bar{\alpha}_{ij}^{sp} = \frac{\sum_{lm} \int_0^{2\pi} \int_0^{\pi} \alpha_{lm}^{mol} (\boldsymbol{x}_i \cdot \boldsymbol{x}_l) \ (\boldsymbol{x}_j \cdot \boldsymbol{x}_m) e^{-W/kT} \sin\theta d\theta d\phi}{\int_0^{2\pi} \int_0^{\pi} e^{-W/kT} \sin\theta d\theta d\phi} \tag{3.41}$$

where θ and ϕ are the standard polar and azimuthal angles used in spherical coordi-

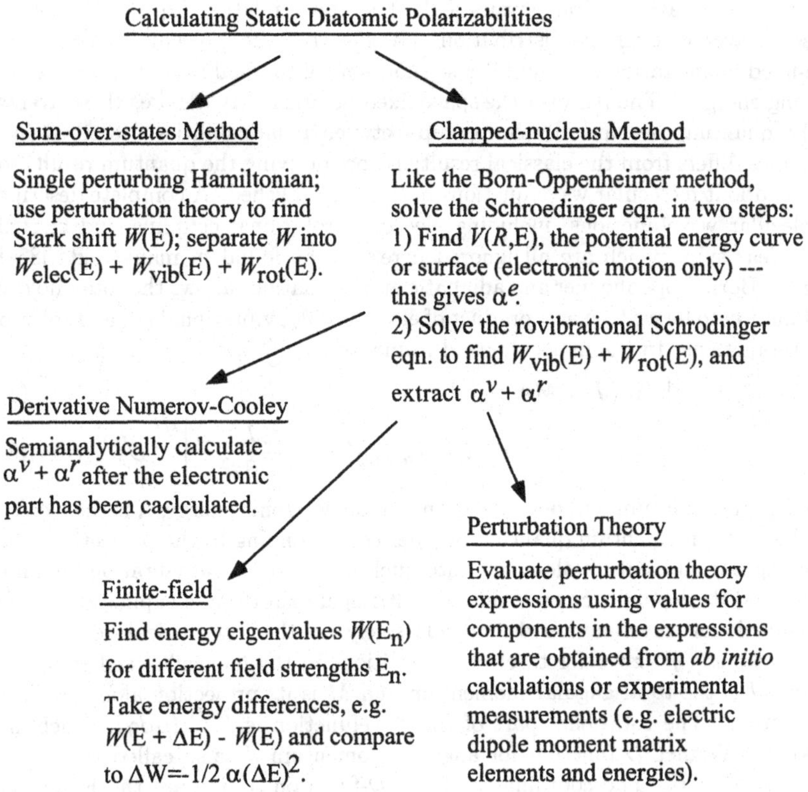

Calculating Static Diatomic Polarizabilities

Sum-over-states Method

Single perturbing Hamiltonian; use perturbation theory to find Stark shift W(E); separate W into W_{elec}(E) + W_{vib}(E) + W_{rot}(E).

Clamped-nucleus Method

Like the Born-Oppenheimer method, solve the Schroedinger eqn. in two steps:
1) Find $V(R,$E), the potential energy curve or surface (electronic motion only) --- this gives α^e.
2) Solve the rovibrational Schrodinger eqn. to find W_{vib}(E) + W_{rot}(E), and extract $\alpha^v + \alpha^r$

Derivative Numerov-Cooley

Semianalytically calculate $\alpha^v + \alpha^r$ after the electronic part has been caclculated.

Finite-field

Find energy eigenvalues $W(\mathrm{E_n})$ for different field strengths $\mathrm{E_n}$. Take energy differences, e.g. W(E + ΔE) - W(E) and compare to $\Delta W = -1/2\ \alpha(\Delta E)^2$.

Perturbation Theory

Evaluate perturbation theory expressions using values for components in the expressions that are obtained from *ab initio* calculations or experimental measurements (e.g. electric dipole moment matrix elements and energies).

nates that describe the orientation of the molecule with respect to the space-fixed axes, $W = W(\theta, \phi)$ is the energy of the molecule in the applied field, and $(\boldsymbol{x}_i \cdot \boldsymbol{x}_l), (\boldsymbol{x}_j \cdot \boldsymbol{x}_m)$ are the direction cosines of the angles between the respective axes (molecule and space). The direction cosines are simply the components of a rotation matrix R_{ij} that connects the two coordinate systems to each other. Any second rank Cartesian tensor, like the polarizability α_{ij}^{sp}, can be transformed from one coordinate system to another by the operation

$$\alpha_{ij}^{sp} = R_{il}\alpha_{lm}^{mol}R_{mj}^{-1}. \tag{3.42}$$

Details about transformations of vectors and tensors under rotations are covered in many texts on mathematical physics [84]. The above relation Eq. (3.41) represents a classical average over the distribution of angles that the molecules make with the space-fixed quantization axis, and it also represents a thermal average over the corresponding energies. The bar over the space-fixed polarizability denotes these averages.

The quantum-mechanical relationship between polarizabilities described in the two frames differs from the classical result. Before deriving the quantum result, some notation about molecular wavefunctions must be established. A complete description of molecular wavefunctions, including electronic-rotational coupling and electronic and nuclear spin, which are all ignored here, can be found in many texts [85–87]. Using the Born-Oppenheimer and adiabatic approximations allows the total molecular wavefunction to be written as a product of electronic ψ_e, vibrational ψ_v, and rotational ψ_r wavefunctions. In the molecule-fixed frame we have [87]

$$\psi_e\psi_v\psi_r = |n\Lambda\nu(J), JM\Lambda> \tag{3.43}$$

$$= \frac{1}{R}\Psi_{n\Lambda}(x^m, y^m, z^m; R)\ \Phi_{n\Lambda\nu(J)}(R)\ \left[\frac{(2J+1)}{8\pi^2}\right]^{1/2} D_{M,\Lambda}^{(J)}(\alpha\beta\gamma) \tag{3.44}$$

where n is the quantum number corresponding to electronic energy, Λ is the electronic orbital angular momentum quantum number corresponding to the projection of electronic angular momentum along the internuclear axis, ν is the vibrational quantum number and its dependence on J due to centrifugal distortion is explicitly displayed. The vibrational wavefunction Φ depends solely on the internuclear separation R. Coordinates x^m, y^m, z^m refer to electron positions in the molecule-fixed frame. The quantity J is the total angular momentum and M is its projection along the axis of quantization. The rotational part of the wavefunction is $D_{M,\Lambda}^{(J)}(\alpha\beta\gamma)$, which is the well-known Wigner D-function for angular momentum J (also called rotation matrices — see [67]). The coordinates of the D-function α, β, γ are the Euler angles associated with rotation from one coordinate system to another [67, 84, 87].

To derive a rigorous expression for the polarizability of a diatomic molecule, consider the energy shift in some level $|n\Lambda\nu(J), JM\Lambda> \equiv |n>$ due to the dipole induced by a linearly-polarized ac field at frequency ω [see Eq. (2.99) with $\hat{\varepsilon} = \hat{z}$]. The energy shift is [see Eq. (2.101)]

$$W_{n\Lambda\nu JM} = <n|\delta H_{shift}|n> = -\frac{|\mathcal{E}|^2}{4} <n|\hat{z}\cdot\boldsymbol{\alpha}\cdot\hat{z}|n> \tag{3.45}$$

$$= -\frac{|\mathcal{E}|^2}{4}(\hat{\alpha}_{zz})_{nn}. \tag{3.46}$$

The transition polarizability $(\hat{\alpha}_{zz})_{nn}$ in the last expression can be written using Eq. (2.45)

$$(\hat{\alpha}_{zz})_{nn} = \sum_{n'(\neq n)} \left[\frac{2\omega_{n'n}}{\hbar(\omega_{n'n}^2 - \omega^2)} \right] (\hat{\mathrm{p}}_z)_{nn'} \, (\hat{\mathrm{p}}_z)_{n'n}. \tag{3.47}$$

Note that the dipole matrix elements in Eq. (3.47) are evaluated in the space-fixed frame. To transform them so the expression contains dipole matrix elements in the molecule-fixed frame p^{mol}, we need to rotate the dipole operator from the space-fixed to the molecule-fixed frame. As already mentioned, this standard rotation procedure is discussed in many texts [67, 84, 87] and the relation between the components of the dipole-moment operator in the space-fixed frame ($\hat{\mathrm{p}}_q$) and the components in the molecule-fixed frame ($\hat{\mathrm{p}}_r^{mol}$) is

$$\hat{\mathrm{p}}_q = \sum_r (-1)^{r-q} D_{qr}^{(1)}(\alpha\beta\gamma) \; \mathrm{p}_r^{mol}. \tag{3.48}$$

Here, the indices q, r run over the values $1, 0, -1$ indicating vector components in spherical coordinates. Since the light is linearly-polarized along $\hat{z}$, $q = 0$ here. From Eq. (3.44) and Eq. (3.48) it can be seen that the dipole moment matrix element $(\hat{\mathrm{p}}_z)_{n'n} = (\hat{\mathrm{p}}_0)_{n'n}$ will involve integrals over D-functions. These integrals can be simplified using the result [67, 87]

$$\begin{aligned} &\tfrac{1}{8\pi^2} \textstyle\int_0^{2\pi} \int_0^{\pi} \int_0^{2\pi} D_{M'\Lambda'}^{(J')*}(\alpha\beta\gamma) D_{qr}^{(j)}(\alpha\beta\gamma) D_{M\Lambda}^{(J)}(\alpha\beta\gamma) d\alpha \sin\beta d\beta d\gamma \\ &= (-1)^{J'-M'} \begin{pmatrix} J' & j & J \\ -M' & q & M \end{pmatrix} (-1)^{J'-\Lambda'} \begin{pmatrix} J' & j & J \\ -\Lambda' & r & \Lambda \end{pmatrix}. \end{aligned} \tag{3.49}$$

Using Eq. (3.49) and the fact that here $j = 1, q = 0$, the dipole matrix element becomes

$$(\hat{\mathrm{p}}_z)_{n'n} = \; < n'\Lambda'\nu'(J'), J'M'\Lambda' | \hat{\mathrm{p}}_0 | n\Lambda\nu(J), JM\Lambda > \tag{3.50}$$

$$\begin{aligned} &= \delta_{M'M} \;\; < n'\Lambda'\nu'(J') | \mathrm{p}_{\Lambda'-\Lambda}^{mol} | n\Lambda\nu(J) > \;\; [(2J'+1)(2J+1)]^{1/2} \\ &\times \;(-1)^{J'-M'} \begin{pmatrix} J' & 1 & J \\ -M' & 0 & M \end{pmatrix} (-1)^{J'-\Lambda'} \begin{pmatrix} J' & 1 & J \\ -\Lambda' & \Lambda'-\Lambda & \Lambda \end{pmatrix}. \end{aligned} \tag{3.51}$$

The remaining matrix element $< n'\Lambda'\nu'(J') | \hat{\mathrm{p}}_{\Lambda'-\Lambda}^{mol} | n\Lambda\nu(J) >$ involves integration over the electronic coordinates x^m, y^m, z^m relative to the molecule-fixed axes and an integration over the internuclear coordinate R. The final result for the energy shift is

$$\begin{aligned} W_{n\Lambda\nu JM} &= -\frac{|\mathcal{E}|^2}{4} \sum_{n'} \sum_{\Lambda'} \sum_{\nu'(J')} \sum_{J'} \left[\frac{2\omega_{n'n}}{\hbar(\omega_{n'n}^2 - \omega^2)} \right] | < n'\Lambda'\nu'(J') | \hat{\mathrm{p}}_{\Lambda'-\Lambda}^{mol} | n\Lambda\nu(J) > |^2 \\ &\times \;(2J'+1)(2J+1) \begin{pmatrix} J' & 1 & J \\ -M & 0 & M \end{pmatrix}^2 \begin{pmatrix} J' & 1 & J \\ -\Lambda' & \Lambda'-\Lambda & \Lambda \end{pmatrix}^2. \end{aligned} \tag{3.52}$$

Note that the frequencies here correspond to the unperturbed energies, i.e.

$$\omega_{n'n} = \frac{W^0_{n'\Lambda'\nu'J'} - W^0_{n\Lambda\nu J}}{\hbar}, \tag{3.53}$$

where

$$W^0_{n\Lambda\nu J} =< n\Lambda\nu(J)J|H^0|n\Lambda\nu(J)J >, \tag{3.54}$$

and H_0 is the field-free (unperturbed) Hamiltonian.

Insight into the physical significance of the terms in Eq. (3.53) can be gained by separating the sums out into four parts:

(a) the terms diagonal in $n\Lambda\nu$, corresponding to the rotational polarizability α^r,

(b) the terms diagonal in $n\Lambda$ and off-diagonal in ν, corresponding to the vibrational polarizability α^v,

(c) the terms diagonal in Λ and off-diagonal in n, corresponding to the electronic polarizability parallel to the internuclear axis $\alpha^e_{||}$,

(c) the terms off-diagonal in Λ, corresponding to the electronic polarizability perpendicular to the internuclear axis $\alpha^e_{\perp}$,

Rewriting Eq. (3.53) to explicitly display these four separate contributions in the order listed above gives

$$\begin{aligned}
W_{n\Lambda\nu JM} &= -\frac{|\mathcal{E}|^2}{4}\sum_{J'}(2J'+1)(2J+1)\begin{pmatrix} J' & J & 1 \\ M & -M & 0 \end{pmatrix}^2 \times \\
&\left[\begin{pmatrix} J' & J & 1 \\ -\Lambda & \Lambda & 0 \end{pmatrix}^2 \left\{ |< n\Lambda\nu(J)|\hat{\mathrm{p}}_0^{mol}|n\Lambda\nu(J') >|^2 \frac{2\omega_{J'J}}{\hbar(\omega^2_{J'J} - \omega^2)} \right.\right. \\
&+ \sum_{\nu'(\neq\nu)} |< n\Lambda\nu(J)|\hat{\mathrm{p}}_0^{mol}|n\Lambda\nu'(J') >|^2 \frac{2\omega_{\nu'J',\nu J}}{\hbar(\omega^2_{\nu'J',\nu J} - \omega^2)} \\
&\left. + \sum_{n'(\neq n)}\sum_{\nu'} |< n\Lambda\nu(J)|\hat{\mathrm{p}}_0^{mol}|n'\Lambda\nu'(J') >|^2 \frac{2\omega_{n'\nu'J',n\nu J}}{\hbar(\omega^2_{n'\nu'J',n\nu J} - \omega^2)} \right\} \\
&+ \sum_{\Lambda'(\neq\Lambda)} \begin{pmatrix} J' & J & 1 \\ -\Lambda' & \Lambda & \Lambda'-\Lambda \end{pmatrix}^2 \sum_{n'}\sum_{\nu'} |< n\Lambda\nu(J)|\hat{\mathrm{p}}^{mol}_{\Lambda'-\Lambda}|n'\Lambda'\nu'(J') >|^2 \\
&\left. \times \frac{2\omega_{n'\Lambda'\nu'J',n\Lambda\nu J}}{\hbar(\omega^2_{n'\Lambda'\nu'J',n\Lambda\nu J} - \omega^2)} \right].
\end{aligned} \tag{3.55}$$

In the last term of Eq. (3.55), with terms off-diagonal in Λ, the dipole operator only has two components that occur: $\mathrm{p}^{mol}_{\pm 1}$. These two components are both perpendicular to the internuclear axis and so it is easy to see that this last term will correspond to $\alpha^e_{\perp}$. Note that the rotational and vibrational contributions to the polarizability are both parallel to the internuclear axis, i.e. only dipole moment operators of the form p^{mol}_0 occur.

A special case of Eq. (3.55) that is of frequent interest is that for a dc field (ω=0) applied to a diatomic molecule in a $^1\Sigma$ ground state ($\Lambda = 0$). This case was first worked out in detail by Breiger for LiH [74]. Briefly discussing this special case will further elucidate the physical contributions of the different motions: rotational, vibrational, and electronic. Here Eq. (3.55) simplifies to

$$\begin{aligned} W_{n^1\Sigma\nu JM} &= -\frac{|\mathrm{E}|^2}{2}\sum_{J'}(2J'+1)(2J+1)\begin{pmatrix} J' & J & 1 \\ M & -M & 0 \end{pmatrix}^2 \\ &\times \left[\begin{pmatrix} J' & J & 1 \\ 0 & 0 & 0 \end{pmatrix}^2 \left\{\alpha^r_{||}(J,J') + \alpha^v_{||}(J,J') + \alpha^e_{||}(J,J')\right\}\right. \\ &+ \left.\begin{pmatrix} J' & J & 1 \\ -1 & 0 & 1 \end{pmatrix}^2 2\alpha^e_{\perp}(J,J')\right], \end{aligned} \tag{3.56}$$

where

$$\begin{aligned} \alpha^r_{||}(J,J') &= \frac{2|<n^1\Sigma\nu(J)|\hat{\mathrm{p}}_0^{mol}|n^1\Sigma\nu(J')>|^2}{W^0_{J'} - W^0_J}, \\ \alpha^v_{||}(J,J') &= \sum_{\nu'(\neq\nu)}\frac{2|<n^1\Sigma\nu(J)|\hat{\mathrm{p}}_0^{mol}|n^1\Sigma\nu'(J')>|^2}{W^0_{\nu'J'} - W^0_{\nu J}}, \\ \alpha^e_{||}(J,J') &= \sum_{n'(\neq n)}\sum_{\nu'}\frac{2|<n^1\Sigma\nu(J)|\hat{\mathrm{p}}_0^{mol}|n'\,^1\Sigma\nu'(J')>|^2}{W^0_{n'\nu'J'} - W^0_{n\nu J}}, \\ \alpha^e_{\perp}(J,J') &= \sum_{n'}\sum_{\nu'}\frac{|<n^1\Sigma\nu(J)|\hat{\mathrm{p}}_{-1}^{mol}|n'\,^1\Pi\nu'(J')>|^2}{W^0_{n'\,^1\Pi\nu'J'} - W^0_{n^1\Sigma\nu J}}. \end{aligned} \tag{3.57}$$

In Eq. (3.56) and Eq. (3.57) the polarizability contributions due to the rotational, vibrational, and electronic motion have become quite clear. To compare the expressions here for diatomic molecules with those found earlier for atoms, we will concentrate on the typically larger electronic contributions to the polarizability (and energy shift). Rewriting the electronic contribution to the energy, Eq. (3.56), using the irreducible representations of the polarizabilities gives [74]

$$W_{n^1\Sigma\nu JM} = -\frac{|\mathrm{E}|^2}{2}\left(\alpha^{sc}(J) - \alpha^t(J)\frac{3M^2 - J(J+1)}{(2J-1)(2J+3)}\right) + \text{smaller terms.} \tag{3.58}$$

Here the polarizability, corresponding to the terms in large parentheses, depends on a scalar part α^{sc} and a tensor part α^t, and is almost identical to that for atoms [see Eq. (3.19) in Sec. 3.2.1]. The relationship between the irreducible representation components of the polarizability and the molecular-frame polarizabilities is

$$\begin{aligned} \alpha^{sc}(J) &= \frac{1}{3}[\alpha_{||}(J) + \alpha_{\perp}(J)] \\ \alpha^t(J) &= \frac{2}{3}[\alpha_{||}(J) - \alpha_{\perp}(J)], \end{aligned} \tag{3.59}$$

where

$$\begin{aligned}\alpha_{||}(J) &= \frac{1}{2}[\alpha_{||}(J,J-1)+\alpha_{||}(J,J+1)] \\ \alpha_{\perp}(J) &= \frac{1}{3}[\alpha_{\perp}(J,J-1)+\alpha_{\perp}(J,J)+\alpha_{\perp}(J,J+1)]. \end{aligned} \tag{3.60}$$

The smaller terms in Eq. (3.58) consist of expressions depending on molecular polarizability differences such as $\alpha_{||}(J,J-1)-\alpha_{||}(J,J+1)$, which are much smaller than the terms that have been retained. The reason for this is that the electronic polarizabilities do not vary much with J since the transition moments are only weakly dependent on J and R. For more details, such as explicit expressions for the smaller terms, consult the work by Breiger [74].

The derivation above for the polarizability of a diatomic molecule is called the sum-over-states technique [75]. There are other approximations and resulting simplifications that one can make to the expressions derived here for the polarizabilities. An extensive discussion of these is presented in the review article by Bishop [75]. In addition, there are other techniques for calculating polarizabilities that will be discussed in Chapter 4.

We will finish this section by discussing rovibrational and thermal averaging of the quantum results so that we will have given both classical and quantum expressions for such averaging. Due to the weak dependence of the electronic part of the polarizability on the rovibrational quantum numbers, we will only retain the *diagonal* parts in our expressions. This approximation was assumed in the early expressions of van Vleck [88] and in those reported by Bishop [75]. The results here are useful in that experimentally measured polarizabilities are commonly isotropic and thermally averaged. The theoretical expression for the thermally-averaged isotropic polarizability $\bar{\alpha}(\omega)$ is the sum of three parts

$$\bar{\alpha}(\omega)=\bar{\alpha}^e+\bar{\alpha}^v+\bar{\alpha}^r. \tag{3.61}$$

For a molecule in a $^1\Sigma$ ground state, the electronic component is

$$\bar{\alpha}^e=\frac{1}{3}\sum_J \rho(\nu,J) < \nu(J)|\alpha_{||}+2\alpha_{\perp}|\nu(J) >, \tag{3.62}$$

where

$$\begin{aligned}\alpha_{||} &= \sum_{n'(\neq n)} \frac{2\omega_{n'n}}{\hbar(\omega_{n'n}^2-\omega^2)}|< n^1\Sigma|\hat{p}_0^{mol}|n'\,^1\Sigma >|^2, \\ \alpha_{\perp} &= \sum_{n'(\neq n)} \frac{2\omega_{n'n}}{\hbar(\omega_{n'n}^2-\omega^2)}|< n^1\Sigma|\hat{p}_{-1}^{mol}|n'\,^1\Pi >|^2. \end{aligned} \tag{3.63}$$

The vibrational part is

$$\bar{\alpha}^v=\frac{1}{3}\sum_J \rho(\nu,J)\frac{[(J+1)X+JY]}{(2J+1)}, \tag{3.64}$$

where

$$X = \sum_{\nu'(\neq\nu)} \frac{2\omega_{\nu' J+1,\nu J}}{\hbar(\omega^2_{\nu' J+1,\nu J} - \omega^2)} |< n^1\Sigma\nu(J)|p_0^{mol}|n^1\Sigma\nu'(J+1) >|^2,$$

$$Y = \sum_{\nu'(\neq\nu)} \frac{2\omega_{\nu' J-1,\nu J}}{\hbar(\omega^2_{\nu' J-1,\nu J} - \omega^2)} |< n^1\Sigma\nu(J)|p_0^{mol}|n^1\Sigma\nu'(J-1) >|^2. \quad (3.65)$$

Finally the rotational part is

$$\bar{\alpha}^r = \frac{1}{3}\sum_J \rho(\nu, J)\frac{[(J+1)R + JS]}{(2J+1)}, \quad (3.66)$$

where

$$R = \frac{2\omega_{\nu J+1,\nu J}}{\hbar(\omega^2_{\nu J+1,\nu J} - \omega^2)} |< n^1\Sigma\nu(J)|p_0^{mol}|n^1\Sigma\nu(J+1) >|^2,$$

$$S = \frac{2\omega_{\nu J-1,\nu J}}{\hbar(\omega^2_{\nu J-1,\nu J} - \omega^2)} |< n^1\Sigma\nu(J)|p_0^{mol}|n^1\Sigma\nu(J-1) >|^2. \quad (3.67)$$

The density matrix ρ is given by

$$\rho(\nu, J) = \frac{(2J+1)g_J e^{-(W_{\nu J} - W_{\nu 0})/kT}}{\sum_J (2J+1)g_J e^{-(W_{\nu J} - W_{\nu 0})/kT}}, \quad (3.68)$$

where g_J is the nuclear-spin degeneracy, which is only present for homonuclear diatomics. The quantity T is the temperature and k is Boltzmann's constant. One has to be careful when applying these results to a molecular beam, where the different degrees of freedom have different temperatures, e.g. $T_{\text{electr}} \neq T_{\text{vib}} \neq T_{\text{rot}}$.

3.3.3 Polyatomic molecules

Most species whose polarizabilities have been measured are polyatomic molecules. For example, of the 423 species whose polarizabilities are listed in the 1992 CRC Handbook of Chemistry and Physics: 102 are atoms, 30 are diatomic molecules, and the other 291 are polyatomic molecules. For polyatomics, this number does not include all the possible isomers of a given chemical compound! Despite their large numbers, no systematic reviews of polarizabilities of polyatomic molecules have been published. However, the reviews of Buckingham on Stark effects [27] and intermolecular forces [28], and the nice review of molecular and atomic polarizabilities by Bogaard and Orr [29] all contain some information on polyatomic polarizabilities.

Many polyatomic molecules are well-approximated by the simple models of a linear rotor, a symmetric rotor, or a spherical rotor. The more complex asymmetric rotor will be discussed shortly. Figure (3.4) shows geometric sketches of each case and provides specific examples of molecules that are members of each category. Molecular rotors are classified according to their principal moments of inertia (see Herzberg [89] for a discussion of principal moments). Defining the origin of coordinates to be the

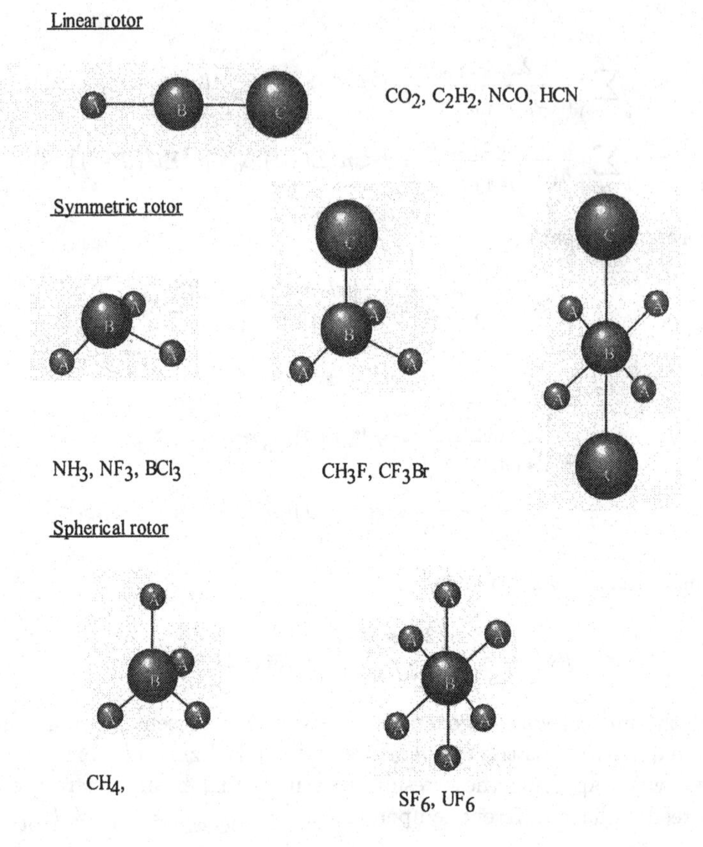

Figure 3.4: Sketches of generic molecules that exhibit the structures of a linear rotor, a symmetric rotor and a spherical rotor. Samples of molecules with these structures are also given.

center of mass of the molecule, the moment of inertia about an axis through the center of mass is $I = \sum_i m_i r_i^2$, where m_i is the mass of the ith particle in the molecule and r_i is the distance of this mass from the axis of rotation. In terms of the principal moments of inertia, the rotors are

1. Linear rotor: one principal moment of inertia, $I_{\|} = 0, I = I_{\perp}$,
2. Spherical rotor: three equal principal moments, $I_1 = I_2 = I_3$,
3. Symmetric rotor: two principal moments, $I_{\perp} = I_{\|}$,
4. Asymmetric rotor: three different principal moments, $I_1 = I_2 = I_3$.

Wavefunctions for the linear and spherical rotors can be written as special cases of the wavefunction for the symmetric rotor,

$$\Psi_n = \psi_n \Theta_{JKM}(\theta) e^{iK\chi} e^{iM\phi}. \tag{3.69}$$

The quantities θ, ϕ, and χ are Euler angles that relate the rotation of molecule-fixed axes to space-fixed axes. The quantity $J = 0, 1, 2, ...$ is the total angular momentum of the molecule, $K = -J, -J+1, ..., J-1, J$ (except for linear rotors, where $K = 0$) is the projection of the angular momentum on the symmetry axis of the molecule, and $M = 0, \pm 1, ..., \pm J$ is the projection of angular momentum along the quantization axis (a space-fixed axis). As in the case of diatomics, the electronic part ψ_n depends weakly on J and K and we ignore the electronic and rovibronic coupling here. Details of the energy, structure, and spectra of these rotors can be found in many texts on molecules [86, 89].

Here we are interested in the behavior of such rotors when placed in an electric field. Specifically, we are interested in the second order response of the energy W to the electric field (E) since this involves the polarizability. The corresponding interaction Hamiltonian for a fixed symmetric rotor in state ψ_n in an electric field E is

$$H_{shift} = -\frac{1}{2}\alpha_n \mathrm{E}^2 - \frac{1}{3}\Delta\alpha_n(\frac{3}{2}\cos^2\theta - \frac{1}{2})\mathrm{E}^2, \tag{3.70}$$

where

$$\begin{aligned} \alpha_n &= \frac{1}{3}(\alpha_{\parallel} + 2\alpha_{\perp}) \\ \Delta\alpha_n &= \alpha_{\parallel} - \alpha_{\perp}. \end{aligned} \tag{3.71}$$

Here $\alpha_{\parallel}$ and $\alpha_{\perp}$ are the polarizabilities parallel to and perpendicular to the symmetry axis of the molecule. The angle θ corresponds to the angle between the axis of symmetry and the direction of the applied electric field.

The shift in energy of a symmetric rotor molecule in the specific state Ψ_n [see Eq. (3.69)] is

$$W_{nJKM} = -\frac{1}{2}\alpha_n \mathrm{E}^2 - \frac{1}{3}\Delta\alpha_n \mathrm{E}^2 \frac{(J^2 + J - 3K^2)}{J(J+1)} \frac{(J^2 + J - 3M^2)}{(2J-1)(2J+3)}, \tag{3.72}$$

where $J \geq 1$ and $\alpha_n, \Delta\alpha_n$ are defined above.

In a static field, symmetric rotor molecules have, at most, three independent components of the transition polarizability tensor $(\alpha_{ij})_{mk}$. Symmetric rotor molecules have C_3 or higher symmetry and the transition polarizability can be written

$$(\alpha_{ij})_{mk} = (\alpha_{ij})_{mk}\delta_{ij} + [(\alpha_{\perp ij})_{mk} - (\alpha_{\parallel ij})_{mk}]l_i l_j + \frac{1}{2}[(\alpha_{xy})_{mk} - (\alpha_{yx})_{mk}]\varepsilon_{ijh}l_h, \tag{3.73}$$

where $l_i = \cos(\hat{x}_i \cdot \hat{z}')$ is the direction cosine between the $\hat{x}_i$ axis in the space frame and the symmetry axis $(\hat{z}')$ of the molecule. This expression is particularly useful for those interested in electric-field induced molecular lines. For more details see the review of the Stark effect, including Stark spectroscopy, by Buckingham [27].

Finally, the most general case that has been analytically treated is that of the *asymmetric* rotor. Simple molecules that come under this category are H_2O, SO_2, and CH_2D_2. The theory of the Stark effect for this case, including second-order effects involving the polarizability, was worked out by Golden and Wilson [90]. In a static field, the energy shift due to the polarizability of the molecule in state $\Psi_{J\tau M}$ is

$$W^{ii}_{J\tau M} = \frac{1}{2}\alpha_{ii}\mathrm{E}^2[C_{J\tau}(\kappa) + M^2 D_{J\tau}(\kappa)], \tag{3.74}$$

where α_{ii} is the polarizability along the ith principal axis of the molecule, M is the magnetic quantum number, τ is the asymmetric-rotor quantum number [91], and $-J \leq \tau \leq J$. Here κ is Ray's asymmetry parameter [92]

$$\kappa = \frac{2B - A - C}{A - C}, \tag{3.75}$$

where $A, B,$ and C are inversely related to the three principal moments of inertia. For example, the energy of an unperturbed asymmetric rotor can be written

$$\begin{aligned} H_0 &= \frac{\hat{J}_x^2}{2I_x} + \frac{\hat{J}_y^2}{2I_y} + \frac{\hat{J}_z^2}{2I_z} \\ &= \frac{2\pi A}{\hbar}\hat{J}_x^2 + \frac{2\pi B}{\hbar}\hat{J}_y^2 + \frac{2\pi C}{\hbar}\hat{J}_z^2. \end{aligned} \tag{3.76}$$

Thus $A = \hbar/4\pi I_x$, $B = \hbar/4\pi I_y$ $C = \hbar/4\pi I_z$. For a prolate (cigar-shaped) symmetric rotor, $\kappa = -1$ $(B = C)$, and for an oblate (disk-shaped) symmetric rotor, $\kappa = 1$ $(A = B)$. Although the quantities $C_{J\tau}(\kappa)$ and $D_{J\tau}(\kappa)$ in Eq. (3.74) are not conveniently tabulated in the literature, they can be readily calculated from published data in the following way [90]

$$\begin{aligned} C_{J\tau}(\kappa) &= \left[\frac{J}{(4J^2-1)}\sum_{\tau'} S_{J\tau,J-1\tau'}(\kappa) + \frac{(J+1)}{(2J+1)(2J+3)}\sum_{\tau'} S_{J\tau,J+1\tau'}(\kappa)\right] \\ D_{J\tau}(\kappa) &= \left[\frac{-1}{J(4J^2-1)}\sum_{\tau'} S_{J\tau,J-1\tau'}(\kappa) + \frac{1}{J(J+1)(2J+1)}\sum_{\tau'} S_{J\tau,J\tau'}(\kappa)\right. \\ &\quad \left. - \frac{1}{(J+1)(2J+1)(2J+3)}\sum_{\tau'} S_{J\tau,J+1\tau'}(\kappa)\right]. \end{aligned} \tag{3.77}$$

The *line strength* $S_{J\tau,J'\tau'}(\kappa)$, corresponding to the transition $J\tau \rightarrow J'\tau'$, is discussed in Golden and Wilson [90], Cross, Hanier, and King [93], and in the classic book by Townes and Schawlow [91]. Line strengths for the asymmetric rotor are conveniently tabulated by Cross, Hanier, and King [93]. Their table is complete for angular momenta $J < 13$. These tables are also reproduced in Appendix V of Townes and Schawlow [91].

Recently, significant success has been achieved in measuring and calculating intensity spectra of rotational and vibration-rotation Raman transitions for asymmetric tops. Progress has been slow due to the difficulty of the measurements and due to the complexity of the calculations [94]. Raman transitions depend directly on reduced matrix elements of the polarizability operator. The relevant reduced matrix elements can be readily computed using expressions derived by Murphy [94]. The following discussion is based on this source. Here the polarizability is most conveniently represented as a spherical tensor (in spherical coordinates) with components α_Q^L, where $L = 0, 1, 2$ is the spherical tensor rank and, for a tensor of rank L, Q is in the range $-L \leq Q \leq L$. To find an expression for the reduced-matrix elements of the polarizability operator $< \nu J\tau||\alpha^L||\nu' J'\tau' >$ that can be evaluated using standard numerical techniques, we need to first find acceptable expressions for the eigenstates of the asymmetric rotor. These are well-known [91, 95, 96] and are given by

$$\Psi(J, \tau, M) = \sum_{K\gamma} a_{J\tau K\gamma} S(J, K, M, \gamma), \tag{3.78}$$

where the Wang functions $S(J, K, M, \gamma)$ are expressed as linear combinations of pairs of symmetric rotor eigenfunctions ψ_{JKM}

$$S(J, K, M, \gamma) = \frac{1}{\sqrt{2}} \left[\psi_{JKM} + (-1)^\gamma \psi_{J-KM}\right], \quad K > 0 \tag{3.79}$$

and

$$S(J, 0, M, \gamma) = \psi_{J0M}, \quad K = 0. \tag{3.80}$$

Here γ corresponds to the parity so $\gamma = 0$ yields an even state and $\gamma = 1$ yields an odd state. Symmetric-top wavefunctions ψ_{JKM} have already been discussed [see Sec. (3.3.2)]. The linear coefficients $a_{J\tau K\gamma}$ in Eq. (3.78) represent the weight of the different Wang functions to the total eigenstate of the asymmetric rotor. Using these results, the reduced-matrix element $RME \equiv \; < \nu J\tau||\alpha^L||\nu' J'\tau' >$ can be expressed as [94]

$$\begin{aligned}
RME &= (-1)^J[(2J+1)(2J'+1)]^{1/2} \\
&\times \sum_{\gamma,\gamma'} \Bigg\{ \sum_{K=0}^{J} \sum_{K'=0}^{J'} \frac{(-1)^{K'}}{2} \alpha_0^L [1 + (-1)^{J+\gamma+J'+\gamma'+L}] a_{J\tau K\gamma} a'_{J'\tau' K'\gamma'} \begin{pmatrix} J & L & J' \\ -K & 0 & K' \end{pmatrix} \\
&+ \sum_{K=0}^{J} \sum_{K'=0}^{J'} \sum_{Q=1}^{L} [\alpha_Q^L + (-1)^{J+\gamma+J'+\gamma'+L} \alpha_{-Q}^L] \\
&\times \Bigg[\frac{(-1)^{\gamma'+K'}}{\sqrt{2}} a_{J\tau 00} a'_{J'\tau' K'\gamma'} \begin{pmatrix} J & L & J' \\ 0 & Q & -K' \end{pmatrix} + \frac{1}{\sqrt{2}} a_{J\tau K\gamma} a'_{J'\tau' 00} \begin{pmatrix} J & L & J' \\ -K & Q & 0 \end{pmatrix} \\
&+ \frac{(-1)^{K'}}{2} a_{J\tau K\gamma} a'_{J'\tau' K'\gamma'} \Bigg[\Bigg(\begin{pmatrix} J & L & J' \\ -K & Q & K' \end{pmatrix} \\
&+ (-1)^{\gamma'} \begin{pmatrix} J & L & J' \\ -K & Q & -K' \end{pmatrix} + (-1)^{\gamma+\gamma'} \begin{pmatrix} J & L & J' \\ K & Q & -K' \end{pmatrix} \Bigg) \Bigg] \Bigg] \Bigg\}.
\end{aligned} \tag{3.81}$$

Reduced matrix elements calculated using Eq. (3.81) have been reported in [94] and [97] and they give excellent agreement between theoretical spectral curves and experimental data for rovibrational Raman transitions.

3.4 Organic and Biological Molecules

Electrical forces drive the biochemical processes that sustain life. Beyond this basic fact about the nature of most biological and chemical phenomena that govern life, there are specific reasons to study the polarizability (an important electrical property) of organic and other biologically-significant molecules. In particular,

1. In the last decade, there has been a great interest in the nonlinear optical properties of π-conjugated polymeric molecules due to their large, nonresonant third-order susceptibility. This has spurred research, both experimental [98–100] and theoretical [101], into the linear and nonlinear optical properties of these systems. Linear and nonlinear properties of a molecule are usually related: large linear polarizabilities are frequently accompanied by very large hyperpolarizabilities [101].

2. Polarizabilities are sensitive to the conformational structure of chain-polymer systems. Hence, polarizability studies can increase our understanding of the structure of these molecules. In addition, conformational energies of DNA and protein molecules depend on dispersive interactions and fragment-fragment polarization. The size of dispersive interactions can be found if the polarizability is known.

3. It has been observed [102] that strong fields are present across membranes of biological systems that result in large polarizations and these polarizations are important in nerve conduction. Cross-membrane potentials of up to 100 mV have been measured which, for a membrane of wall thickness 100 Å, creates a field of 10^7 V/m!

4. Protein molecules generate electric fields which may be significant in understanding their properties [103]. Also, electrostatic interactions play an important role in determining protein structure [104].

Experimentally, the polarizability of organic molecules or biomolecules is usually determined from light scattering (either refractive index measurements or absorption, but not off-axis scattering). Refractive indices are measured using (i) a thin film containing the molecules or (ii) a liquid solvent containing the molecules. In addition, use is sometimes made of the absorption spectrum and the Kramers-Krönig relation (see Sec. 2.5). These bulk methods are described in the chapter on experimental techniques — see Ch. 5. Bulk techniques rely on a phenomenological model to go from the bulk quantities (extinction coefficient, refractive index) to the microscopic quantity (the polarizability). One potential problem with such techniques is that the presence of the solvent or substrate molecules changes, through intermolecular forces, the polarizability of the biomolecules from their "free" values. Many investigators use the linear additivity approximation, in which the solvent or substrate molecules and biomolecules are assumed to be noninteracting, and the solute concentration is small so the solute-solute interactions are negligible. However, Coulombic interactions of nearest-neighbor molecules are incorporated into the Lorentz-Lorenz relation for the dielectric permitivity or refractive index, in an attempt to account for the bulk polarization of condensed material by treating nearest neighbors as randomly-oriented electric dipoles (see [56]). The Lorentz-Lorenz relations are discussed in the section on refractive index measurements of polarizabilities in Ch. 5. A classical treatment of the absorption and refractive index of molecular aggregates in terms of the microscopic polarizability that accounts for the dipole-dipole Coulomb interactions of solute-solvent molecules is given by DeVoe [105, 106]. This classical, coupled-oscillator approach can be applied to aggregates of monomer units of varying sizes, e.g. polymers, molecular crystals, and molecular aggregates. It was subsequently shown [107] that the classical theory of DeVoe is equivalent to time-dependent Hartree theory ("within approximations common to all perturbation treatments of polymer optical properties" [108]).

Several theoretical approaches are used to calculate polarizabilities for large organic and biological molecules. The most common techniques include the sum-over-states method based on perturbation theory, coupled-perturbed Hartree-Fock, and finite-field methods. In these cases, Gaussian basis sets are generally used. However, in some cases [101], the medium-size polarized basis set of Sadlej [109, 110] is used since it was purposefully designed to provide maximum accuracy in the calculation of electric dipole moments and polarizabilities. Frequently, excellent values for α of large molecules can be obtained by inverting accurate experimental data on a large number of smaller species to obtain atomic polarizabilities (bond, group, atomic-hybrid polarizabilities). There is an extensive and useful literature on these methods [73, 111–116], which are called additivity methods (see Sec. 4.8). Additivity methods can also be used to obtain tensor polarizabilities [73]. However, application of additivity methods to aromatic species is not as accurate, and although more subtle application of experimental data can be used to obtain average polarizabilities of aromatic species, the use of spherical atomic components limits their accuracy in calculating anisotropies [73, 117].

Since structure is so deeply related to function for organic molecules and biomolecules, it is worthwhile to discuss how the polarizability and structure of these molecules are closely linked. We will describe two specific examples: oligothiophenes and DNA.

The relatively large values for optical properties of large π-conjugated polymers (like oligothiophenes) have made them popular candidates for both theoretical and experimental studies [118, 119]. Oligothiophenes consist of polythiophene chains like those shown in Fig. 3.5. A *conjugated* molecule is one having alternating single and double bonds, like benzene. A π-conjugated polymer is simply a molecule with repeating unit cells that are π-conjugated [120]. We can explain π-conjugated bonds by using the example of benzene. The structure of benzene is conveniently described using plane, triangular sp^2 molecular orbitals and a perpendicular $2p$ orbital for carbon – see Fig. 3.6. For each C atom, the sp^2 hybrid orbitals form σ bonds with two other carbons and one hydrogen – see Fig. 3.7. Additional π-bonds form between the $2p$ orbitals of the six carbon atoms that form the benzene ring. However, it is incorrect to think of the 3 π-bonds being formed so each carbon atom participates in a single double-bond with one of its neighboring carbon atoms, forming 3 pairs of distinct double bonds and three distinct single bonds. Instead, the electrons in the perpendicular p-orbitals of the carbon atom are *delocalized*, sharing in bonds around the ring. This delocalization of π-conjugated systems, such as the oligothiophenes, helps us to understand their observed polarizabilities.

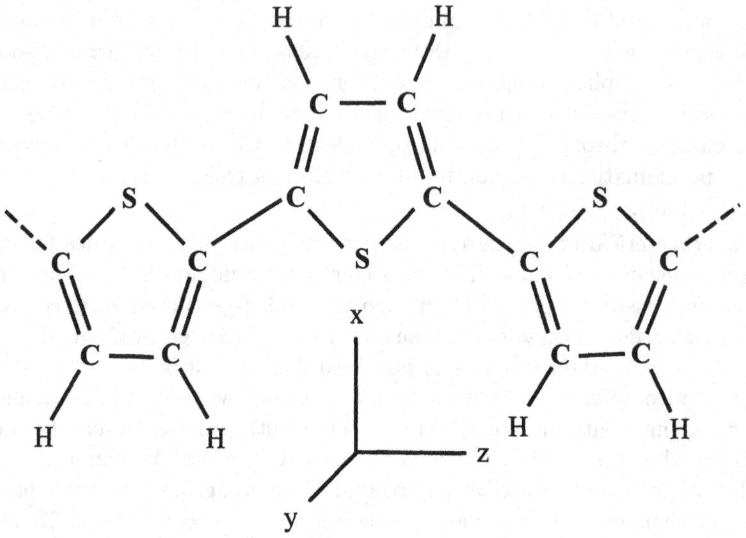

Figure 3.5: A schematic of the chemical structure of oligothiophenes showing several polythiophene chains.

Here we will review experimental and theoretical work on oligamers (small polymers with 2-10 repeated — or 'mer' — units). Consider a polymer of polythiophene chains as shown in Fig. 3.5, and assume the structure is planar (there is some evidence to support this assumption [121–123]). The polarizability is expected to be smallest along y, slightly larger along x, and large along the z-axis due to electron delocalization along the polythiophene backbone. The transversal (α_{xx}), perpendicular (α_{yy}), and longitudinal (α_{zz}) polarizabilities have been computed using uncoupled and coupled Hartree-Fock levels with a Gaussian basis set [101]. These calculations can be checked against the results of a similar calculation on the thiophene monomer using a better basis set for finding polarizabilities — the medium-size, polarized basis set of Sadlej [109, 110]. The calculated results for the average polarizability $\overline{\alpha}$ of the thiophene monomer agree well with experiment [101], and so we infer that the calculations have some merit in calculating the α tensor in the polymer regime.

These results are fascinating and they illustrate the deep relation between the polarizability and structure. The results of the coupled Hartree-Fock calculations on $\bar{\alpha}$ and α_{zz} for polythiophene chains with N unit cells [101] are reproduced in Table (3.2). As the chain length grows (N increases), the polarizability per chain length (α/N) is flat for α_{xx} and α_{yy} (as expected). However, the longitudinal polarizability per chain

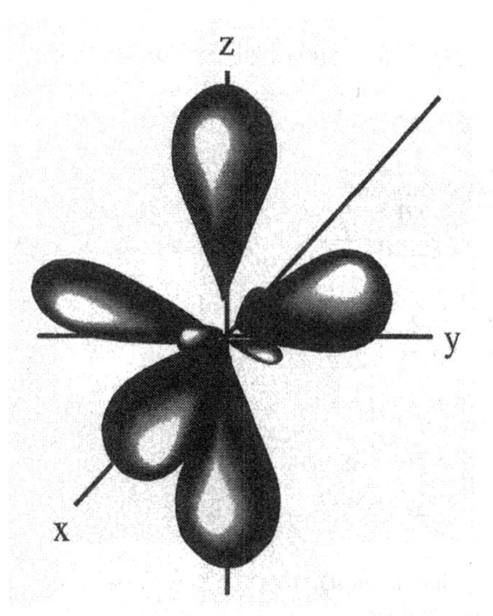

Figure 3.6: Figure showing the plane, triangular sp^2 molecular orbitals (lying in the x-y plane) and a perpendicular $2p$ orbital (along the z-axis) for carbon.

Figure 3.7: Figure illustrating the conjugated π-bonds that form between the $2p$ orbitals of the six carbon atoms that comprise the benzene ring.

length α_{zz}/N should increase with N, as indicated by the calculated results in Table (3.2). To understand the behavior of the three different polarizability components as the chain length is increased, consider the following simple argument. Recall the dipole moment and the polarizability are proportional to each other (p=αE). The dipole moment (p=er) can increase if there is more available charge to separate or if there is a greater separation of the charge. For α_{xx} and α_{yy} an additional unit adds to the available charge but does not increase the possible separation along x or y. Hence, since the available charge is proportional to N, so are α_{xx} and α_{yy}. However, for the z component of the dipole moment, an additional unit increases the available charge and, due to delocalization, it increases the size of the separation of the charge. Hence p_z and therefore α_{zz} grow as N^2. Thus, the α_{zz}/N component is larger than the other two components due to electron delocalization along the polythiophene backbone. The trend for the longitudinal polarizability dominates the behavior of the average polarizability as well. When measured in a THF solution at 589 nm [98, 99], the average polarizability per chain length $\bar{\alpha}/N$ increases with N (see Table 3.2). Both the theoretical and experimental results are summarized in Table (3.2). Moreover, the measured increase in $\overline{\alpha}/N$ has not saturated at $N = 6$, whereas the computed values begin showing saturation. Measurements by Prasad's group [98, 99] were also done by forming a thin film and measuring the refractive index, and hence α, at 632.8 nm. These results were much smaller and they exhibit saturation (as predicted by theory). Measurements by a different group on alkyl-substituted oligothiophenes in a PMMA matrix [100] show a decrease in $\overline{\alpha}/N$ as N increases!

Table 3.2: Experimental data and calculations of $\overline{\alpha}/N$ for oligothiophenes of different chain lengths N. Units here are atomic units (a.u.), where 1 a.u. $= 0.148176$ Å^3 $= 0.148176 \times 10^{-24}$ esu $= 0.164867 \times 10^{-40}$ C^{2}m^2/J.

N	Calculated		Experimental		
	α_{zz}/N	$\bar{\alpha}/N$	$\bar{\alpha}/N^a$	$\bar{\alpha}/N^b$	$\bar{\alpha}/N^c$
1	55.5	44.5	66.9	66.0	
2	75.6	49.6	84.4		
3	94.1	54.8	101.2	51.2	24.7
4	109.0	59.3	168.7	50.7	23.6
5	120.0	62.7	256.4	51.0	35.1
6	129.0	65.5	506.1		
7	136.0	67.6			63.6
11					42.9

[a]Experimental measurements at $\lambda = 589$ nm in THF solution [98, 99].
[b]Experimental measurements at $\lambda = 632.8$ nm in a thin film [98, 99].
[c]Experimental measurements at $\lambda = 632.8$ nm in a thin film matrix of PMMA [100].

As emphasized by Champagne, *et al.* [101], these results highlight that the polarizability is extremely sensitive to "the nature and physical state of the chains". Clearly, much work has been done, but much more is needed.

Interestingly, DNA behaves oppositely to π-conjugated polymers. Here the active chromophores are the nucleic acid base pairs that hydrogen bond to each other in the center of the double helix formed by the sugar-phosphate backbone (see Fig. 3.8). A *chromophore* is a group of atoms in a molecule that are the active absorbers of electromagnetic radiation. The absorption spectrum is largely independent of the particular molecule or environment in which the chromophore is placed. The fascinating result for DNA is that the polarizability perpendicular to the helical axis is much larger than the polarizability parallel to the axis. It is the presence of the base-pair chromophores that makes this so, and the lack of conjugated bonds reduces electron delocalization.

The bases in DNA consist of purines: adenine and guanine, and pyridamines: thymine and cytosine. These bases show strong absorption in the ultraviolet part of the spectrum (200-300 nm). There are several conformations of DNA, but the most significant one biologically is B DNA that forms the double helix structure determined by Watson and Crick. The B form is right-handed, while the Z form is left-handed. The polarizabilities of these structures can be determined indirectly by measuring refractive indices and absorption. DNA, RNA, sugars, and many other organic and biological molecules are *optically active.* Optically-active substances exhibit circular birefringence (CB), or circular dichroism (CD) or both. *Circular birefringence,* also called optical-rotary dispersion (ORD), results from having *refractive indices* that differ for left- and right-circularly polarized light. Circular birefringence (or ORD) will cause linearly-polarized light to rotate its polarization plane as it propagates through the substance. Larger path lengths produce larger rotation angles. *Circular dichroism* results from having *absorption coefficients* that differ for left- and right-circularly polarized light. Circular dichroism will manifest itself by changing linearly-polarized light into elliptically-polarized light, i.e. it changes the degree of circular polarization of the light. Circular birefringence is a measure of the difference in the real part of the polarizability for left- and right-circularly polarized light, while circular dichroism is a measure of the difference in the imaginary part of the polarizability for left- and right-circularly polarized light. Since the polarizability is frequency dependent, both optical activity and circular dichroism depend on the wavelength of light being used. Books covering these topics in detail include those by Caldwell and Eyring [124], Mason [125], and Barron [72]. A brief, and very clear review of these effects and their use in fast, time-resolved spectroscopy can be found in the paper by Goldbeck and Kliger [126].

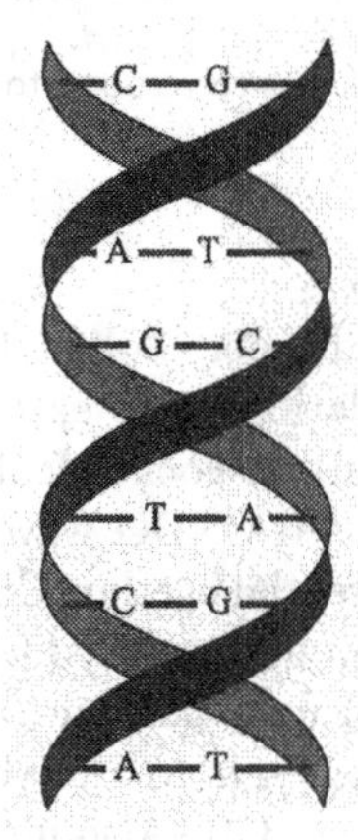

Figure 3.8: A sketch of the structure of B DNA showing the nucleic acid base pairs in planes between the two spiral filaments representing the sugar-phosphate backbone of DNA. The nucleic acid base pairs lie in planes that are nearly perpendicular to the spiral axis of the molecule.

Circular dichroism is a powerful, frequently-used tool to study DNA and other optically active structures [108, 127–129]. Theoretically calculated polarizabilities have produced excellent agreement between calculated absorption spectra and circular dichroism and measured results. For example, applying the classical coupled oscillator theory for molecular aggregates of DeVoe [105, 106], Cech, Hug, and Tinoco [108] calculated the circular dichroism of the polynucleotides ApA and oligoadenylic acid in RNA and B-DNA geometries. Their calculated spectra agree remarkably well with experiment. Subsequently, Cech and Tinoco [130] successfully applied the same technique to double-stranded polynucleotides in RNA, B-DNA, and C-DNA conformations.

A similar calculation that uses a Fano-DeVoe model [131] to account for solvent effects on copolymers was carried out by Ito and I'Haya [127] about a decade after the work by Cech and Tinoco [108, 130]. In this work, Ito and I'Haya calculated the ultraviolet absorption and the circular-dichroism (CD) spectrum of B- and Z-DNA conformations. They calculate the red-shift of the UV absorption spectrum and the inversion of the CD profile of the B-Z transition of poly(dG-dC)·poly(dG-dC), and reasonable qualitative agreement with experiment [102, 132–136] is obtained for their calculated bandshapes.

Here we summarize theoretical calculations of the polarizability of the five nucleic acid bases that occur in RNA or DNA (adenine, cytosine, guanine, thymine, and uracil) and compare the results to experimental measurements. Ab initio calculations of polarizabilities for these nucleic acid bases were done using a coupled perturbed Hartree-Fock (CPHF) technique [137]. This self-consistent field calcu-

Table 3.3: Summary of experimental and calculated average polarizabilities of the nucleic acid bases of RNA and DNA. Units here are atomic units (a.u.), where 1 a.u. = 0.148176 Å^3 = 0.148176 × 10^{-24} esu = 0.164867 × 10^{-40} C^{2}m^2/J.

Base	$\bar{\alpha}$(exp)[a]	additivity[b]	CPHF (CNDO)[c]	CPHF (+MP2)[d]
adenine	88.4	92.7, 101.6	101.0, 91.0	88.7
cytosine	69.5	69.4, 75.0	71.8, 63.4	76.7
guanine	91.8	96.2, 105.8	93.9, 90.0	94.9
thymine	75.8	77.6, 81.7		76.1
uracil				68.1

[a]Experimental measurements cited in Bottcher [139].

[b]Calculations of using additivity methods [116]. The two values correspond to the atomic hybrid component (ahc) method and the atomic hybrid polarizabilities (ahp) method. See the work of Miller [116] for details.

[c]Theoretical calculations of Papadopoulos and Waite [138]. The two values correspond to calculations using an olefinic and aromatic basis set respectively.

[d]The best results of the theoretical calculations of Basch *et al.* [137].

lation also accounted for electron correlation by using second-order Möller-Plesset perturbation theory (see Sec. 4.6). The results of their calculations were extremely good — see Table (3.3) below. Another coupled perturbed Hartree-Fock calculation performed slightly earlier by Papadopoulos and Waite [138] used CNDO wavefunctions (see Sec. 4.3.2 for a discussion of complete-neglect of differential overlap(CNDO) wavefunctions). They used an extended basis set of atomic orbitals optimized for calculating polarizabilities and hyperpolarizabilities of either (i) olefinic molecules or (ii) aromatic molecules. The results of calculations using these basis sets to find the average polarizabilities of the nucleic acid bases are given in Table (3.3). Also included in the table are the polarizability results found when using additivity methods — see Sec. 4.8. All of the results are quite respectable, and the ab initio calculations of Basch, *et al.* [137] are within 5% of the experimental values (except for cytosine).

Finally, we mention that studies of the polarizabilities of nucleosomal DNA fragments have provided insight into the conformational structure and flexibility of DNA and its fragments(see the work of Marion [140–143] and Diekmann and Porschke [144–146] and references therein). The chain-length dependence of the polarizability is studied as well as its variation with base sequence. For more information, consult the literature [140–146].

3.5 Clusters

Atomic clusters occupy the range intermediate between small molecules and bulk matter. A cluster is an aggregate of identical atoms (denoted A_n), or a set of atoms (e.g., A_nB_m). The number of atoms n ranges, roughly speaking, from several to tens of thousands, spanning particle dimensions from a few angstroms up to several tens of nanometers.

Until the 1970s, most microcluster research focused on ensembles of particles condensed on a substrate or captured in a matrix. Such samples manifested a variety of finite-size effects (see, e.g., the reviews [147–149]), but suffered from unavoidable wide distribution of particle sizes and from strong, but poorly understood, particle-substrate interactions. To follow cluster evolution in detail, it was imperative to be able to focus on the intrinsic properties of a particle of well-defined size.

This goal was realized with the advent of molecular beam techniques for the study of clusters (see, e.g., the reviews [150–153]). Bulk material is vaporized (in a hot oven, by laser ablation or sputtering, etc.) and expands through a small nozzle into vacuum, condensing and forming a beam of clusters. Cluster beam formation is facilitated by entrapping the expanding vapor in a flow of inert "carrier" gas. In this way, free clusters of almost any chemical element or compound can be produced. Particle size is determined by passing the beam through a mass spectrometer. By tuning the spectrometer to progressively higher masses, it becomes possible to study continuous sequences of clusters literally atom by atom. (The upper limit of the sequence is set by the range - and cost - of the mass spectrometer; free clusters with as many as 20,000 atoms have been produced and mass-selected [154].)

Thus in cluster science the size n of the particle forms a new physical variable. By following the step-by-step evolution of cluster properties one can map out the transition from molecular to solid-state behavior. In addition, microclusters display interesting and useful size-dependent properties of their own. Recent general reviews of the rapid and robust progress in the field can be found, e.g., in books and review articles [152, 155–160], compendia [153, 161–163], and conference proceedings [164–169].

It should be noted that many common points are becoming apparent between the studies of finite clusters and those of semiconductor nanostructures and quantum dots, single-electron tunneling contacts, and other confined systems operating in the quantum size regime. Advances in microfabrication and lithography techniques have resulted in nanostructure research beginning to approach from above the size range that cluster science has been reaching from below.

While, as described above, it is now possible to produce clusters of a variety of sizes and materials, the same cannot yet be said about our ability to interpret their behavior. The situation is different from that in solid-state physics where metals, insulators, and semiconductors can be understood from a unified point of view. Thus for describing the behavior of, say, noble gas clusters one appeals to models quite different from those for metal clusters. There are both theoretical and practical reasons for this predicament. From the experimental point of view, the control over and knowledge of the internal state and structure of clusters are still limited. On the theoretical side, the situation reflects both the diversity of chemical bonding and the still incomplete understanding of it. Nevertheless, within some specific cluster families it has become possible to formulate "unified viewpoints" which are very successful at explaining a wide body of experimental data. This represents a dramatic shift from the earlier days when clusters were regarded as either individual overgrown molecules or small chunks of bulk materials.

We now discuss some specific atomic cluster families and examine how their structure and evolution are manifested in the size dependence of the polarizability. We begin by looking at simple metal clusters, a category which in just over ten years has developed a solid conceptual foundation and has given rise to a variety of fundamental questions.

3.5.1 Metal clusters

(a) Delocalized electrons and shell structure. The family of simple metals includes such materials as the alkalis, aluminum, copper, etc. In the bulk, these form so-called free (or nearly-free) electron metals. It is found that the outer-shell valence electrons leave their parent atoms and form a delocalized sea of conduction electrons. Their interaction with the lattice of the remaining ion cores can be adequately taken into account by means of a residual pseudopotential (see, e.g., [170, 171]).

It turns out that when just "a handful" of atoms of these elements form a cluster, the same phenomenon occurs: the atoms lose their weakly bound valence electrons which then form a delocalized cloud. There is thus a strong analogy with the physics of atoms and nuclei, in that we are dealing with a finite-sized system of identical Fermi particles. As in those cases, the energy levels of valence electrons in metal clusters organize into shells [152, 172–174] (Fig. 3.9). Since the spacing between valence electron energy shells in small clusters can be quite large (as much as several tenths of an electron volt), the shell structure is not washed out by temperature or surface effects and leads to observable consequences. This phenomenon is referred to as the quantum size effect.

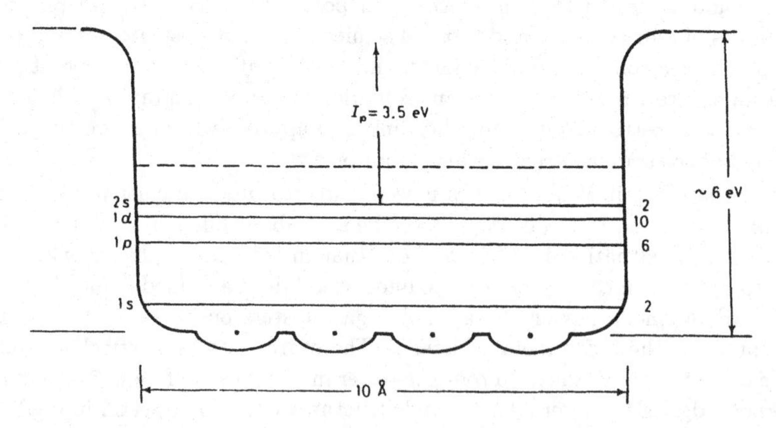

Figure 3.9: A schematic potential well for electrons in the Na_{20} cluster. The filled shells, indicated by solid horizontal lines, accommodate 20 valence electrons. The positive ionic cores are screened, producing a rather smooth well. The effect of the cores can be taken into account by introducing pseudopotentials, see Section 4.3.2. (Reprinted with permission from Ref. [175].)

Clusters with closed energy shells of delocalized electrons (e.g., those with $8, 20, 40\ldots$ valence electrons) possess spherical symmetry and are especially stable; clusters with open shells find it energetically favorable to acquire distorted shapes [176] (see Fig. 3.10). A clear manifestation of this effect is observed in the mass spectra of cluster beams. Closed-shell metal clusters, known as "magic number" clusters, are dramatically more abundant than their neighbors [177] (Fig. 3.11). This effect has by now been seen in mass spectra extending to hundreds and even thousands of atoms [178, 179]. Cluster properties as diverse as abundances, cohesive energies [180], ionization potentials [152, 172, 181], photoelectron spectra [182, 183], fission patterns [184] and optical properties [152, 158, 185–187] and polarizabilities (discussed in this book) can be understood and analyzed on the basis of the picture of a delocalized valence electron cloud exhibiting shell structure.

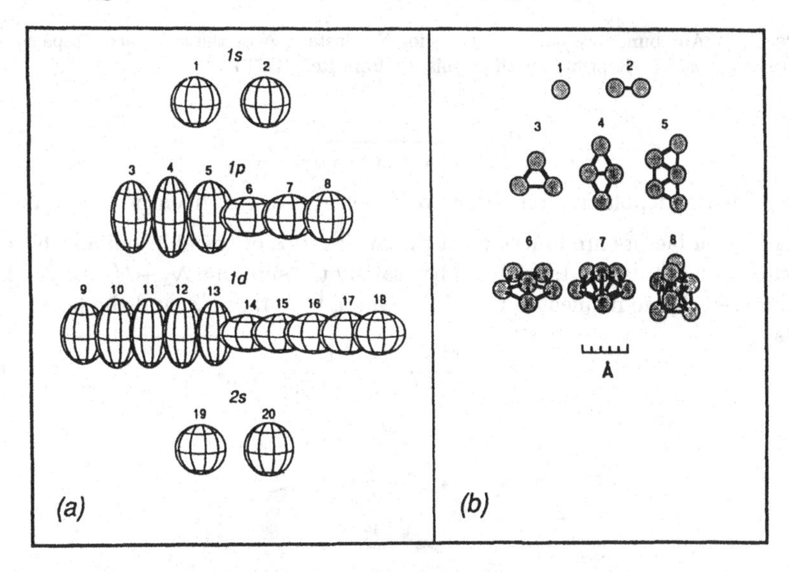

Figure 3.10: (a) The shapes of simple metal clusters according to the Clemenger-Nilsson model. (b) For comparison, some structures predicted by *ab initio* calculations [188] are shown. (Reprinted with permission from Ref. [152].)

(b) Polarizability of metal clusters: high or low? Since in the lowest approximation a cluster may be visualized as a microscopic metallic droplet, it is reasonable to begin by considering the classical problem of the polarizability of a conducting particle [189].

It is a famous result of classical electrostatics that when a dielectric ellipsoid (semiaxes a, b, c; dielectric constant ϵ) is placed in a uniform external field, a uniform depolarization field arises in the interior [Fig. 3.12]. If the external field E is directed along a principal axis of the ellipsoid, the induced electric polarization (dipole moment per unit volume) is given by

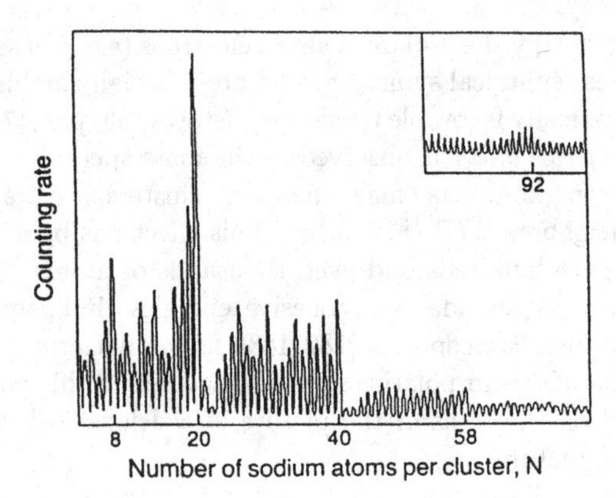

Figure 3.11: An abundance mass spectrum for Na clusters. Note the prominent steps associated with electronic shells. (Reprinted with permission from Ref. [177].)

$$\mathrm{P}_i = \frac{1}{4\pi}\frac{\epsilon - 1}{1 + (\epsilon - 1)\mathcal{N}_i}\mathrm{E}, \tag{3.82}$$

where $\mathcal{N}_i$ is the depolarization factor corresponding to the given principal axis. The depolarization factors are functions of the axial ratios of the ellipsoid and have been computed and tabulated [190, 191]. They satisfy the sum rule $\mathcal{N}_x + \mathcal{N}_y + \mathcal{N}_z = 1$. The total induced dipole moment is $\mathrm{p}_i = \mathrm{P}_i V$, where $V = 4\pi abc/3$, and the polarizability is thus

$$\alpha_i = \frac{abc}{3}\frac{\epsilon - 1}{1 + (\epsilon - 1)\mathcal{N}_i}. \tag{3.83}$$

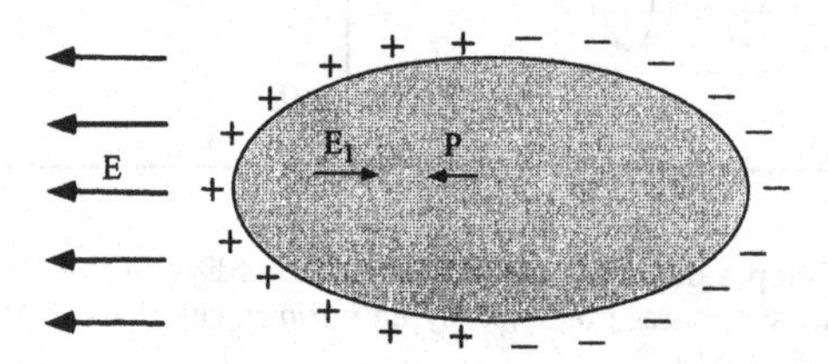

Figure 3.12: Depolarization field in a dielectric ellipsoid. The external field **E** gives rise to the polarization $\mathbf{P}_0$ and the depolarization field $\mathbf{E}_1$.

For an ideal conductor $\epsilon \to \infty$, and $\alpha_i = abc/(3\mathcal{N}_i)$. In particular, for a sphere $\mathcal{N}_i = 1/3$ and $a = b = c \equiv R$, leading to the well-known result

$$\alpha = R^3. \tag{3.84}$$

To what extent are these expressions applicable to microscopic clusters? It turns out that they provide correct order-of-magnitude estimates of cluster polarizabilities; nevertheless, there are noticeable deviations from the classical values. These size-dependent deviations are precisely what one wishes to understand, for they reflect quantum behavior of the finite Fermi system formed by the delocalized valence electrons.

The issue of the static electric polarizability of size-quantized systems has an interesting history. It first attracted a lot of attention when it was predicted [192] (see also [193]) that very small metal particles would exhibit an orders-of-magnitude increase in polarizability compared to the classical value Eq. (3.84). The calculation assumed that for applied electric fields E$< \Delta/(eR)$ (where Δ is the average spacing of the quantized energy levels) each cluster electron responds independently to the applied field E. Subsequent experiments [194, 195] showed that no such effect was present. Soon thereafter, it was pointed out [196, 197] that even inside a finite system the external field is strongly screened by the delocalized electrons. These newer calculations considered the polarizability of N electrons confined to an infinite potential well of radius R, and concluded that screening would actually lead to the polarizability of metal particles being *smaller* than the classical value.

Finally, accurate size-resolved experiments on small sodium and potassium clusters [198, 199] demonstrated that their static electric polarizabilities are enhanced after all, albeit moderately, reaching values of up to $\simeq 2R^3$ and slowly decreasing to the bulk value (here R is the radius of the positive ion background in the cluster, see below). These results are shown in Fig. 3.13. The experiment was performed by deflecting a cluster beam from a supersonic oven in a strong inhomogeneous electric field.

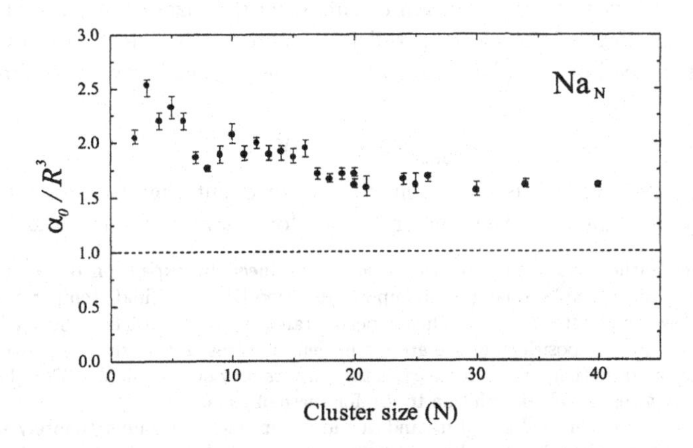

Figure 3.13: Experimental static electric polarizabilities of sodium clusters [198, 199] expressed in terms of the classical value for a sodium sphere (R is the radius of the positive cluster background). Dashed line denotes the classical limit.

To interpret the observed behavior we must keep in mind that the size-dependent polarizability of the cluster electrons is determined by the interplay of the following factors: *(1) Electron screening*, (2) *Size quantization*, and (3) *Surface effects* (i.e., inhomogeneity of the electron density distribution or "spill-out"). For a quantitative analysis of these effects we first need to define a tractable model of a microcluster.

(c) Jellium picture of a metallic cluster. Since clusters with as few as half a dozen or so electrons already display clear shell structure, and the ionic pseudo-potentials are frequently weak, one may conclude that many cluster properties, including their response, is determined primarily by the delocalized electron cloud. A simple approximation, and one that has proven its usefulness, is the so-called jellium model. In this model, extrapolated from the theory of nearly-free-electron metals (see, e.g., Refs. [171, 200, 201]), the positive background formed by the ions is represented by a uniformly charged sphere of radius $R = a_0 r_s N_e^{1/3}$ and charge number density $\rho_+ = 3/(4\pi r_s^3 a_0^3)$. Here N_e is the total number of valence electrons, r_s is the Wigner-Seitz density parameter, and a_0 is the Bohr radius.[1] It is also commonly assumed that the r_s parameter in a cluster does not differ significantly from its bulk value [172].[2]

In a large metallic particle, the conduction electrons are essentially all located in the interior (except for a thin surface layer [207]). In a finite cluster, on the other hand, a sizable fraction of the valence electrons are located near the surface, and as a result a significant portion of the electron cloud spills out beyond the edge of the jellium background. The cluster profile then looks like that shown in Fig. 3.14. This spill-out is an essential attribute of metal clusters and is responsible for many of their size-dependent characteristics. [3]

(d) Cluster structure and polarizability. Recognizing the importance of the spill-out effect, we can understand the origin of enhanced polarizability of metal clusters (Fig. 3.13). The effect of the electron tail is to increase the effective radius of the particle. Thus instead of Eq. (3.84) we can write the polarizability approximately as [209]

$$\alpha_{\text{cluster}} \cong (R + \delta)^3, \tag{3.85}$$

where δ is the effective radius enhancement. It turns out that the magnitude of δ is only weakly dependent on the cluster radius; for example, for the alkali clusters

[1] As was noted earlier, cluster beams are typically produced by expansion of hot vapor into vacuum. Although considerable cooling occurs upon expansion [151], the final "temperature" of the cluster is not known (estimates for alkali cluster beams range from $\sim$ 100K to $\sim$ 500 K [202–205], and neither are the precise positions of the atomic nuclei. It seems certain that the latter are in a state of highly agitated motion, and in some cases the clusters may even be liquid [206]. The jellium picture is therefore a reasonable description to use for these objects.

[2] The concepts of electron shell structure and of the jellium model are unfortunately sometimes used interchangeably. It should be emphasized that there is a fundamental difference between the two. While jellium is a convenient approximation, shell structure is an experimental fact of far greater generality and applicability. In other words, while the former is only a model, the existence of electron shells in metal clusters has been unequivocally demonstrated and must be reflected by whatever theoretical approach one chooses.

[3] In practice, the surface density profile of the cluster electrons is usually calculated by Thomas-Fermi, extended Thomas-Fermi, variational, and density-functional methods (see the reviews [185–187]).

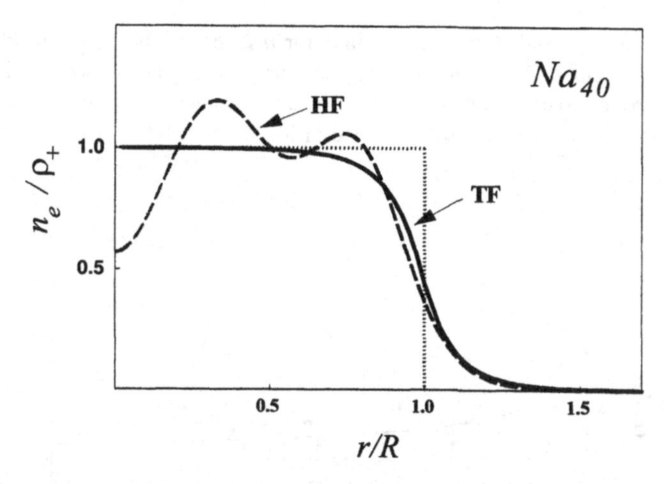

Figure 3.14: Radial electronic density of the Na_{20} cluster, normalized to the density of the positive core, ρ_+ (depicted by the dotted line). The radial distance is expressed in terms of the core radius. Solid line: Hartree-Fock calculation (courtesy C. Guet, after [208]); dashed line: Thomas-Fermi approximation [185].

shown in Fig. 3.13 a reasonable fit can be obtained with $\delta \simeq 1.5$ a.u. [198].

Another qualitative way to interpret the increased polarizability is that the spill-out profile reduces the restoring force the electron cloud feels when displaced by an external field.

These observations confirm that even in small clusters the delocalized electrons are mobile and provide efficient screening. Of course, Eq. (3.85) is phenomenological, and a full quantum-mechanical analysis is needed to understand the behavior of the polarizability in more details. Such approaches are discussed in Chapter 4. However, it is appropriate to mention here that the theoretical calculations of alkali cluster polarizabilities tend to yield values lower (by about 20%) than the experimentally measured values shown in Fig. 3.13. The origin of this discrepancy has not yet been satisfactorily explained. This issue will be discussed further in Section 4.10.4.

So far we have considered spherical particles. In general the polarizability is, of course, a tensor[4] , but particle rotation in the beam averages out anisotropic effects. The experimentally measured quantity is thus the averaged scalar polarizability. As a result, shape effects and shell structure are not as easily resolved in the polarizability data as they are in the mass spectra. Nevertheless, Fig. 3.13 displays a clear dip near the magic number $n = 8$, and shell effects are responsible for the oscillations in α. The new technique of light force deflection (see Sec. 5.2.4) has the potential of resolving tensor components of the polarizability. The only other metal clusters for which polarizabilities have been directly measured are $Al_n(n = 1, 2, 16 - 60)$ [210],

[4]For simple metal clusters it has been shown [185] that the classical ratios $\alpha_i \sim 1/\mathcal{N}_i$ should remain valid to an accuracy of ≈10% even in the presence of electron spill-out.

see Fig. 3.15. The overall trend of the data for $n \gtrsim 20$ can be understood within the jellium picture, and $\alpha(Al_2)$ is in good agreement with a pseudo-potential calculation [211]. The strongly oscillating behavior of α has not been explained, but may be due to non-jellium effects, i.e., to the influence of ion cores [210].

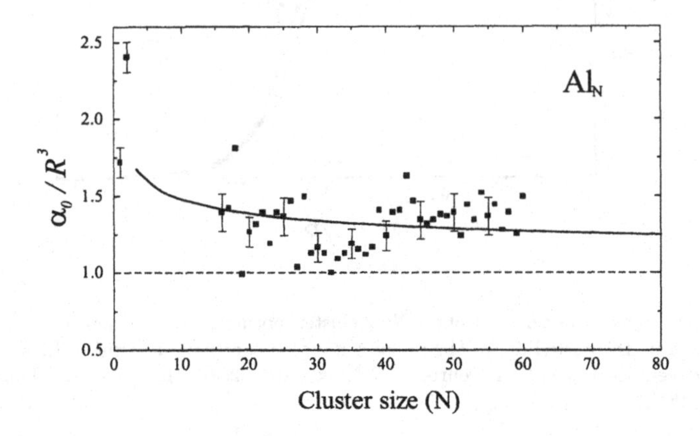

Figure 3.15: Static electric polarizabilities of Al_n clusters. The experimental data are from Ref. [210]. The solid line is the result of an analytical many-body (RPA) calculation (see Section 4.10.4). It has been proposed that the low polarizabilities of some of the smaller clusters are due to non-jellium effects, i.e., to the influence of ionic cores [210].

(e) Connection with giant dipole resonances. High polarizability of metallic clusters has its counterpoint in another response phenomenon: strong photoabsorption resonances seen in the cluster optical spectra (see the reviews in [152, 158, 178, 185, 186]). The high mobility of the valence electrons gives rise to collective motion whereby the electron cloud oscillates coherently under the influence of an external time-varying field.

In bulk metals, on surfaces, in thin films, and in colloidal particles these oscillations are known as plasma waves or plasmons. Remarkably, experiments on size-resolved metal clusters have revealed that analogous "giant resonances" begin to dominate photoabsorption spectra already in particles with as few as half a dozen atoms. Some such spectra are shown in Fig. 3.16.

In parallel with macroscopic terminology, these prominent resonances are referred to as cluster surface plasmons. The origin of this term can be visualized by considering collective electron motion in a classical conducting particle (see e.g., [214]), cf. Fig. 3.12. Back-and-forth oscillations of the electron cloud give rise to a large dipole moment and are equivalent to a charge wave running on the surface. In the quasi-static approximation (neglecting retardation effects), the classical dipole resonance frequency is given by the pole of Eq. (3.82). Of course, in this dynamic case the frequency dependence of the dielectric constant must be included. The resonance

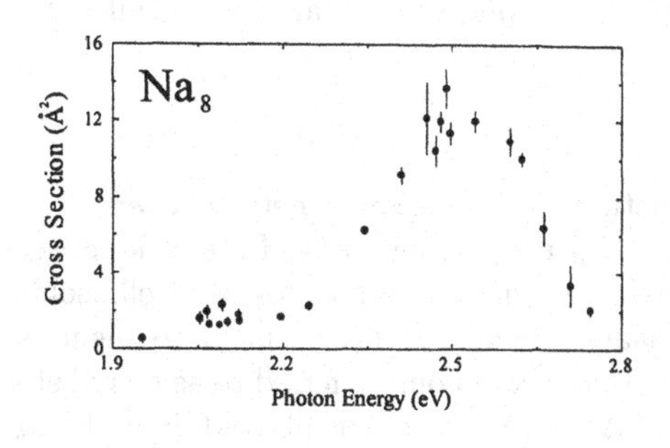

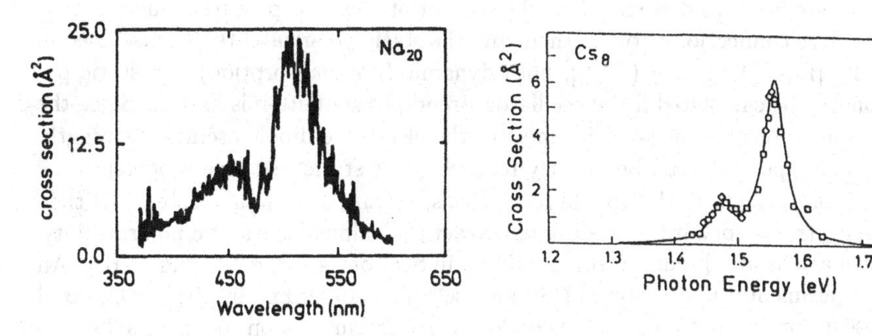

Figure 3.16: Examples of strong photoabsorption resonances in neutral metal clusters. These resonances dominate the dipole response of the delocalized valence electrons. "Top": Na_8 (reprinted with permission from [204]). A collective resonance is seen at 2.5 eV, taking up 70% of the dipole oscillator strength of the valence electrons. The classical surface plasma resonance of a sodium sphere is at 3.4 eV. "Bottom-left": Na_{20} (reprinted with permission from [212]). "Bottom-right": Cs_8 (reprinted from Ref. [213] with the kind permission of Elsevier Science Publishers). The classical surface plasma resonance of a cesium sphere is at 2.0 eV. The red shift of the resonances is directly connected with the enhanced static polarizabilities of small clusters.

condition reads

$$\epsilon(\omega) = 1 - 1/\mathcal{N}_i. \tag{3.86}$$

For a sphere one finds the well-known condition $\epsilon(\omega) = -2$. This is a special case of the general Mie theory [215] which describes multipole scattering of light by spheres of arbitrary dielectric composition. If we use the Drude expression for a free-electron gas:

$$\epsilon(\omega) = 1 - \omega_p^2/\omega^2, \tag{3.87}$$

where

$$\omega_p = (4\pi n e^2/m)^{1/2} \tag{3.88}$$

is the bulk plasma frequency (n is the number density of conduction electrons in the particle), the resonance frequency of a classical ellipsoid is found to be

$$\omega_i = \omega_p\sqrt{\mathcal{N}_i}. \tag{3.89}$$

In particular, for a conducting sphere we obtain the so-called Mie frequency

$$\omega_M = \frac{\omega_p}{\sqrt{3}}. \tag{3.90}$$

The resonance structure in microscopic metal clusters is a bit more intricate. Instead of a resonance at ω_M exhausting 100% of the conduction electrons' oscillator strength (or two or three resonances in a spheroidal or ellipsoidal particle), in small clusters the surface plasma resonances: (1) occur at frequencies up to ~25% lower than the classical value (this is commonly referred to as a "red shift"); (2) are missing up to ~25% of the oscillator strength; (3) display additional fragmentation.

The reason we have digressed into the subject of cluster dipole resonances is that there is a close connection between them and the static polarizability. As discussed in Section 4.2 [Eqs. (4.11) and (4.18)], both dynamic (photoabsorption) and static polarization can be calculated if the oscillator strength distribution is known. Since the cluster giant resonances account for most of the electronic dipole oscillator strength, the resonance spectrum can be directly related to the static response properties.

Thus if one calculates the optical resonance spectrum of a metal cluster, it is then a relatively straightforward procedure to extract the value of the static polarizability. Such calculations are discussed in more detail in Sections 4.7.3, 4.7.4, and 4.10.4. An elegant experimental realization of the same scheme is described in [216]: integrated photoabsorption cross sections of size-selected silver clusters on an inert substrate were converted to cluster polarizabilities. The results are shown in Fig. 3.17. Note the qualitative similarities to the alkali-cluster data in Fig. 3.13, including the apparent electronic shell edge at $n = 8$.

The direct connection between the enhanced polarizability of small metal clusters and the aforementioned red shift of their optical resonances is most easily seen within the so-called plasma-pole approximation. This approximation assigns the entire dipole oscillator strength of the system to the surface plasma peak. Then from very general sum rules (Sec.4.2.2) one can show that for a spherical particle

$$\alpha_{\text{cluster}}/R^3 = (\omega_M/\omega_{\text{cluster}})^2 \tag{3.91}$$

$$= (1 - \Delta N/N)^{-1} \tag{3.92}$$

where $\Delta N/N$ is the fraction of the valence electrons spilled out beyond the cluster edge.[5] This approximation illustrates the essence of the spill-out effect. In Eq. (3.85) we argued that spill-out increases the effective radius of the particle. Equivalently, we can say that it reduces the average electron density, in which case Eq. (3.88) shows that the plasma frequency is decreased, leading to a red shift of the dipole resonance. The magnitude of both shifts is seen to be essentially governed by $\Delta N/N$. More

[5]Physically, the result Eq. (3.91) can be understood by visualizing the collective response of the electron cloud as that of a driven harmonic oscillator (the Drude model). In an applied static field, the displacement of the oscillator, and hence its polarizability, are inversely proportional to the restoring constant k. The resonance frequency, on the other hand, is proportional to $k^{1/2}$. Hence $\alpha \sim \omega^{-2}$.

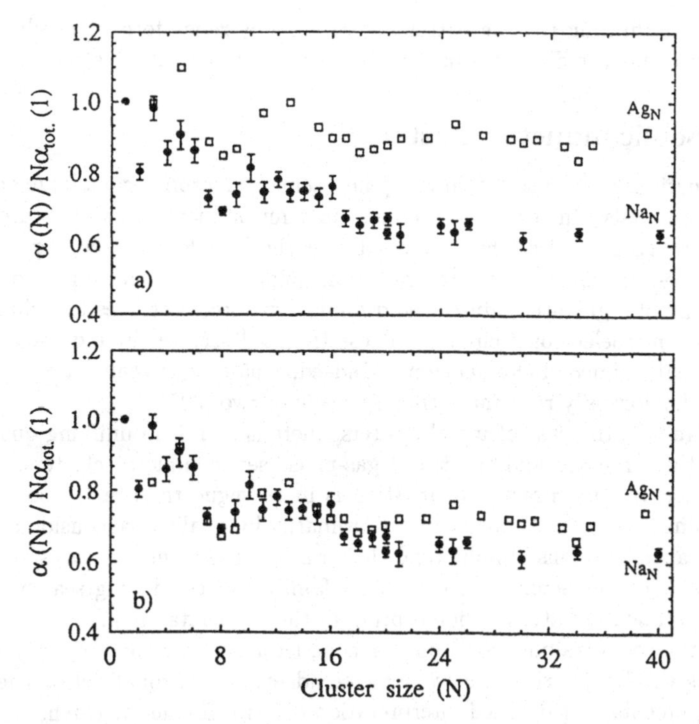

Figure 3.17: Static dipole polarizability for deposited Ag_n clusters as extracted from the absorption spectra (open squares): (a) Normalized to the theoretical value for Ag_1, (b) same as (a) but shifted by -0.2. The solid circles are the experimental data for Na_n (Fig. 3.13). Note the strong similarity in the dependence of the polarizability on the number of valence electrons. (Reprinted from Ref. [216] with the kind permission of Elsevier Science Publishers.)

accurate treatments (Sec. 4.10.4) refine the size dependence of the polarizability, but reflect the same physics.

(f) Cluster polarizabilities and long-range interactions. A vivid manifestation of the high polarizability of metallic clusters is provided by the strength of their long-range interaction potentials. For example, an electron external to a neutral cluster will feel a "polarization potential" $V_p = -\alpha e^2/(2r^4)$ (equivalent to the "image charge" attraction familiar from classical electrostatics [56]), while a neutral particle will be attracted by the van der Waals potential which is also proportional to α (see Section 6.2 for a discussion of these long-range potentials). The polarization potential was found to be responsible for near-threshold bound states of Au_6^- [217] and for the high capture cross sections of slow electrons by sodium clusters [218]. The van der Waals interaction has been investigated in cluster-atom and cluster-molecule scattering experiments [219, 220], which showed very large elastic scattering cross sections. For example, cross sections for Na_n-C_{60} collisions (see Fig. 6.14) exceed the geometric dimensions of the collision partners by well over an order of

magnitude, confirming that an extremely strong long-range force is involved. All of these experiments are discussed in more detail in Sections 6.2.1(b,e) and 6.2.2(c).

3.5.2 Semiconductor clusters

(a) Introduction. From a scientific viewpoint, semiconductors form a large and varied family of materials. In particular, interest in semiconductor nanostructures (quantum dots, wells, and wires) has exploded over the last decade (see, e.g., [221–226] and references therein). From a technological point of view, semiconductors are the underpinning of a gigantic industry, and miniaturization continues to reduce the dimensions of microelectronic circuits. These trends illustrate the motivation for the study of small semiconductor clusters. Indeed, cluster size scales are expected to become technologically relevant within a decade or two [227].

In contrast to the case of metal clusters, there is so far no unifying quantitative thread to the structure and spectra of gas-phase semiconductor clusters. As mentioned earlier, in this regard the situation is more vague than in the solid-state or atomic realms. At present, one is guided primarily by qualitative considerations and by numerical calculations of individual cluster structures. There is less spectroscopic data available on the semiconductor cluster family, but continuing search for reproducible experimental patterns should bring further understanding.

Interesting polarizability data on semiconductor clusters have recently begun to appear. As will be described below, some trends can be rationalized on the basis of general arguments, but detailed interpretation calls for further research.

(b) Silicon clusters. The normal bulk structure of silicon is a diamond lattice with tetrahedrally coordinated directional bonds arising from sp^3 hybridization (Fig.3.18). By analogy with metals, one might be tempted to speculate that silicon clusters resemble small fragments of the bulk phase to a fair degree. However, this would not be a favorable state of affairs. Indeed, Si likes to be tetrahedrally coordinated, while carving out a small lattice fragment would result in a large number of dangling bonds. Thus, one should expect that small Si_n clusters will undergo reconstruction, akin to that found on bulk crystal surfaces. Indeed, for $n \leq 100$ most of the atoms are at the surface, so that reconstruction should be very extensive.

Many theoretical calculations have been carried out to search for optimum cluster geometries (see, e.g., the references quoted in [23, 228–232]). The qualitative conclusion is that small clusters want to acquire markedly more compact structures with high coordination numbers (see, e.g., Fig. 3.19). This suggestion is confirmed by chemical reactivity measurements which indicate that Si_n^+ and Si_n^- cluster ions are less reactive than bulk silicon surfaces [227, 233]. It appears probable that in such reconstructed clusters the average electronic binding and transition energies are higher than in the bulk. For example, photoelectron spectroscopy has shown that the HOMO-LUMO (Highest Occupied - Lowest Unoccupied Molecular Orbital) gap of the free Si_{10} cluster is especially high [234].

Even leaving aside the influence of structural changes, it is known that the optical spectra of semiconductor nanocrystals shift to the blue with decreasing size [224,

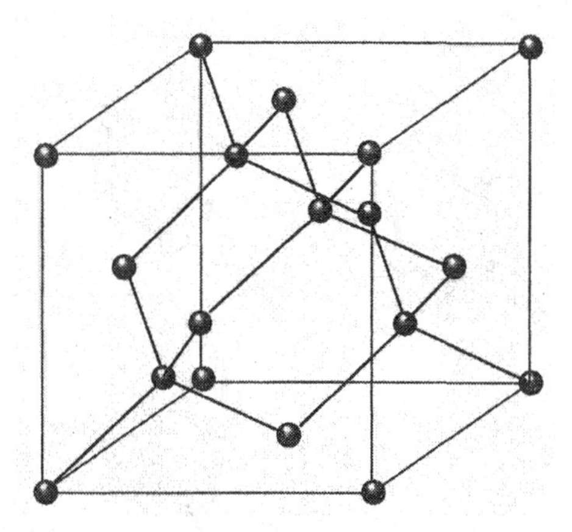

Figure 3.18: Crystal structure of silicon, showing the tetrahedral bond arrangement of the diamond lattice.

225]. This phenomenon has been ascribed to the *quantum confinement effect*: as the particle size gets smaller, the carrier energy levels break up into discrete states and (as in a particle-in-a-box picture) the separation of the valence and conduction bands increases. Numerical calculations on hydrogen-passivated silicon quantum dots with between 87 and 1315 Si atoms in Ref. [235] suggested that the size scaling of the of the band gap change could be fit to $\Delta W_{gap} \sim R^{-n}$ with $n \approx 1.4$, where R is the dot radius.

Since the static polarizability of a system is inversely proportional to the square of the characteristic transition energy W_{eff} [cf. Eqs. (4.7), (4.14)]:

$$\alpha \sim W_{\text{eff}}^{-2}, \tag{3.93}$$

a "blue shift" in the spectrum should manifest itself in a decreased cluster polarizability per atom. Such a trend was indeed observed in beam-deflection experiments [23, 236]. The results of this work, which represents the first measurement for a semiconductor microcluster, are shown in Fig. 3.20. The "silicon bulk" limit was calculated as the polarizability per atom of a dielectric sphere:

$$\alpha_a = \frac{\epsilon - 1}{\epsilon + 2} R_{WS}^3, \tag{3.94}$$

where $\epsilon = 11.8$ is the bulk dielectric constant and $R_{WS} = 1.68$ Å is the Wigner-Seitz radius of the Si diamond lattice. The mean cluster polarizability is indeed below the bulk value, corresponding to an effective transition energy, Eq. (3.93), of ~ 15 eV, as opposed to $W_{\text{eff}} \sim 11$ eV for the bulk limit.

This overall shift is consistent with the R^{-n} effect of quantum confinement described above. The strong oscillations of α with cluster size cannot, however, be explained by the gradual band gap shift. The authors [236] suggest that these variations can be modeled by an energy scheme shown in Fig. 3.21(a) which proposes that

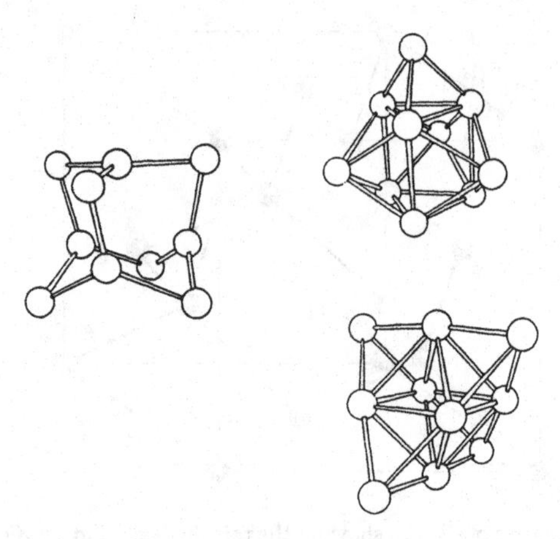

Figure 3.19: Possible structures for Si_{10}. The structure on the left is a bulk fragment; the more compact ones on the right are around 5 eV more stable. (Reprinted with permission from Ref. [227]. Copyright 1991 American Association for the Advancement of Science.)

in clusters there may exist electronic states similar to defect states in the band gap of semiconductors. Clearly, the concept of defect-like states in pure small semiconductor clusters is but another way of visualizing the effect of structural rearrangements. Thus it appears that silicon cluster polarizabilities reflect a combined strong influence of both electronic confinement and structural reconstruction.

Since the polarizability is sensitive to the cluster energy spectrum, measurements of α may contribute to the debate over the existence of structural isomers of semiconductor clusters. Chemical reactivity studies of silicon clusters in ion traps [233] and drift tubes [227, 237], as well as drift tube mobility studies [237] suggest that clusters of a given size may exist in a number of geometric incarnations. The abundances of different isomers are, however, governed by their stabilities and appear to be strongly affected by "annealing" via collisions or laser heating. The shapes of beam deflection profiles in the polarizability experiment [23] showed no evidence for the presence of structural isomers. It is interesting to speculate whether a higher-resolution measurement could resolve different isomeric polarizabilities.

A set of theoretical calculations for Si_n tensor dipole polarizabilities has been published [238], based on the tight-binding model. These range from $\alpha_{xx} = \alpha_{yy} = 80$ Å^3 and $\alpha_{zz} = 60$ Å^3 for Si_7 to $\alpha_{xx} = \alpha_{yy} = \alpha_{zz} = 140$ Å^3 for Si_{13}. As can be seen from Fig. 3.20, the experimental points are quite a bit lower than the predicted values. Local-density functional calculations on Si_4 and Si_6 [243] gave average polarizabilities per atom of 5.0, 4.7, 4.3, and 4.4 Å^3 for Si_4 (D_{2h}, T_d) and Si_6 (D_{4h}, C_{2v}), respectively. While these values concern clusters smaller than those studied experimentally, they are of the same general magnitude. There are clearly both room and need for continued studies.

(c) Gallium arsenide clusters. While silicon is the backbone of today's microelectronics, GaAs has a lot of technological promise and has also been extensively studied

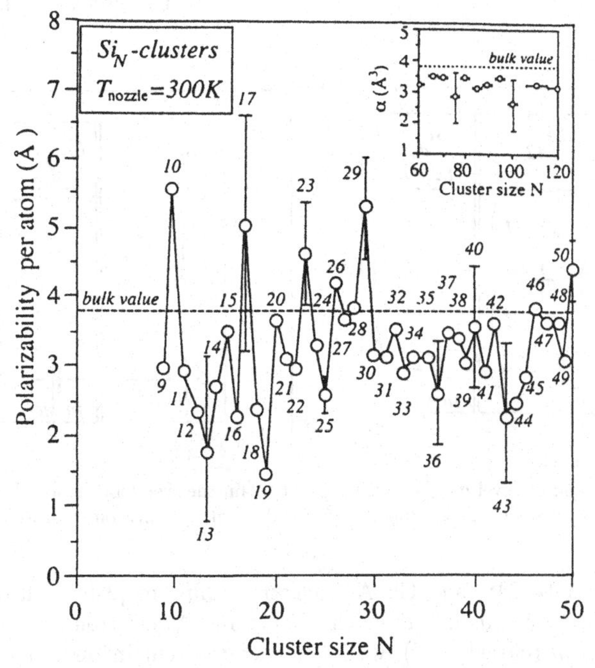

Figure 3.20: Polarizabilities per atom of free Si_N clusters. (Reprinted with permission from Ref. [236]).

[239]. The properties of Ga_nAs_m clusters are likewise of interest, both on their own and for comparison with the structure of Si_n and Ge_n. Experimental work has involved mass spectrometry, photoelectron and photofragmentation spectroscopy and chemical reactivity studies, while theoretical efforts have concentrated on molecular structure calculations limited to small ($m+n \leq 10$) systems (see, e.g., [240–242], and references therein).

Fig. 3.22 shows the results of a beam-deflection measurement of the polarizabilities of GaAs clusters [236, 244]. There are two conspicuous differences from the silicon data in Fig. 3.20: (a) The polarizabilities are primarily enhanced (rather than suppressed) relative to the limit of a bulk dielectric sphere, Eq. (3.94); and (b) An odd-even oscillation pattern is observed. [6]

The first difference may be related to the fact that the bonds in GaAs are noticeably more polar than in Si [171, 245], and the crystal structure is more compact (zincblende). As a result, even in a cluster fragment the bonds will be more saturated and there will be less tendency toward reconstruction and compactification. This reasoning is consistent with the fact that both electron affinities [246] and chemisorption rates [247] of Ga_nAs_m clusters quickly converge to the bulk surface limit.

The odd-even variation of polarizabilities parallels the oscillations of Ga_nAs_m ion-

[6]Due to mass spectrometer limitations, the data points for a given $x = n+m$ contain a distribution of n:m compositions. However, the signal is dominated by the stoichiometries $n = m$ and $n = m \pm 1$.

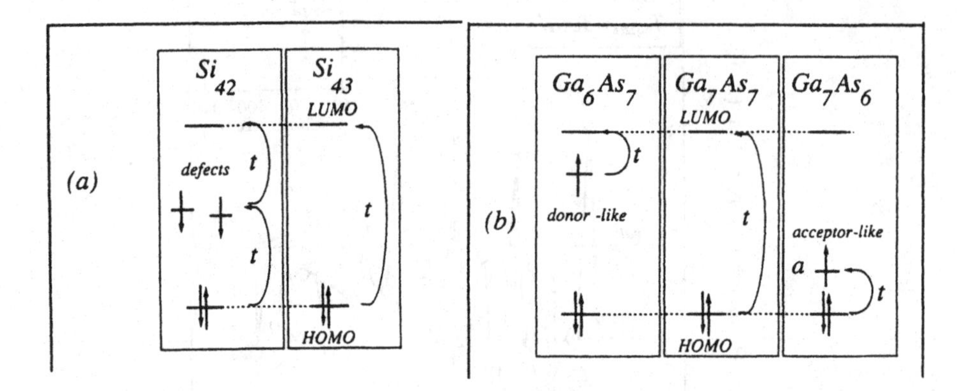

Figure 3.21: Energy level models proposed to explain the size dependence of the polarizability of (a)Si_n clusters, and (b) Ga_nAs_m clusters. (Reprinted with permission from Ref. [236].)

ization potentials [248] and $Ga_nAs_m^-$ electron affinities [246]. All of these patterns are consistent with a model in which the Ga and As electrons are paired up in clusters with $n = m$ (closed shell), while one electron remains unpaired in clusters with $n = m \pm 1$ (open shell). A schematic energy scheme is shown in Fig. 3.21(b). The energy required to remove an unpaired electron is relatively low, thus odd-numbered clusters will have lower ionization potentials. Similar reasoning works for electron affinities. The picture resembles that of a doped bulk semiconductor where additional energy levels are introduced by donors or acceptors. The donor-like electron level [d in Fig. 3.21(b)] will be close to the LUMO, while the acceptor-like level [a in Fig. 3.21(b)] will be close to the HOMO of the closed-shell system. The diagram is also consistent with the fact that electron affinities decrease as clusters become more As-rich [246].

Both the donor- and acceptor-like configurations allow low-energy transitions, indicated by t in Fig. 3.21(b). By virtue of the dependence Eq. (3.93), these transitions make a significant contribution to the cluster polarizability. This provides a qualitative explanation for the odd-even behavior of the latter.

It should be noted, however, that the odd-even polarizability pattern appears to break down near $n + m \cong 17$. This size range also shows a curious flip in the dependence of α on cluster stoichiometry (Fig. 3.23). For $n+m < 13$, Ga-rich clusters have higher polarizabilities, which is consistent with the fact that $\alpha_{\mathrm{Ga\ atom}} > \alpha_{\mathrm{As\ atom}}$. However, for larger sizes the trend reverses. The origin of this switch is not known, but it may imply a readjustment in the energy level scheme in Fig. 3.21(b).

A further intriguing aspect of the Ga_nAs_m data [236] is the temperature dependence of the cluster polarizability. As shown in Fig. 3.22, when the nozzle of the laser vaporization source is heated up from 38 K to 300 K, the polarizability of several smaller clusters is slightly increased, while that of several larger clusters goes down. These changes suggest an important contribution of ionic (vibronic) effects.

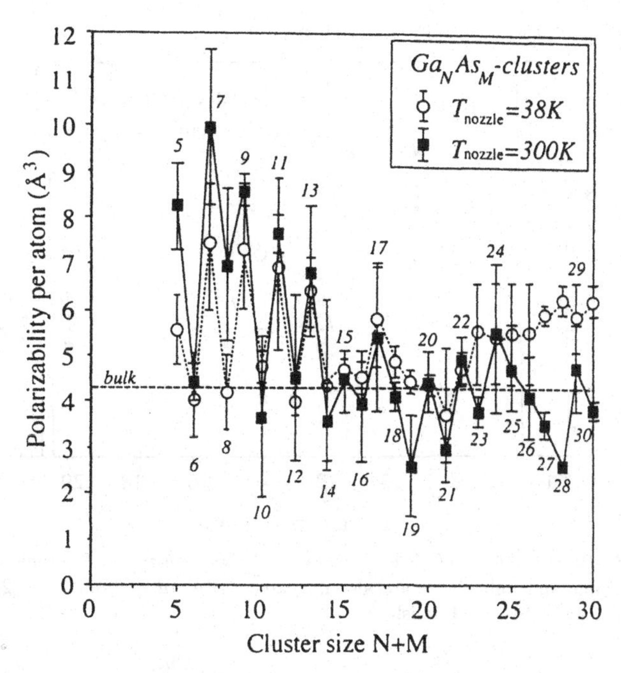

Figure 3.22: Polarizabilities per atom of free Ga_NAs_M clusters measured with two different nozzle temperatures. (Reprinted with permission from Ref. [236].)

An even more spectacular shift was reported in the same paper for Ge_nTe_m clusters: the polarizability of some species more than doubled when the nozzle temperature was similarly raised, while others were only weakly affected. The authors hypothesize that a melting or ferroelectric transition may be responsible for such a dramatic effect. As remarked repeatedly throughout this section, continued research is required for a thorough understanding of the polarization properties of semiconductor clusters.

3.5.3 Fullerenes

(a) C_{60} molecule. Fullerenes are cage-like carbon structures which include the beautifully symmetric C_{60} and its larger relatives (see Fig. 3.24) that have generated a lot of excitement since they were first spotted in carbon cluster beam mass spectra [249] and then isolated in bulk quantities [250]. An abundance of reviews and compendia have appeared since then; see, e.g. [159, 251] and references therein. The C_{60} cluster has icosahedral symmetry, with carbon atoms located at the vertices of twelve pentagons and twenty hexagons. It is essentially a conjugated system, with σ and π bonds forming the molecular framework.

In interpreting the electronic properties of the fullerenes, and in particular their polarizabilities, one is tempted to start with some rudimentary models. One such model treats each vertex atom as an individual dipole and calculates the total polarizability as the self-consistent response of a system of interacting dipoles. The

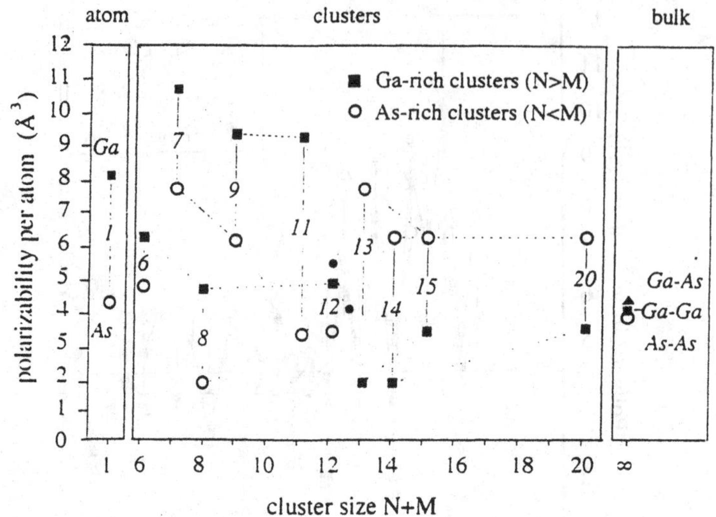

Figure 3.23: Polarizabilities of Ga_NAs_M clusters for gallium-rich and arsenic-rich compositions. Calculated atomic and bulk limits are also indicated. (Reprinted from Ref. [244] with the kind permission of Elsevier Science Publishers.)

polarizabilities of the discrete dipoles are adjusted to fit the data for planar aromatic molecules [252] or graphite [253]. These calculations predict polarizability values for C_{60} of 61 Å^3 and 77 Å^3 respectively. This can be compared with the experimental estimates of about 90 Å^3 based on bulk crystal dielectric properties (see below).

An opposite approach is based on the fact that C_{60} latticework is covered with mobile delocalized electrons. The simplest image is that of a spherical shell containing 60 free electrons. Their wave functions can be expressed as spherical harmonics, and a calculation for this non-interacting spherical electron gas yields [254] a polarizability of 390 Å^3, a considerable overestimate. What is missing in the above picture is, of course, screening due to the electron-electron interaction. Like the delocalized electrons in metal clusters (see Sec. 3.5.1), the mobile electrons are efficient in screening out the applied electric field. The crude model of a classical conducting shell of outer radius $R = 4.3$ Å (we added the 3.5 Å radius of the ionic frame of C_{60} [251] and one-half of the 1.5 Å thickness of its electron cloud [255]) has a polarizability of $\alpha = 80$ Å^3, a value strongly reduced due to the screening effect and one not far from the experimental values. A qualitatively similar result is obtained if C_{60} is modeled as a dielectric shell [3]. The same dramatic influence of screening by the active electrons is seen in more involved many-body calculations (see., e.g., [256] and references therein). In this respect, screening in the fullerene balls is dramatically more efficient than in planar hydrocarbon molecules [257].

A number of calculations of $\alpha(C_{60})$ have appeared in the literature. Generally speaking, in accordance with the scheme sketched above, they start either with a tight-binding picture, or with one based on a set of delocalized electrons superimposed upon the proper ionic potential of the fullerene cage. The polarizabilities are either

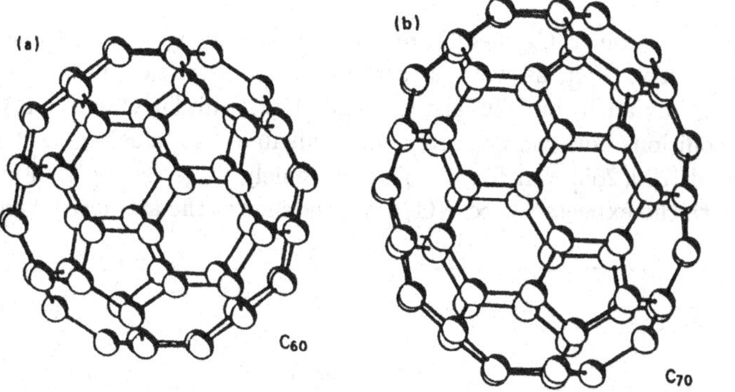

Figure 3.24: Representations of C_{60} and C_{70}. These are the first two members of the fullerene family of cage-like carbon molecules.

computed directly, or are derived from the dipole oscillator strength distribution as a by-product of collective electron resonance calculations (somewhat analogously to metal clusters, fullerenes also display strong photoabsorption resonances [258] which can be interpreted as collective oscillations of the σ and π electrons, see, e.g., [259, 260] and references therein). Some theories treat all 240 σ and π electrons fully, while others concentrate on the mobile electrons and assign the rest to a relatively rigid background. In general, the various calculated values of pure C_{60} polarizability fall within a factor of two of each other. Section 4.10.5 contains a representative listing of theoretical results.

To date, no direct measurements of polarizabilities of isolated fullerenes have been reported. Experimental results for the polarizability have been derived primarily from the dielectric constant of fullerite, the molecular solid of C_{60}. Since fullerite is a collection of van der Waals bonded molecules, its dielectric response can be visualized as arising from a lattice of point dipoles, and the Clausius-Mossotti relation [cf. Sec. 5.1.1 and Eq. (3.94)]

$$\alpha = \frac{3}{4\pi n}\frac{\varepsilon_0 - 1}{\varepsilon_0 + 2}, \tag{3.95}$$

can be used to determine the polarizability from the low-frequency dielectric constant ε_0 of the crystal. (Here $n = 1.40 \times 10^{-3}$ Å^{-3} is the number density of dipoles in the crystal [251].) A review of the experimental measurements of the low-frequency dielectric function $\varepsilon(\omega)$ of C_{60} by optical and electron-energy-loss methods can be found in [159]; an example is shown in Fig. 3.25. The experimental values of ε_0 range from 3.6 to 4.4, corresponding to $\alpha \simeq$ 80-90Å^3, with a scatter on the order of 20% (see the listing in Sec. 4.10.5). Intermolecular interactions and higher-order moments may introduce corrections to the Clausius-Mossotti relation, but the importance of such corrections in fullerite crystals has not yet been assessed.

Similar to the case of metallic clusters, Section 3.5.1(f), the high polarizability of the fullerene molecule manifests itself in strong long-range interactions with external electrons, molecules and atoms. It has been proposed that the electron-cluster polarization potential $V_p = -\alpha e^2/(2r^4)$ makes an important contribution to the photodetachment cross section of C_{60}^- near threshold [261] (see Fig. 6.12), to slow electron capture by C_{60} [262–264], as well as to the on-site interaction parameter U of the Hubbard Hamiltonian in bulk fullerite crystals[159]. Similarly, van der Waals cross sections for collisions between C_{60} and neutral alkali atoms, clusters, and atoms have been measured [220, 265] and found to be extremely large (see, e.g., Fig. 6.14), in agreement with the expected $V \propto \alpha(C_{60})$ dependence of the attractive van der Waals potential.

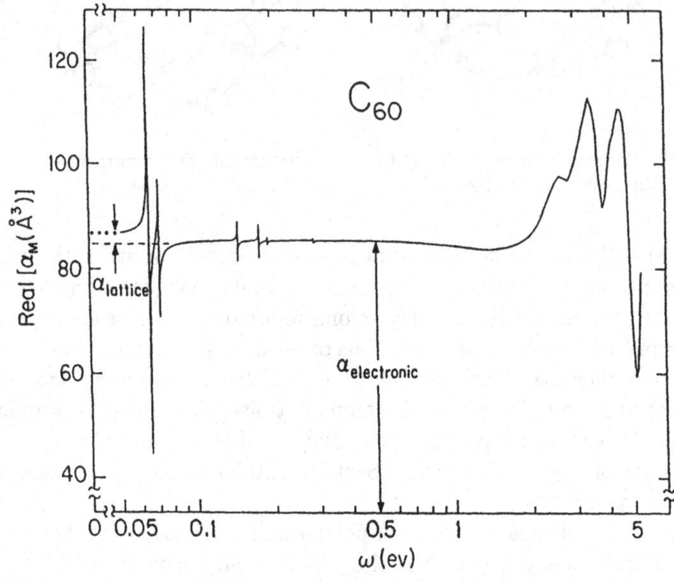

Figure 3.25: The real part of the complex polarizability $\alpha(\omega)$ for solid C_{60} obtained from the dielectric function $\varepsilon(\omega)$ of solid fullerite films using the Clausius-Mossotti relation Eq. (3.94). From these data, the electronic and vibrational contributions to the static polarizability α_0 were estimated to be $\alpha_{\text{electronic}} \approx 83$ Å^3 and $\alpha_{\text{lattice}} \approx 2$ Å^3. (Reprinted from [266] with the kind permission of Elsevier Science Publishers.)

(b) Higher fullerenes. So far there exists no experimental information on the polarizabilities of larger members of the fullerene family. Since the higher-order fullerenes such as C_{70}, C_{76}, ... are expected to be ellipsoidal in shape, a tensor polarizability is needed to describe the electronic response in these species. An interesting question concerns the dependence of the polarizability on the number n of atoms in the fullerene. An atomic additivity picture argues that $\alpha \sim n$, while if the polarizability resembles that of the corresponding sphere or ellipsoid, then α should scale with $R^3 \propto n^{3/2}$. *Ab initio* calculations [267] found $\alpha(C_{70})/\alpha(C_{60}) \simeq (7/6)^{1.25}$, while an extended Hubbard Hamiltonian treatment yielded an exponent of 1.66 for the same ratio [268]. The semiempirical model of point-atom arrays interacting by means of their induced monopole and dipole moments [252], applied to 46 fullerenes in the range C_{20}-C_{240} predicts a middle-of-the road value $\alpha \sim n^{1.30}$.

(c) Endofullerenes and other derivatives. A number of interesting structures have

been prepared on the basis of fullerene molecules: endofullerenes (atoms, ions, or small molecules encapsulated inside the carbon cage) [159], clusters of fullerenes $(C_{60})_n$ [269, 270], "buckyonions" (concentric fullerene shells) [159, 271], and fullerenes surface-coated by alkali atoms [272]. All of the above can be expected to display interesting polarizability properties, and although no experimental data are yet available, some theoretical predictions have been published. Metal atoms encased inside C_{60} molecules are partially ionized due to charge transfer to the cage; at the same time the inside confining potential is rather flat. This implies that an external field can strongly displace the encapsulated ion and give rise to a large dipole moment, corresponding to a very high polarizability. A classical estimate [80], taking into account the shielding of the ionic charge by the carbon cage [273], concludes that an extremely large vibrational contribution to the polarizability results, with a magnitude inversely proportional to the temperature of the system. For example, the $[Mg@C_{60}]^{2+}$ complex (by convention, the @ sign denotes an endofullerene structure) is supposed to have a vibrational polarizability as high as 1400 $Å^3$ at T=20K, exceeding that of pure C_{60} by well over an order of magnitude. In the "opposite" picture of metal atoms coating the outer surface of C_{60}, a gradual evolution from fullerene to metallic-like polarizability will take place as the metal coverage increases. An estimate for the $[C_{60}Na_n]^+$ cluster [274] suggests that the rise in polarizability per atom is gradual, and, for example, for n=93 the carbon-sodium interface still has a role in the response.

Chapter 4

Theory

4.1 Overview

In evaluating the dipole polarizability of a many-particle system, one is faced with the following general problem. In the absence of an external field, the Hamiltonian for M particles is, neglecting spin

$$\hat{H}_0 = -\sum_{l=1}^{M} \frac{\hbar^2}{2m_l} \nabla_l^2 + \frac{1}{2} \sum_{l \neq m} \frac{q_l q_m}{|\boldsymbol{r}_l - \boldsymbol{r}_m|}. \tag{4.1}$$

The summation applies to all electrons and nuclei in the system. For an electron-only calculation one would write

$$\hat{H}_0^{(N\ electrons)} = -\sum_{i=1}^{N} \frac{\hbar^2}{2m_e} \nabla_i^2 - e \sum_{i=1}^{N} \Phi(\boldsymbol{r}_i) + \frac{1}{2} \sum_{i \neq j}^{N} \frac{e^2}{|\boldsymbol{r}_i - \boldsymbol{r}_j|}, \tag{4.2}$$

where Φ is the electrostatic potential due to the nuclei (or inert ions).

Let the unperturbed many-particle wavefunction of the ground state of the system be $\Psi^{(0)}(\{\boldsymbol{r}_l\})$ and the corresponding energy $W^{(0)}$. When a uniform external electric field $\mathbf{E}$ is applied to the system, the interaction of the field with the system is given by the perturbing Hamiltonian

$$\hat{H}' = -\mathbf{E} \cdot \sum_{l=1}^{M} \mathbf{p}_l = -\mathbf{E} \cdot \sum_{l=1}^{M} q_l \boldsymbol{r}_l. \tag{4.3}$$

Hence, it is important to calculate the resultant dipole moment, i.e., the expectation value

$$\mathbf{p} \equiv \langle \Psi | \sum_{l=1}^{M} \mathbf{p}_l | \Psi \rangle, \tag{4.4}$$

where Ψ is the new wavefunction of the full Hamiltonian. Since the focus here is on *linear* response, one is interested in the lowest-order result for the polarization. This quantity can be evaluated in a variety of ways. While all of them must be equivalent in principle, in practice they lend themselves to different implementations and approximations, and in the end one is faced with a potpourri of methods. Without attempting completeness, we will start with an overview of the basic lines of attack, and then proceed to comment on some of the theoretical tools.

Many calculations of atomic and molecular polarizabilities seek to evaluate the expectation value of the dipole moment by means of the first-order form

$$\langle \Psi^{(1)} | \mathbf{p} | \Psi^{(0)} \rangle. \tag{4.5}$$

Here $\Psi^{(1)}$ is the wavefunction of the system in the first-order of perturbation theory. It is usually determined by Hartree-Fock and related techniques.

Instead of using wavefunctions directly, one may calculate the second-order energy shift $W^{(2)}$ of the system and derive the linear polarizability as

$$\alpha_0 = -\frac{d^2 W^{(2)}}{d\mathrm{E}^2}. \tag{4.6}$$

Although Eq. (4.6) is written for the case of isotropic polarizability, it can be readily extended to the tensor case. For very small systems, such as the He atom or the H_2 molecule, the energy shift, i.e., the Stark effect, can be accurately determined by the variational method with a trial wavefunction (see, e.g., [275, 276] for an early review). For systems containing many particles, the energy shift is calculated within the self-consistent field framework: the Hartree-Fock approximation and its generalizations, extended Thomas-Fermi and density functional theory, and field-theoretical (diagrammatic) many-body techniques.

As described in Sections 2.1 and 3.2.1, second-order perturbation theory also allows the polarizability of the kth state to be expressed in terms of the dipole oscillator strength distribution

$$\alpha_0 = \frac{e^2}{m_e} \sum_{l \neq k} \frac{f_{kl}}{\omega_{kl}^2}, \tag{4.7}$$

where

$$f_{kl} = \frac{2 m_e \omega_{lk}}{3 \hbar g_k} \left| \langle k | \sum x_i \, | l \rangle \right|^2 \tag{4.8}$$

This elegant formula relates the dynamic and static properties of any quantum-mechanical system: if the frequencies and strengths of its principal transitions are known, the response to a static field can be derived. Recall that the summation includes all electronic states, both discrete and continuum. The oscillator strengths of the dominant transitions may be taken from experiment, from a calculation of the excitation spectrum, or from estimates based on general sum rules.

Another possible route is to calculate the field-induced density $\delta n(\boldsymbol{r})$ instead of the perturbed wavefunction. The induced dipole moment then follows as

$$\mathbf{p} = \int \delta n(\boldsymbol{r}) \boldsymbol{r} d^3 r. \tag{4.9}$$

The self-consistent field methods mentioned above are used to calculate the induced charge density.

Finally, for some classes of molecules one can appeal to semi-empirical additivity methods which deduce the polarizability of the molecule from the properties of its individual subgroups and their bonds.

All of the approaches listed above have been extensively explored. Their development and application to atoms, molecules and clusters are discussed in detail in a number of reviews [1, 25, 29, 172, 185, 277, 278]. Additional general references can be found in the sections to follow.

4.2 Oscillator Strengths and Sum Rules

4.2.1 Summation of oscillator strengths

If oscillator strengths of the dominant dipole transitions are known, either from experiment or a calculation, then the static polarizability $\alpha(0)$ can be found from the summation Eq. (4.7). An early application of this formula to alkali atoms resulted in values accurate to an estimated 5%-10% [277, 279]. A calculated number that agrees with experiment to better than 1% can be obtained for rubidium using newer measurements of the oscillator strengths [280].

Even if it is not feasible to obtain all f_{kl}, Eq. (4.7) is still extremely useful for estimating bounds on the polarizability. Its utility is based on the fact that oscillator strengths satisfy the famous Thomas-Reiche-Kuhn (TRK) sum rule [281]

$$\sum_{l\neq k}\!\!\!\!\!\!\!\int f_{kl} = N, \tag{4.10}$$

where N is the total number of electrons. Thus if only a subset l' of transitions have known oscillator strengths, then a lower bound on α_0 is given by

$$\alpha_0 \geq \alpha_{\min} = \frac{e^2}{m_e} \sum_{l'\neq k}\!\!\!\!\!\!\!\int \frac{f_{kl'}}{\omega_{kl}^2}, \tag{4.11}$$

while an upper bound is found by assigning the remaining oscillator strength, $N - \sum_{l'\neq k} f_{kl'}$, to the smallest possible energy difference ω_{min} [25]:

$$\alpha_0 \leq \alpha_{\min} + \frac{e^2}{m_e}\left[\frac{N - \sum_{l'\neq k} f_{kl'}}{\omega_{\min}^2}\right]. \tag{4.12}$$

For most alkali metal atoms nearly the entire contribution comes from the strong $n_0S \rightarrow n_0P$ resonance transition of the valence electron (n_0 is the principal quantum number), e.g., the well-known sodium D-line[1]. Since valence electron transitions dominate the polarizability, one may take $N \approx 1$ and count only these one-electron states. High accuracy can be achieved by the introduction of a model potential for the motion of the valence electron in the presence of a frozen core [282].

[1]Among group I elements, this approximation is poorest for atomic lithium and hydrogen. For the latter, the contribution of higher states to the $1s$ ground state polarizability is [29]: $2p$-66%, $3p$-9%, $4p$-3%, $5p$-1%, continuum states - 19%.

4.2.2 Sum rules

The TRK sum rule is the most famous one in a series of so-called energy-weighted sum rules satisfied by quantum-mechanical oscillator strengths. These identities are very general and are useful as controls on model calculations. In cases where most of the dipole transition strength is concentrated within a relatively narrow spectral region or within a small number of regions, they can provide direct estimates of response parameters. Sum rules have been used extensively in nuclear physics [283–285] and more recently in the study of clusters [185–187, 286]. The power of sum rules comes from the fact that certain energy-weighted moments of the spectral function can be expressed purely in terms of ground-state properties of the system.

The pth moment of the dipole oscillator strength is defined as

$$I_p = \sum_{l \neq k} f_{kl} \omega_{kl}^p. \tag{4.13}$$

These dipole oscillator strength sums I_p are also called Cauchy moments for reasons we shall see at the end of this section. Many fascinating theorems can be derived for various values of p. Here let us consider only p=0, -2, and 2. The first sum is just the TRK case Eq. (4.10), and the second sum is related to the static polarizability via Eq. (4.7). Thus $I_0 = N, I_2 = (m_e/e^2)\alpha_0$.

If the dipole transition strength is concentrated around a strong resonance at frequency $\tilde{\omega}$, then we can write $I_{-2} \approx I_0/\tilde{\omega}^2$, which brings us to a simple estimate analogous to that discussed above:

$$\alpha_0 \approx \frac{Ne^2}{m_e \tilde{\omega}^2}. \tag{4.14}$$

This estimate works well, e.g., for alkali atoms, simple metal clusters (see Section 3.5.1), and fullerenes (Section 3.5.3). The characteristic resonance frequency may be taken from experiment or from a theoretical calculation.

If $|k>$ is the ground state, the p=2 moment turns out to be related to the spatial derivative $\partial^2\Phi(\vec{r}_i)/\partial x_i^2$ of the nuclear potential in Eq. (4.2) and to the ground-state electron density $\rho_e(\vec{r})$ [284, 287]. For the special case of a spherically symmetric ground state one can simplify this further by making use of the Poisson equation for Φ, obtaining

$$I_2 = \frac{4\pi}{3} \frac{e^2\hbar^4}{m_e} \int \rho_+(\vec{r})\rho_e(\vec{r})d^3r, \tag{4.15}$$

where ρ_+ and ρ_e are the nuclear and electronic densities, respectively. It is interesting to look at a couple of special cases. For an atom with a point-like nucleus of charge Z this reduces to

$$I_2 = \frac{4\pi}{3} \left(\frac{Ze^2\hbar^4}{m_e} \right) \rho_e(0). \tag{4.16}$$

For a spherical cluster with a uniform positive background one obtains

$$I_2 = N\omega_M^2(1 - \frac{\Delta N}{N}), \tag{4.17}$$

where ω_M is the classical Mie resonance frequency for a conducting sphere defined in Eq. (3.90) and ΔN is the amount of valence electron spill-out beyond the cluster edge.

If we again assume that the dipole photoabsorption spectrum is dominated by one resonance, then we can form the estimate $\tilde{\omega}^2 \approx I_2/I_0 = I_2/N$. This estimate for $\tilde{\omega}^2$ can then be inserted into Eq. (4.14), yielding an interesting connection between the static, ground-state dipole polarizability and the ground-state charge distribution, Eq. (4.15). We already commented on the significance of this result for metal clusters in Section 3.5.1(e); an application to fullerenes is discussed in Ref. [288], and the more general form of I_2 is used for non-spherical clusters in [185, 287].

In general, the frequency-dependent polarizability can be described in terms of dipole oscillator strength sums I_p, and it is particularly useful when the frequency is far away from an absorption band. A sum-over-states expression similar to Eq. (4.11) can be written as

$$\alpha_0(\omega) = \frac{e^2}{m_e} \sum\!\!\!\!\!\!\int_{l \neq k} \frac{f_{kl}}{(\omega_{kl}^2 - \omega^2)}, \tag{4.18}$$

For frequencies less than the first transition frequency ω_1, this expression can be expanded to give the Cauchy formula

$$\alpha_0(\omega) = \sum_{p=1}^{\infty} I_{-2p} \omega^{2p-2}, \tag{4.19}$$

where I_{-2p} are the Cauchy moments defined in Eq. (4.13). Cauchy moments are discussed in [289]. Besides their usefulness in expressing polarizabilities, Cauchy moments are important in other basic properties including the dipole-dipole interaction constant C_6 [290], second hyperpolarizabilities [291], and pair polarizabilities [292]. Recent interest has been focused on metal atoms [24, 282, 293–296].

4.2.3 Expression for the anisotropic case

The discussion above dealt with the scalar, or average, polarizability α_0; the more general expressions corresponding to Eq. (4.7) are:

$$\alpha_i(\omega) = \frac{e^2}{m_e} \sum\!\!\!\!\!\!\int_{l \neq k} \frac{f_{kl}}{(\omega_{lk}^2 - \omega^2)} \phi_i(J_k, J_l) \quad i = 0, 2, \tag{4.20}$$

$$\alpha_1(\omega) = \frac{3}{2} \frac{e^2}{m_e} \sum\!\!\!\!\!\!\int_{l \neq k} \frac{\omega f_{kl}}{\omega_{lk}(\omega_{lk}^2 - \omega^2)} \phi_1(J_k, J_l). \tag{4.21}$$

The factor ϕ_2 that occurs in the tensor polarizability α_2 (see Eq. (3.13)) and the factor ϕ_1 that occurs in the vector polarizability (see Eq. (3.13)) change sign depending on J. Hence, accurate values of oscillator strengths are especially critical when computing α_1 or α_2.

4.3 Hartree-Fock Method and Perturbation Theory

4.3.1 Hartree-Fock Equations

The most famous and widely used procedure for determining the wavefunctions of a many-electron system is the Hartree-Fock (HF) method. It assumes that the total wavefunction Ψ is an antisymmetrized product (Slater determinant) of single-electron wavefunctions (orbitals), i.e.,

$$\Psi(1,2\cdots,N)=\frac{1}{\sqrt{N}}\begin{vmatrix}\psi_1(1) & \psi_2(1) & \cdots & \psi_N(1)\\ \psi_1(2) & \psi_2(2) & \cdots & \psi_N(2)\\ \vdots & \vdots & \cdots & \vdots\\ \psi_1(N) & \psi_2(N) & \cdots & \psi_N(N)\end{vmatrix}=\frac{1}{\sqrt{N}}\det\{\psi_1(1)\psi_2(2)\cdots\psi_N(N)\} \tag{4.22}$$

where the single-electron orbitals are

$$\psi_l(j)=\phi_l(\mathbf{r}_j)\chi_l(m_j). \tag{4.23}$$

Here $\phi_l(\mathbf{r}_j)$ are spatial orbitals and the spin orbitals $\chi_l(m_j)$ are either α or β spin functions depending on whether $m_j = \pm 1/2$. The HF equations are derived by minimizing the total energy of the system, given by [170, 281, 297]

$$\begin{aligned}\left\langle \hat{H}_0^{(N\ electrons)}\right\rangle &= \sum_i \int d^3r\phi_i^*(\boldsymbol{r})\left[-\frac{\hbar^2}{2m}\nabla^2+\Phi(\boldsymbol{r})\right]\phi_i(\boldsymbol{r})\\ &+\frac{1}{2}\sum_{i,j}\int d^3rd^3r'\frac{e^2}{|\boldsymbol{r}-\boldsymbol{r}'|}\left|\phi_i(\boldsymbol{r})\right|^2\left|\phi_j(\boldsymbol{r}')\right|^2 \\ &-\frac{1}{2}\sum_{i,j}\int d^3rd^3r'\frac{e^2}{|\boldsymbol{r}-\boldsymbol{r}'|}\phi_i^*(\boldsymbol{r})\phi_i(\boldsymbol{r}')\phi_j^*(\boldsymbol{r}')\phi_j(\boldsymbol{r}')\delta_{m_im_j},\end{aligned} \tag{4.24}$$

with respect to the orbitals ϕ_i . Here $\Phi(\boldsymbol{r})$ is the ionic potential as in Eq. (4.2) . The result is

$$\hat{f}(\boldsymbol{r})\phi_l(\boldsymbol{r})=\varepsilon_l\phi_l(\boldsymbol{r}), \tag{4.25}$$

where $\hat{f}(\boldsymbol{r})$ is the Fock operator:

$$\hat{f}(\boldsymbol{r})=-\frac{\hbar^2}{2m}\nabla^2+\Phi(\boldsymbol{r})+v^{HF}(\boldsymbol{r}). \tag{4.26}$$

Here v^{HF} is the effective self-consistent single-particle potential made up of the direct and exchange Coulomb terms originating in the last two lines of Eq. (4.25). The eigenvalues ε_l are the Lagrange multipliers of the variational problem. In general they cannot be assumed to represent the energies of single-particle orbitals, but may give a good approximation to the removal energy of the ith electron("Koopmans theorem" [281]).

The essence of the HF method is that it defines an effective single-electron Hamiltonian which depends on all the other electrons through their *average* Coulomb and exchange fields. The entire set of single-particle wavefunctions must be solved simultaneously until self-consistency is achieved. This type of approach is termed a Self-Consistent Field (SCF) technique. It is an important feature of HF equations that for closed-shell systems they lead rigorously to a spherically symmetric self-consistent field [281]. This justifies the use of the central-field concept in closed-shell atoms and metallic clusters, affording a great simplification in the calculation of polarizabilities and other properties. In atoms with incomplete shells the central-field approximation still can be defined by appropriate averaging and remains quite accurate [281], while in clusters this is no longer possible due to strong Jahn-Teller deformations (see Section 3.5.1(a)). Systems which can be described by pairs of spin up/down electrons in closed-shell orbitals (closed-shell ground states) can be treated most compactly; this case is called *closed-shell* Hartree-Fock. In dealing with open-shell situations, one encounters two procedures: *restricted* HF (whereby all electrons, except those specifically assigned to open-shell orbitals, occupy closed-shell ones), and the most general *unrestricted* HF method.

For atoms (see, e.g., [298, 299] and references therein) and spherically symmetrical clusters [208] the wavefunctions can be computed from a direct solution of the HF equations. However, this frontal approach may quickly turn intractable, in which case approximation methods become necessary. These typically involve expanding the single-particle orbitals in terms of a convenient basis set.

4.3.2 Basis set expansions of Hartree-Fock wave functions

Solving the HF equations is commonly done using a method introduced independently by Hall [300] and Roothaan [301]. The basic idea is to express each molecular orbital (MO) $\phi_l(\mathbf{r}_k)$ as a linear combination of atomic orbitals (LCAO), i.e.

$$\phi_l(\mathbf{r}_k) = \sum_t^M c_{tl}\varphi_t, \tag{4.27}$$

where the wavefunctions φ_t are from some suitably chosen set. The quantity M is the number of basis wavefunctions chosen to represent the molecular orbitals $\phi_l(\mathbf{r}_k)$. The key to the method is to then find the coefficients c_{tl}. The molecular orbital $\phi_l(\mathbf{r}_k)$ obviously depends critically on the size and choice of the basis set φ_t. Calculations require evaluation of Coulomb direct and exchange integrals of the form

$$V_{ijkl} = \int d^3\mathbf{r} d^3\mathbf{r}' \varphi_i^*(\mathbf{r})\varphi_j(\mathbf{r}')\frac{e^2}{|\mathbf{r}-\mathbf{r}'|}\varphi_k^*(\mathbf{r}')\varphi_l(\mathbf{r}). \tag{4.28}$$

Since these integrals contain four molecular orbitals, the computational time goes as M^4. Hence, limiting the size of the basis set can significantly increase the speed of the computation (at the cost of potential loss of accuracy).

For atoms and small molecules the basis set frequently consists of Slater-type orbitals (STO) of the form [297, 299, 302–304]

$$\varphi_{nlm}(r,\theta,\phi) = N r^{n-1} e^{-\gamma r} Y_{lm}(\theta,\phi). \tag{4.29}$$

where N and γ are constants and $Y_{lm}(\theta,\phi)$ is the mth-component of the lth spherical harmonic. Use of STO's has been limited to atoms and diatomic molecules because it is difficult and time-consuming to evaluate multi-center electrostatic integrals of the form given in Eq. (4.28), where the different φ_l are STO basis functions centered at locations i, j, k, l.

Polyatomic molecules are evaluated using Gaussian-type orbitals (GTO) first suggested by Boys [305]. Cartesian GTO's are of the form

$$\varphi_{tuv}(\mathbf{r}) = N x^t y^u z^v e^{-\gamma r^2} \qquad t,u,v,=0,1,2,3.... \tag{4.30}$$

For example, the GTO exponents for a $2p_y$-orbital will have the values $t = v = 0$ and $u = 1$. The primary advantage of GTO's in calculating polarizabilities is that all integrals like Eq. (4.28) can be evaluated analytically. A useful property of GTO's is that the product of two or more Gaussian functions is another Gaussian function. A disadvantage is that large basis sets are needed since atomic orbitals (which can have nodes) are not well-approximated by Gaussian functions(where nodes are absent). Far an in depth discussion of Gaussian basis sets, we refer the reader to one of the many books on the subject, e.g. see Ref. [306, 307].

A review of the HF method and the basis sets used in applying the method can be found in Ref. [297, 308] and references therein. An extensive discussion of the application of LCAO theory to conjugated organic molecules is given in [120]. For a more complete review of basis sets see Ref. [309]. For tabulations of basis sets see the book [310]. There are a number of simplified SCF schemes that are frequently applied in the literature to atoms and small molecules. For the *valence* electrons it is sometimes assumed that the basis set functions do not overlap in $\mathbf{r}$-space. Mathematically, such semi-empirical zero-differential overlap (ZDO) approximations are expressed as

$$\varphi_k(\mathbf{r})\varphi_l(\mathbf{r}) = |\varphi_l(\mathbf{r})|^2 \delta_{kl}. \tag{4.31}$$

STO's are used and the basis set is minimal, that is the number M of basis set functions equals the number N of electrons: $M = N$. In this case the Coulomb matrix elements V_{ijkl} do not all survive. Only the direct integrals are nonzero, i.e. only $V_{iikk} = 0$. This is the complete neglect of differential overlap (CNDO) approximation. One other variant that occurs is the modified neglect of differential overlap (MNDO) that assumes the ZDO approximation applies only to functions centered at *different* atoms. The key attraction of these approximations is that the number of computations goes as M^2 instead of M^4.

Another important approximation scheme is the pseudo-potential method [171, 311, 312]. Here the full problem of the valence electrons interaction with the atomic core and with the nucleus is replaced by an effective potential for the valence electrons. Hence another term for this approach is the "effective core potential" method.

The pseudopotential accounts for the screening of the nucleus by the core electrons and generates valence orbitals which have the correct form in the outer atomic regions. The usual approach is to take a model pseudopotential (which in general may be non-local) with several adjustable parameters, and the art is in deriving the best and sufficiently simple forms that have a wide range of applicability. Recently pseudopotentials have been extensively used in calculations of cluster response.

4.3.3 Perturbation theory in the Hartree-Fock approximation

Thus far we have been concerned with the determination of HF wavefunctions for the unperturbed system. Now consider the situation when a weak uniform external field is applied to the system, as in Eq. (4.3). In the spirit of perturbation theory, one wishes to find the first-order correction to the total wave function to use in Eq. (4.5). Within the HF approximation, the correction to the total wavefunction is built up of corrections to the single-particle orbitals Eq. (4.23) which make up the Slater determinant Eq. (4.22) :

$$\phi_l(\mathbf{r}) = \phi_l^{(0)}(\mathbf{r}) + \phi_l^{(1)}(\mathbf{r}) + O(\mathrm{E}^2), \tag{4.32}$$

accompanied by a corresponding change in the HF eigenvalues:

$$\varepsilon_l = \varepsilon_l^{(0)} + \varepsilon_l^{(1)} + O(\mathrm{E}^2). \tag{4.33}$$

The most consistent course of action is to insert these corrected functions into the Slater determinant and, using the perturbed Hamiltonian $\hat{H}_0 + \hat{H}'$ defined in Eqs. (4.2) and (4.3), work out a new set of self-consistent single-particle equations to complement the original Eqs. (4.25). This approach, first presented by Dalgarno [313], Kaneko [314], and Allen [276], is known as the "coupled Hartree-Fock approximation." It results in a set of coupled integro-differential equations for the orbital wave functions. In the radially-symmetric case it reduces to a coupled set of radial equations, but even here the solution requires quite laborious computations and is not practical for systems other than moderate-sized atoms. It does, however, represent the full HF solution correct to first order. Further details and references can be found, e.g., in [25, 277, 315, 316].

The so-called "uncoupled Hartree-Fock approximation" [277, 315] represents a less precise but time-saving shortcut. Instead of substituting the perturbed Slater determinant into the full many-body Schrödinger equation, one simply augments the single-particle HF equation with an external-field term:

$$[\hat{f}(\boldsymbol{r}) + e\mathbf{E}\cdot\boldsymbol{r}]\phi_l(\boldsymbol{r}) = \varepsilon_l\phi_l(\boldsymbol{r}). \tag{4.34}$$

This formulation allows one to calculate the corrected orbitals efficiently, e.g., by means of conventional single-particle perturbation theory (making use of the fact that the HF wavefunctions ϕ_l form a complete set). The price one pays is that the mutual influence of the orbitals on each other is no longer accounted for, or in other words the phenomenon of "correlation" between electrons is omitted. Correlation effects, also known as screening, can be extremely important in systems with mobile electrons: many-electron atoms, metallic clusters, conjugated organic molecules, etc. Here the electron cloud redistributes itself in response to the external field, and it is no longer proper to assume that each orbital sees the same bare applied field as implied by Eq. (4.34) . In cases where screening cannot be neglected, correlation effects must be taken into account. Within the boundaries of the HF approximation this implies use of the fully-coupled formalism. Ultimately, one has to proceed to many-body perturbation theory, including diagrammatic methods, discussed below.

4.3.4 Shielding factors

It is desirable to have the ability to assess the quality of a perturbation calculation against a rigorous standard. Earlier we saw how energy-weighted sum rules define one such benchmark. For verifying the precision of a perturbation HF calculation, dipole shielding factors [277, 317] serve a similar purpose. The concept is based on a simple and elegant idea. Imagine an isolated atom placed in a uniform external electric field. If the atom is neutral, it experiences no net force and will not move anywhere. This implies that the total force on the nucleus is zero (otherwise it would start moving), i.e., the electrons must completely shield the electric field at the center of the atom.

Now, the electric field at the nucleus due to the dipole-polarized electron cloud is

$$\mathrm{E}_{ind}(0) = \int \frac{\delta n(\boldsymbol{r})}{r^2} \cos\theta d^3r, \tag{4.35}$$

where δn is the electron charge density induced by the field. This quantity can be immediately calculated from the perturbed wavefunctions. The accuracy of a perturbation calculation can be assessed by whether the dipole shielding factor $\beta \equiv \mathrm{E}_{ind}(0)/\mathrm{E}_{external}$ is reasonably close to 1. It is straightforward to generalize this argument to show that for ions of net charge $(Z-N)$ the shielding factor must equal N/Z.

4.4 Configuration Interaction

4.4.1 Configurations

Even in the ground state, all independent-electron approximations suffer from the disadvantage that electron correlation is not fully accounted for. Each electron's motion is correlated with the location of other electrons in a direct and local way, not only on an average basis. Qualitatively, one can say that there is a so-called correlation hole associated with each electron that prevents the other electrons from coming too close. It is frequently necessary to go beyond the HF approximation to obtain a more accurate description of the ground-state and, especially, of the excited states.

A conceptually (but not computationally!) simple way to generalize the self-consistent field method is by means of the *configuration interaction* (CI) approach. This technique has been extensively reviewed in the literature; detailed discussions can be found, e.g., in [297, 308, 318, 319] and references therein. The basic idea is to represent the exact wavefunction not just as a single Slater determinant, but as a linear combination of N-electron trial functions:

$$\Psi(1,2,\cdots,N) = \sum_k d_k \Phi_k(1,2,\cdots,N), \tag{4.36}$$

where the set d_k are constant parameters that are varied in finding a solution and Φ_k is the kth *configuration*. Each configuration is a determinantal wavefunction like

$$\Phi_k(1,2,\cdots,N) = \frac{1}{\sqrt{N}} \det\{\psi_{1k}(1)\psi_{2k}(2)\cdots\psi_{Nk}(N)\}. \tag{4.37}$$

Actually a single configuration may consist of a sum of Slater determinants like that in Eq. (4.37).

To understand some of the notation used when referring to CI calculations, suppose an HF calculation has been performed on a system and $M > N$ molecular orbitals ψ_i calculated (note: M is the number of orbitals in the set of basis functions and N is the number of electrons). A total of $M - N$ unoccupied orbitals in the HF scheme, called virtual orbitals, will result. Again note that the Hartree-Fock ground-state configuration is [cf. Eq. (4.22)]

$$\Phi_0(1,2,\cdots,N) = \frac{1}{\sqrt{N}} \det\{\psi_1(1)\psi_2(2)\cdots\psi_N(N)\} \tag{4.38}$$

A more general CI calculation incorporates other configurations that have electrons in the virtual MO's. For example, a *singly-excited* configuration, in which an electron has been promoted from level N to $N+1$ is

$$\Phi_1(1,2,\cdots,N) = \frac{1}{\sqrt{N}} \det\{\psi_1(1)\psi_2(2)\cdots\psi_{N-1}(N-1)\psi_{N+1}(N)\}. \tag{4.39}$$

Doubly-excited (*doubly-excited* CI, or DCI) and higher configurations are usually also included. When such excited configurations are combined as in Eq. (4.36), the set of coefficients d_k is not completely arbitrary but must preserve all the good quantum numbers of the system. For example, it is necessary to ensure that the wavefunction $\Psi(1,2,\cdots,N)$ be an eigenstate of the total spin (spin-adapted configuration), etc.

The CI procedure consists of solving the time-independent Schrödinger equation in the Hilbert space spanned by the constructed configurations, i.e., in determining the lowest eigenvalues and eigenvectors of the Hamiltonian matrix $\langle \Psi_A | \hat{H} | \Psi_B \rangle$. For example, in the variational approach the coefficients d_k of the Slater determinants can be allowed to vary to minimize the energy. In multiconfiguration self-consistent field (MC-SCF) theory the orbitals comprising each configuration are also allowed to vary. Since it is possible to identify lowest-energy states with different symmetry properties, the variational procedure yields the ground-level as well as excited-state eigenvalues and eigenfunctions. Modern SCF calculations are frequently carried out using widely disseminated software packages (e.g. Gaussian94 [320], MOPAC 93 [321], GAMESS [322], etc.).

4.4.2 Size Limitations

In principle, CI provides an exact solution of the many-electron problem. However, this would be strictly true only if complete, i.e. infinite, sets of trial wavefunctions could be employed. This ideal would represent a full CI calculation. In practice, one has to employ finite basis sets, but even for small molecules and moderately-sized basis sets the number of N-electron determinants is enormous. Thus the space of virtual orbitals is further reduced by truncating the trial wavefunctions at some excitation level. Still, CI is a very computationally demanding method (accurate calculations may include up to 10^6 configurations [323]); the art of selecting the most appropriate basis set and optimally truncated configurations is discussed in the literature, including the references cited above.

CI ceases to be feasible when a large number of electrons have to be correlated and a large number of excited states included. Furthermore, any CI expansion truncated to low-order excitations may be affected by errors due to so-called *size inconsistency.* For example, consider a supermolecule made up of two individual subunits [308]. If we attempt to describe the subsystems as well as the full unit by means of, say, DCI, we run into an inconsistency: the situation when the subunits are both simultaneously doubly excited represents a *quadruple* excitation of the supermolecule. Put another way, the energy of an N-particle system does not become proportional to N in the limit $N \to \infty$. This indicates that — even apart from ever-increasing computational complexity — there is an intrinsic limit to the applicability of CI to larger systems. (Interestingly, the simpler Hartree-Fock approximation does not suffer from this deficiency.) In practice, CI including quadruple excitations may be effectively size consistent for molecules with less than a few tens of electrons. Metal clusters, in which the energy spectrum becomes progressively denser as cluster size N increases and it becomes necessary to include more and more excited configurations, have been analyzed by CI methods for up to $N \simeq 10 - 20$ [323, 324].

An extensive family of approximations designed to fortify the CI approach is known as *pair* and *coupled-pair* theories. For example, the idea of the Coupled-Cluster Approximation is to truncate the full CI hierarchy by approximating the coefficients of quadruple configurations as sums of products of the coefficients of doubles. An introduction to this vast area can be found, e.g., in Ref. [308] and references therein.

A consistent description of the quantum states of an arbitrary-sized interacting many-particle system is given by *many-body perturbation theory,* which also provides a systematic recipe for deriving the linear response properties. A brief account of this formalism and some approximate forms is given in Section 4.7.

4.4.3 Calculation of Polarizabilities

The calculation of static dipole polarizabilities in the CI framework is usually done by the *finite field method.* It was first employed by Cohen and Roothaan for atoms within the Hartree-Fock approximation [325]. The perturbation of a small but finite external field, Eq.(4.3), is added to the Hamiltonian, and the change in energy or the induced dipole moment computed for a series of electric field strengths. The polarizability is then evaluated from numerical differentiation via Eqs. (4.4) or (4.6). The crucial factor in this procedure is the choice of the basis set; in particular, the response of the valence regions of the electronic cloud must be adequately described. This issue is reviewed, e.g., in Ref. [326], which illustrates the point by showing how the calculated polarizability and its anisotropy for N_2 can vary by as much as a factor of 2 or 3 depending on the basis set used. Instead of augmenting the Hamiltonian with a uniform-field perturbation, one may also simulate a finite external field by introducing one or several point charges outside the molecule [327, 328]. This method can be particularly useful for the calculation of higher-order polarizabilities.

4.5 Density-Functional Theory

4.5.1 Basic theorem

The self-consistent field methods discussed so far have at their core the calculation of wave functions that can be used to determine response properties. In contrast, density-functional theory (DFT) concentrates on the ground state electronic *density* distribution $n(\mathbf{r})$. A detailed treatment of this approach can be found in a number of recent books and reviews, see, e.g., [278, 329–331].

DFT is built upon the fundamental theorem by Hohenberg and Kohn, which states that all the *ground state* properties of an electron system are fully determined by $n(\mathbf{r}) = |\Psi_0|^2$ (here Ψ_0 is the many-particle wavefunction of the ground state). Note that this is a stronger statement than the general quantum mechanical description which requires a knowledge of the full wave function and not just its modulus. There is thus a one-to-one connection between $n(\mathbf{r})$ and the ground state energy W; in other words, W is a unique functional of the density:

$$W[n] = T[n] + \int [v_i(\mathbf{r}) + v_{ext}(\mathbf{r})]n(\mathbf{r})d^3r + \frac{1}{2}\int \frac{n(\mathbf{r})n(\mathbf{r}')}{|\mathbf{r}-\mathbf{r}'|}d^3rd^3r' + W_{xc}[n]. \quad (4.40)$$

This equation divides the total energy into several specific components: the kinetic energy functional T, the energy due to the ionic $v_i(\mathbf{r})$ and externally applied $v_{ext}(\mathbf{r})$ fields, the direct electron-electron Coulomb interaction term, and finally the so-called exchange-correlation functional. The latter is frequently split up into two parts: $W_{xc} = W_x + W_c$, where W_x accounts for the exchange interaction ("exchange energy") while W_c sweeps all the remaining quantum corrections under the blanket term of many-body "correlation energy." The density distribution $n(\mathbf{r})$ is to be found variationally as that which minimizes the ground state energy W.

While it is gratifying that the ground state can be characterized completely in terms of the electron density, the real problem is to understand the functionals in Eq. (4.40). The Hohenberg-Kohn theorem states that such functionals must exist, but provides no information about their specific forms; in general these will be complicated and non-local operators. Therefore the main aim of DFT is to design workable and reasonably accurate approximations to the functionals.

In practice, there are two main schemes for working with Eq. (4.40). One is to employ the density n directly; this leads to the Extended Thomas-Fermi (ETF) model. The other is to expand the density in terms of a set of single particle wave functions. This formalism is known as the Kohn-Sham (KS) theory. We briefly outline both approaches.

4.5.2 Extended Thomas-Fermi model

In the ETF model, one writes

$$T[n] \approx \frac{\hbar^2}{2m} \int \left[\frac{3}{5}(3\pi^2)^{2/3} n^{5/3} + \frac{\lambda}{4} \frac{(\nabla n)^2}{n} + \ldots \right] d^3r, \tag{4.41}$$

and

$$W_x \approx -\frac{3}{4} \left(\frac{3}{\pi} \right)^{1/3} e^2 \int n^{4/3} d^3r. \tag{4.42}$$

If we keep only the first term in the integrand of Eq. (4.41) and neglect both W_x and W_c, we recover the famous Thomas-Fermi (TF) semiclassical equation [281, 329, 332, 333]). The exchange energy estimate Eq. (4.42) is due to Dirac; adding this term we obtain the Thomas-Fermi-Dirac equation. The so-called gradient-correction term in the integral for $T[n]$ represents the first correction to the TF kinetic energy density for inhomogeneous systems. It was first proposed by von Weizsäcker, who calculated a value of 1 for the constant λ. Subsequent refinements showed that it is part of a systematic expansion in terms of gradients and higher powers of the density. They also corrected the value of the constant to $\lambda = 1/9$. In practice, it appears that different values of λ have different regions of validity [329], and this constant is sometimes treated as an adjustable parameter in calculations.

Several approximate forms for the correlation energy density have been proposed. Typically, they are taken over from the theory of the homogeneous electron gas and represent analytical interpolations between low-density and high-density limits. The original and still frequently used formula is due to Wigner:

$$e_c(n) \approx -\frac{0.88}{r_s(n) + 7.8}, \tag{4.43}$$

where e_c is in atomic units and is written in terms of the dimensionless electronic Wigner-Seitz radius $r_s(n) = (3/4\pi n a_0^3)^{1/3}$, and a_0 is the Bohr radius. Alternative expressions for $e_c(n)$ have been proposed by Gunnarsson and Lundqvist, Ceperley and Alder, and others. A thorough discussion of these correlation densities can be found in the sources quoted above. While differing in detail, they lead to qualitatively similar results.

In all cases, the correlation energy is written in terms of the energy density as

$$W_c = \int n e_c(n) d^3r, \tag{4.44}$$

signifying an assumed local relationship between the density and the correlation energy. This assumption forms the basis of the *Local Density Approximation* (LDA) version of the theory. Being simpler computationally, LDA is prevalent in applications. It is frequently found to yield good results even in strongly inhomogeneous systems such as atoms, molecules, and surfaces, although no systematic criterion for its applicability has been established.

4.5.3 Kohn-Sham formalism

Kohn and Sham proposed to write the electron density in terms of a complete set of single-particle "wavefunctions":

$$n(\mathbf{r}) = \sum_{i=1}^{N} |\phi_i(\mathbf{r})|^2, \tag{4.45}$$

where the sum runs over the N lowest states. The advantage of expressing the density in this way is that an exact expression for the kinetic energy term can now be written. In fact, it can be shown (see, e.g., [329]) that if the energy functional Eq. (4.40) is minimized with the density written in the form Eq. (4.45), the single-particle functions are found to satisfy a set of Schrödinger-like equations:

$$\left[-\frac{\hbar^2}{2m}\nabla^2 + v_{eff}(\mathbf{r})\right]\phi_i(\mathbf{r}) = \varepsilon_i \phi_i(\mathbf{r}), \tag{4.46}$$

where

$$v_{eff}(\mathbf{r}) = v_i(\mathbf{r}) + v_{ext}(\mathbf{r}) + \frac{\delta W_{xc}[n]}{\delta n(\mathbf{r})}, \tag{4.47}$$

the last term being the functional derivative of the exchange-correlation functional. These are called the *Kohn-Sham (KS) equations*, and the functions $\phi_i(\vec{r})$ are commonly referred to as Kohn-Sham orbitals.

While in principle the potential v_{eff} is a non-local operator, in practice one usually works within the LDA and makes use of the exchange and correlation functionals described in the preceding sections. This reduces Eqs. (4.46) to a set of local coupled differential equations which are solved self-consistently (either computed directly or expanded in terms of basis sets, see Section 4.3.2). It is important to keep in mind that the eigenvalues ε_i which result from the solution of these equations (corresponding to both occupied and unoccupied levels ϕ_i) *cannot* be identified with physical particle energies.

In the preceding section we pointed out that LDA written in terms of densities is a direct descendant of the extended Thomas-Fermi theory. Similarly, one notices a strong resemblance between the KS equations and those which arise in Hartree-Fock theory. Both represent systems of self-consistent single-particle Schrödinger-like equations. In fact, the KS-LDA formalism has a direct precedent in the Hartree-Fock-Slater theory (also known as the $X\alpha$ method [334]). In the latter, the full HF exchange term is approximated by a local exchange potential of the form Eq. (4.42) and equations identical to Eq. (4.46) result. The one extra component of LDA is the assumption of an additional local correlation potential.

Typically, W_c is only about 10% of the ground state energy [278], but when one is concerned with energy *differences* its contribution may become important, as, for example, in the calculation of single-electron properties or polarizabilities. It should be mentioned that LDA has definite limitations in describing electronic excitations [335]; for example it underestimates band gaps in semiconductors and insulators by as much as 50%. This shows the importance of non-local exchange and correlation effects in describing single-particle energies. As discussed in the references quoted above, a number of refinements of DFT have been attempted: non-local formulations, self-interaction corrections, local spin-density (LSD) theory for open-shell systems, etc.

4.5.4 Calculation of polarizabilities

Since density functional theory yields ground-state energies and electron densities, one can use either Eq. (4.6) or Eq. (4.9) to determine the polarization by an external field.

In the Extended Thomas-Fermi formulation which employs the density directly, a common technique is to use a trial variational function

$$n(\mathbf{r}) = n_0(\mathbf{r}) + \delta n(\mathbf{r}) \tag{4.48}$$

and minimize the total energy in the presence of an external dipole potential v_{ext}.

It is interesting that the ordinary Thomas-Fermi equation leads to infinite dipole polarizabilities and shielding factors [317, 336]. The reason is that the electron density in a neutral TF atom decays to zero too slowly. The same feature is responsible for the inadequacy of the TF model to describe ionization processes. Weakly bound electrons far from the nucleus make the primary contribution to polarization and ionization, and having too many such electrons overestimates the degree of both. It is necessary to account for exchange to obtain a sensible answer. Thus the Thomas-Fermi-Dirac equation improves the situation, although it still considerably overestimates the shielding factors [317]. For a better result, correlation corrections must be included.

In the Kohn-Sham formalism, the calculation of induced dipole moments proceeds analogously to the Hartee-Fock case: either by perturbation theory[2] similar to the

[2]Since the KS theory does not involve any Slater determinants, the perturbation theory equations for single-particle eigenfunctions, while forming an interconnected set, are less complicated than in the coupled HF approximation. They are usually solved by iteration, as discussed in detail in

approach outlined in Section 4.3.3, or by the finite field method already encountered in Section 4.4.3.

Finally, the *Time-Dependent Local Density Approximation* (TDLDA) uses the occupied and vacant KS eigenfunctions and eigenvalues Eq. (4.46) as input for many-body perturbation theory discussed below. Starting with this basis, TDLDA uses many-body linear response techniques (e.g., the random-phase approximation) to compute the static polarizability either directly, or from the correlated dipole excitation spectrum by means of the sum rule Eq. (4.7).

Since, as mentioned earlier, the KS orbitals cannot be identified with a physical particle-hole basis (the Hohenberg-Kohn theorem is strictly limited to the static ground state), TDLDA is essentially an ad hoc procedure. It also carries the danger of double-counting correlation effects [187, 337]. Nevertheless, empirically it is frequently found to produce adequate results (see, e.g., [278, 338], and references therein). In a sense, this is not unreasonable, because we saw that the KS eigenstates are closely related to the Hartree-Fock-Slater wavefunctions.

4.6 Perturbative Corrections to the Hartree-Fock Hamiltonian

We have seen that it is often essential to take into account many-particle correlation effects beyond the HF approximation. One possibility is to proceed to the Configuration Interaction method: in principle, a full CI treatment yields an exact solution. In practice, however, not only is CI very intensive computationally, it also must be limited to a finite number of excitations which results in concerns about size consistency.

An alternative systematic procedure is perturbation theory in terms of corrections to the HF Hamiltonian. We write the exact many-electron Hamiltonian Eq. (4.2) as

$$\hat{H}_0^{(N\ electrons)} = \sum_{i=1}^{N} \hat{f}_i + \hat{V}, \tag{4.49}$$

where $\hat{f}_i$ is the single-particle Fock operator and $\hat{V}$ is an operator corresponding to the difference between the exact and the HF Hamiltonian. The operator $\hat{V}$ is called the *residual* interaction and it will be treated as the perturbation term. Since the HF wavefunctions Ψ are Slater determinants made up of eigenfunctions of $\hat{f}_i$, they are themselves eigenfunctions of the zeroth-order operator $\sum \hat{f}_i$ with eigenvalues $\sum \varepsilon_i$. Then standard time-independent perturbation theory gives corrections to the ground-state energy W_0:

$$W_0 = W_0^{(0)} + W_0^{(1)} + W_0^{(2)} + \ldots$$

Ref.[278]. This book also describes two numerical techniques used to solve such inhomogeneous differential equations of perturbation theory: the *modified Sternheimer method* and the *Green's function method.*

$$= \sum_{i=1}^{N} \varepsilon_i + \langle 0 | \hat{V} | 0 \rangle + \sum_{n} \frac{\left| \langle 0 | \hat{V} | n \rangle \right|^2}{W_0^{(0)} - W_n^{(0)}} + ... \tag{4.50}$$

Here $|0\rangle$ is the ground state HF wavefunction Ψ_0, and $|n\rangle$ are HF states with electrons in excited orbitals. Now, it is easy to show that the ground state HF energy, i.e. the expectation value $\left\langle \hat{H}_0^{(N\ electrons)} \right\rangle$, Eq. (4.25), is actually equal to $W_0^{(0)} + W_0^{(1)}$ [281, 308]. Thus, corrections to the HF energy start with second-order perturbation theory. What states $|n\rangle$ actually contribute to the sums in Eq. (4.50)? A general theorem, due to Brillouin, states that the full Hamiltonian Eq. (4.49), $\hat{H}_0^{(N\ electrons)}$, does not mix singly-excited Slater determinants. On the other hand,

$$\hat{V} = \frac{1}{2} \sum_{i \neq j} \frac{e^2}{r_{ij}} - \sum_{i} v^{HF}(\mathbf{r}_i) \tag{4.51}$$

where v_{HF} is the HF single-particle potential. This is a two-particle perturbation, hence it will not mix triply- and higher-excited states either. It follows [308] that corrections to the energy of an isolated atom or molecule involve summation over doubly-excited states. In fact, from the explicit form of v_{HF} and the definition Eq. (4.51), it is straightforward to show [297, 308] that energy corrections are of the form

$$W_0^{(2)} = \sum_{\substack{a<b \\ \alpha<\beta}} \frac{\langle ab | \sum_{i<j} \frac{e^2}{r_{ij}} | \alpha\beta \rangle}{\varepsilon_a + \varepsilon_b - \varepsilon_\alpha - \varepsilon_\beta}, \tag{4.52}$$

where the virtual transitions from the ground state into excited configurations are labeled by the indices of the filled (a, b) and unfilled (α, β) orbitals.

Perturbation theory with the HF Hamiltonian as its starting point was first used by C. Møller and M.S.Plesset in the early days of quantum mechanics [339] and, as a consequence, it is often referred to as *Møller-Plesset* (MP) perturbation theory. The second-order expansion (MP-2) written out above, Eq. (4.52), is the most commonly used one, and it often recovers a significant portion of the correlation energy at a computational cost drastically reduced compared to CI methods. One can proceed to higher-order terms in the expansion Eq. (4.50) (labeled "MP-4," etc.); these expressions quickly become very bulky and diagrammatic methods have been devised to construct them in a systematic manner for all orders of perturbation theory (e.g., Hugenholtz or Goldstone diagrams [308, 340]). It can be rigorously shown that the perturbation expansion of the energy is size consistent to all orders. This gratifying result is a consequence of the so-called linked-cluster theorem due to Goldstone.

In many cases, e.g., for systems with a large number of strongly correlated electrons (in particular, complex atoms, clusters, conjugated hydrocarbon molecules, etc.) it becomes imperative to account for correlation effects fully, i.e., to all orders in perturbation theory. There exists a rigorous and systematic method to achieve this requirement, and it is based on the technique of many-body Green's functions and Feynman diagrams. We comment on this method in the next section.

Polarizability calculations within MP perturbation theory proceed along the same lines as above. For example, the external field may be incorporated into the zeroth-order HF Hamiltonian from the start, i.e., included in the term $\Phi(\mathbf{r})$ in Eq. (4.2). Then the perturbation-theory expansion will be of the form Eq. (4.52), as derived above. Alternatively, the external field itself may be treated as a perturbation as well, in which case the perturbation expansion will include matrix elements of both the electron-electron Coulomb interaction e^2/r_{ij} and the applied potential:

$$\hat{V} = \frac{1}{2}\sum_{i\neq j}\frac{e^2}{r_{ij}} - \sum_i v^{HF}(\mathbf{r}_i) + \sum_i e\mathbf{E}\cdot\mathbf{r}_i \tag{4.53}$$

This choice is labeled "double perturbation theory." In either case, the finite-field approach is used: the energy must be computed for several small values of the applied field and the polarizability is derived from the energy shift. Further discussion and specific examples can be found, e.g., in Refs. [341] (finite-order MP calculations on molecules), [342, 343] (Goldstone perturbation theory calculations on atoms), and in references quoted in [308].

4.7 Many-Body Perturbation Theory

4.7.1 Many-body Green's functions and diagrams

As mentioned above, the Green's function method provides a systematic technique for calculating the properties of many-particle systems. It can incorporate correlation effects to all orders, and gives rise to the concise language of Feynman diagrams. The technique of Green's functions was originally defined in quantum field theory, and first applied to many-body physics by Migdal and co-workers [344, 345].

The tools of quantum field theory make use of second-quantization operators which reflect the appropriate quantum statistics (Bose or Fermi) at each step. As a result, one can work in the occupation-number representation in contrast to the more cumbersome use of the symmetrized or antisymmetrized products of single-particle orbitals. The Green's function contains all the important physical information about the system: its ground-state energy, the excitation spectrum, and the linear response to external perturbations. Even a superficial overview of this method would be quite beyond the scale and scope of the present book, and we refer the reader to a large number of excellent books on the subject, including [340, 346–349] and others, as well as many review articles. A recent paper contains a thorough description of the Green's function approach to atomic many-body calculations and its application to alkali atoms [350].

Briefly, the one-particle Green's function (GF) is defined as the ground state expectation value

$$G(\mathbf{r},t;\mathbf{r}',t') = -i\left\langle \Psi_0|\hat{T}[\hat{\psi}(\mathbf{r},t)\hat{\psi}^\dagger(\mathbf{r}',t')]|\Psi_0\right\rangle, \tag{4.54}$$

where $\hat{\psi}$ and $\hat{\psi}^\dagger$ are the second-quantization field operators which destroy a particle at coordinates $(\mathbf{r},t)$ and create one at $(\mathbf{r}',t')$, respectively. The field operators, as well as the ground state Ψ_0 are written in the "Heisenberg" time-dependent representation, and $\hat{T}$ is the time-ordering operator. Essentially, the GF measures the probability that if a particle is created in the system, it will be present at some point at another time. Due to this image of the propagation of an excitation, the GF is also frequently referred to as the *propagator*.

We can take advantage of the fact that for an isolated system the GF Eq. (4.54) will depend only on the time difference $t - t'$. This allows us to define the Fourier transform with respect to this difference, $G(\mathbf{r}, \mathbf{r}', \omega)$, where ω now represents the energy of the excitation. In books on condensed matter physics one usually encounters the model of a uniform homogeneous system where G simplifies further by depending only on the coordinate difference $(\mathbf{r} - \mathbf{r}')$; this permits another Fourier transform and the introduction of a momentum- and frequency-dependent GF $G(\mathbf{k}, \omega)$. However in finite systems such as atoms, molecules, clusters, nuclei, etc., momentum is not a good quantum number and we must keep the explicit coordinate dependence.

As was stated earlier, the many-body GF contains much information about the properties of the system. As might be expected, it is as impractical to write down the exact GF of an interacting system as it is to write down the original wavefunctions. The power of the method stems from the fact that G can be evaluated by perturbation theory, efficiently and up to infinite order, aided by the diagrammatic representation.

For example, Fig. 4.1 depicts the first- and second-order Feynman diagrams in the perturbation expansion for G (denoted by the thick solid line) for a system of electrons. The thin solid lines denote G^0, the zeroth-order GF built from the Hartree-Fock basis. The dashed lines correspond to the electron-electron Coulomb repulsion. The number of vertex points gives the order of the expansion. At each vertex the coordinate variables are integrated over, while the energy variable is conserved. The references cited above derive the full set of rules for the evaluation of such diagrams. Note that the perturbation expansion contains only connected diagrams, i.e., no sub-units that are not connected to the rest of the diagram by any lines. This is a manifestation of the aforementioned linked-cluster theorem. The diagrammatic representation has a valuable physical interpretation: it depicts the self-interaction of electrons by the emission of virtual photons and creation of virtual electron-hole excitations. (Note a parallel between this picture and the configuration interaction concept.)

4.7.2 Self-energy and Dyson's equation

Diagrammatic perturbation theory allows one to isolate the most important terms in the expansion and sum their contributions to infinite order. For example, let us focus on the diagrams in Fig. 4.1(a,c,e,l). These are the first members of the series shown in Fig. 4.2. It is shown in the theory of the electron gas and other many-electron systems that this series of "bubble," or "loop," diagrams makes the dominant contribution to the perturbation expansion.[3] Interestingly, the importance of individual terms

[3] This series represents the so-called Random-Phase Approximation, or RPA, which will be discussed in more detail below.

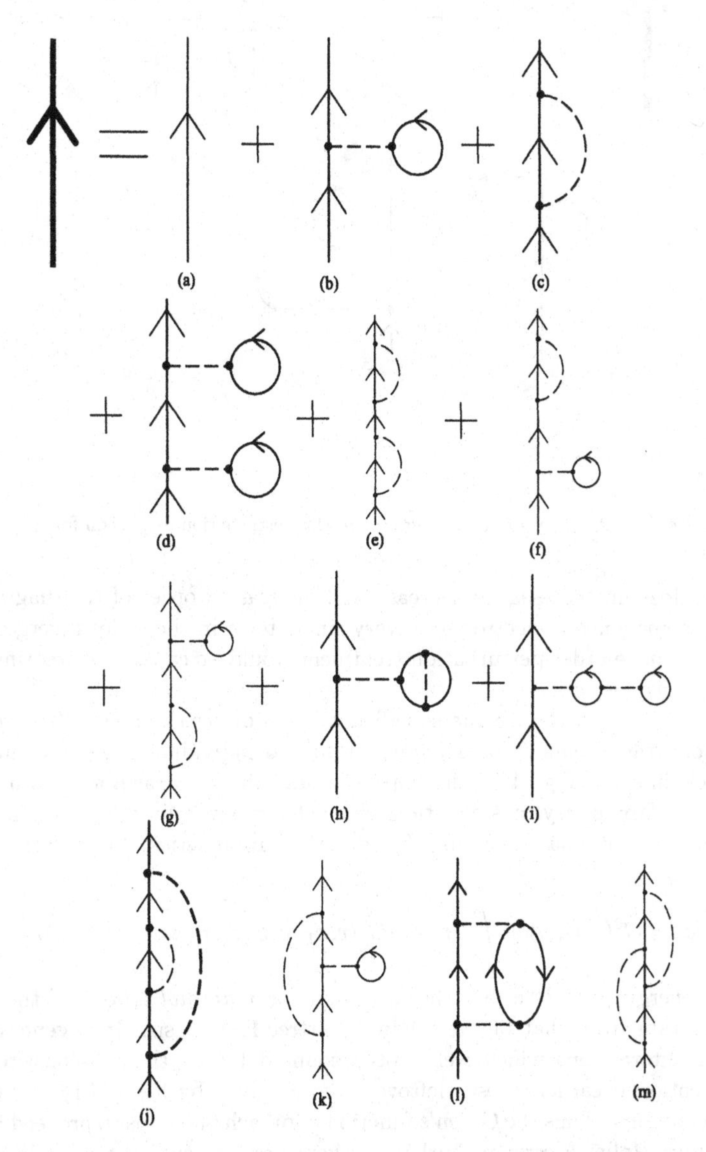

Figure 4.1: All first- and second-order Feynman diagrams for the perturbation expansion of the Green's function for a system of interacting electrons (after [340]).

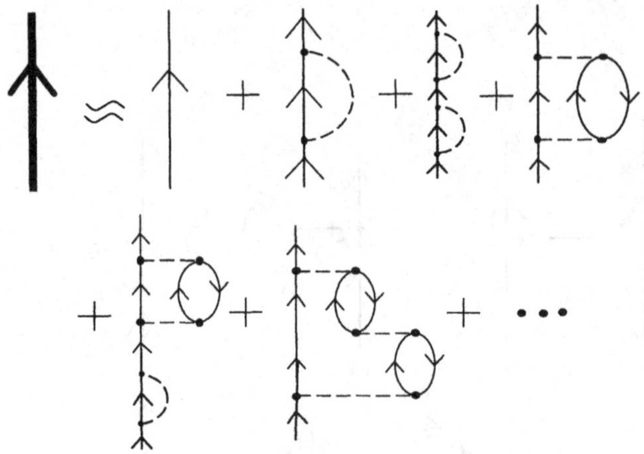

Figure 4.2: A subset of 'loop' diagrams in the perturbation expansion for G.

in the figure does not necessarily decrease with increasing order of the diagram. In fact, for the homogeneous electron gas every single term in the series diverges! This means that no finite-order perturbation treatment would be capable of treating these correlation terms correctly.

Notice that if we introduce the so-called *proper* (or *irreducible*) *self-energy part* $\Sigma(\mathbf{r},\mathbf{r}',\omega)$ (one which cannot be separated into two individual pieces by cutting a single particle line, see Fig. 4.3), the infinite series can be redrawn as shown in Fig. 4.4. This is the famous Dyson's equation which shows how a finite correction can be obtained from the summation of an infinite perturbation series. In analytical form, it reads

$$G(\mathbf{r},\mathbf{r}',\omega) = G^0(\mathbf{r},\mathbf{r}',\omega) + \int d^3r_1 d^3r_2 G^0(\mathbf{r},\mathbf{r}_1,\omega)\Sigma(\mathbf{r}_1,\mathbf{r}_2,\omega)G(\mathbf{r}_2,\mathbf{r},\omega). \tag{4.55}$$

The self-energy part describes the deviation (or *renormalization*) of the many-particle spectrum from that of the original Hartree-Fock basis. It is generated by interelectron interactions which were not accounted for by the self-consistent HF field. As mentioned earlier, these "leftover" terms, given by Eq. (4.51), are termed *residual* interactions. Thus the Green's function approach allows us to proceed beyond the finite-order Møller-Plesset perturbation theory and to sum the most important correlation effects to infinite order.[4]

[4]As mentioned in Sec. 4.6, time-independent perturbation theory expansions can also be described with the help of so-called Goldstone diagrams. Although based on a different grouping and interpretation of the expansion terms, they of course yield fully equivalent results for the energy shifts. However, the Feynman-Dyson method is much more compact: it has the fundamental ad-

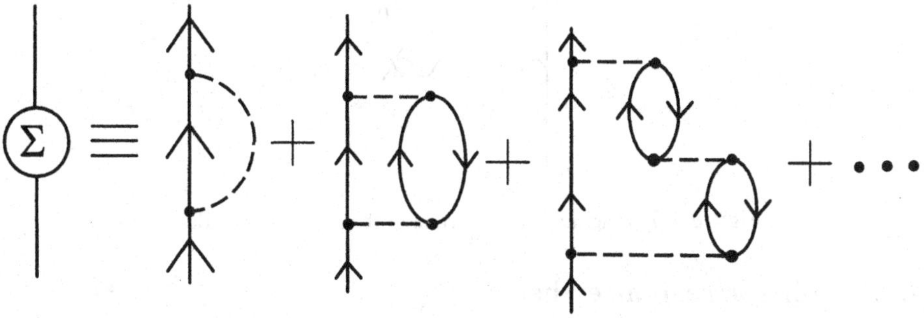

Figure 4.3: Diagrams forming the irreducible self-energy part Σ.

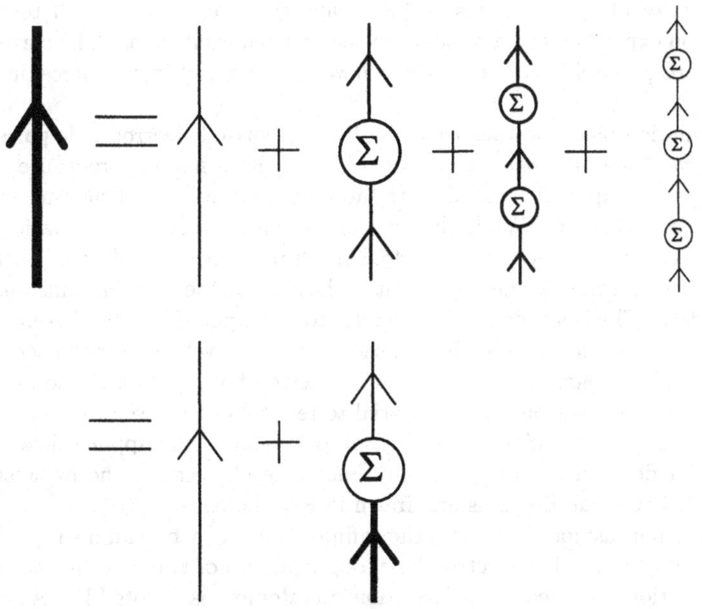

Figure 4.4: Green's function written in terms of the self-energy part: Dyson's equation.

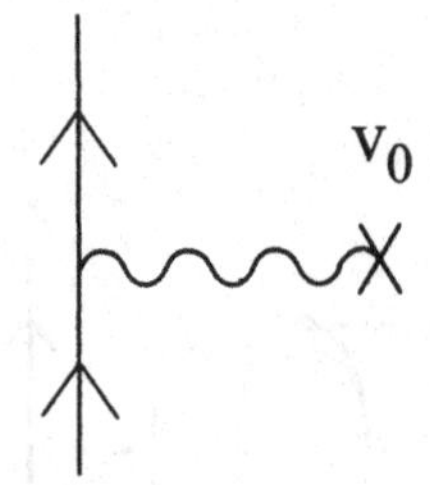

Figure 4.5: Diagrammatic depiction of the external potential.

4.7.3 Linear response theory

The effect of an external potential $v_0(\mathbf{r})\exp(i\omega t)$ on the propagator can be denoted by a line from an external source entering the Feynman diagram, see Fig. 4.5. In systems with mobile electrons this first-order diagram will not remain unchanged: as mentioned, for instance, in Section 4.3.3, the electrons will screen the applied potential and produce an effective field $V(\mathbf{r},\omega)\exp(i\omega t)$. The most important mechanism of screening in systems of this type is by the excitation of electron-hole pairs giving rise to a series of virtual states. This means that we must sum all terms in the perturbation expansion corresponding to such virtual excitations. The corresponding diagrams are precisely the bubbles which we encountered in the preceding section, see Fig. 4.3.

A calculation incorporating screening by such virtual electron-hole pairs is called the *Random-Phase Approximation* (RPA). It is quite a famous procedure which has been extensively applied in solid-state, atomic, molecular, and nuclear physics, to scattering problems, etc. While the final result can be obtained in a variety of ways (e.g., "equation of motion method" [346, 351], "time-dependent Hartree-Fock theory" [283, 352], etc.), a diagrammatic derivation has the virtue of being quick and physically intuitive. The basic idea is that, similar to the expansion of the Green's function, an infinite diagrammatic perturbation series can be developed for the screened field V as well. The expansion will involve the electron-hole loops and the Coulomb interaction between electrons. It is essential to remember that we are interested in the linear response regime (effective field V proportional to the applied field v_0), hence the external field line will appear only once in each term in the expansion. The appropriate Feynman diagrams are drawn in Fig. 4.6(a).

In the same fashion as before, the infinite series can be summed to obtain the result in Fig. 4.6(b). The electron-hole loop made up of two zeroth-order electronic Green's functions is called the *polarization operator* and is denoted Π_0. As mentioned above, it represents screening of the external field by virtual electron-hole excitations.

To translate this equation into an analytical form, we remember that at each vertex we must integrate over the spatial coordinates. Energy is conserved at each vertex, therefore v_0, V, and Π_0 are all evaluated at the same frequency ω. The resulting *linear*

vantage of effectively combining many terms of time-independent perturbation theory into a single Feynman diagram [340].

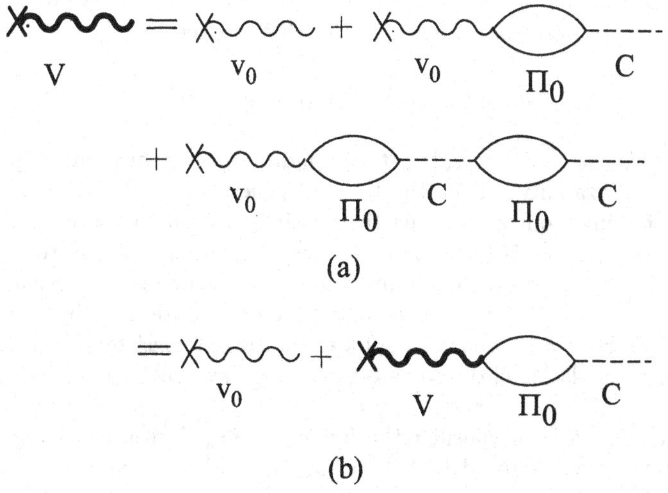

Figure 4.6: Perturbation expansion for the self-consistent field V in the random-phase approximation. (a) The field V originates in the external potential v_0, which is screened by the electron-electron Coulomb interaction (denoted by the dashed lines C) producing virtual electron-hole excitations (polarization operator Π_0). (b) Result of summation of the infinite series (a), leading to the integral linear response equation for V.

response equation has the form

$$V(\mathbf{r},\omega) = v_0(\mathbf{r}) + \iint d^3r_1 d^3r_2 \frac{e^2}{|\mathbf{r}-\mathbf{r}_1|}\Pi_0(\mathbf{r}_1,\mathbf{r}_2,\omega)V(\mathbf{r}_2,\omega). \tag{4.56}$$

The RPA polarization operator can be evaluated from the loop of Green's functions according to the diagram rules, and the result is [345]

$$\Pi_0(\mathbf{r},\mathbf{r}_1,\omega) = \sum_{\lambda\lambda'} \frac{n_\lambda - n_{\lambda'}}{\varepsilon_\lambda - \varepsilon_{\lambda'} + \hbar\omega + i\delta}\varphi_\lambda(\mathbf{r})\varphi^*_{\lambda'}(\mathbf{r})\varphi^*_\lambda(\mathbf{r}_1)\varphi_{\lambda'}(\mathbf{r}_1), \tag{4.57}$$

where ε_λ and φ_λ are the energies and wavefunctions of the single-electron orbitals, and n_λ are the occupation numbers. These zeroth-order orbitals are taken from a Hartree-Fock or another appropriate basis (e.g., TDLDA uses Kohn-Sham orbitals, see the discussion in Section 4.5.4). Thus the integral equation Eq. (4.56) fully defines the RPA linear response of the system to an external perturbation.[5]

[5]The diagrammatic language makes it clear that one can proceed beyond RPA by incorporating additional features in the polarization operator. For example, adding loops bridged across by interaction lines results in the so-called Random-Phase Approximation with Exchange (RPAE) used in atomic and nuclear physics [340, 343, 350]. The most general linear response equation is obtained by replacing Π_0 by the complete polarization operator. In terms of Feynman diagrams, this means that the bubble will contain the exact screened interaction vertex and the full "dressed" electron and hole Green's functions.

The second term on the right-hand side of Eq. (4.56) is just the Coulomb potential due to the field-induced charge density $\delta n(\mathbf{r})$. We see that

$$\delta n(\mathbf{r}) = -e \int \Pi_0(\mathbf{r}, \mathbf{r}_1, \omega) V(\mathbf{r}_1, \omega) d^3 r_1. \quad (4.58)$$

Knowing $\delta n(\mathbf{r})$, one can then calculate the induced dipole moment, Eq. (4.9), and the electric polarizability. Thus the task, in general, is to build the polarization propagator starting from a single-particle basis and then to use it in the integral response equation Eq. (4.56) to determine the polarization of the system.

Static polarizabilities can be obtained from the linear response equation in two complementary ways. One is to focus directly on the static case by setting $\omega = 0$, evaluating $\Pi_0(\mathbf{r}, \mathbf{r}_1, \omega = 0)$, and solving for the static self-consistent field. In this way, for example, the polarizabilities of rare-gas atoms have been calculated in TDLDA [353, 354].

The other choice is to consider the full *dynamic* polarizability. It corresponds to the dipole moment induced by a time-varying field $v_0(\mathbf{r}) \exp(i\omega t)$, with $v_0(\mathbf{r}) = -eEz = -eEr\cos\theta$, and describes the complete dipole excitation spectrum. This yields the distribution of dipole oscillator strengths, whereby the static polarizability can be calculated by means of Eq. (4.7) (see also Section 4.2.1). This approach may be advantageous if the excitation spectrum contains strong high-frequency resonances, in which case the static polarizability can be derived without ever needing to consider the (frequently more difficult) limit $\omega = 0$ explicitly. Such is the case, e.g., for clusters (cf. Sections 4.2.2, 3.5.1, 3.5.3, and 4.10), for which the dynamical response equation Eq. (4.56) can be solved analytically [185] or numerically [259, 338]

4.7.4 Matrix RPA

A formulation of RPAE extensively employed in atomic, molecular, and nuclear physics casts the linear response equations in the form of a matrix eigenvalue problem. In this context, it is usually interpreted as a linearized time-dependent extension of the Hartree-Fock theory (see, e.g., [283, 343, 352]). The physical contents of matrix and diagrammatic RPA are, of course, equivalent [340]. For electron systems, the matrix equations have the form [355]:

$$\begin{pmatrix} A & B \\ B^* & A^* \end{pmatrix} \begin{pmatrix} X^k \\ Y^k \end{pmatrix} = \omega_k \begin{pmatrix} X^k \\ -Y^k \end{pmatrix}, \quad (4.59)$$

where A and B are matrices made up of direct and exchange matrix elements of the residual interaction $\hat{V}$, Eq. (4.51). The matrix A contains matrix elements between particle-hole excitations, while B is composed of matrix elements between the ground-state and two-particle/two-hole excitations. In cases when the ground state has been defined as an exact HF Slater determinant, the matrix elements of the residual interaction reduce to those of the Coulomb potential, similar to Eq. (4.52).

The eigenvalues ω_k are the excitation energies of the system, and the matrix element for the dipole transition from the ground state $|0\rangle$ to the correlated excited

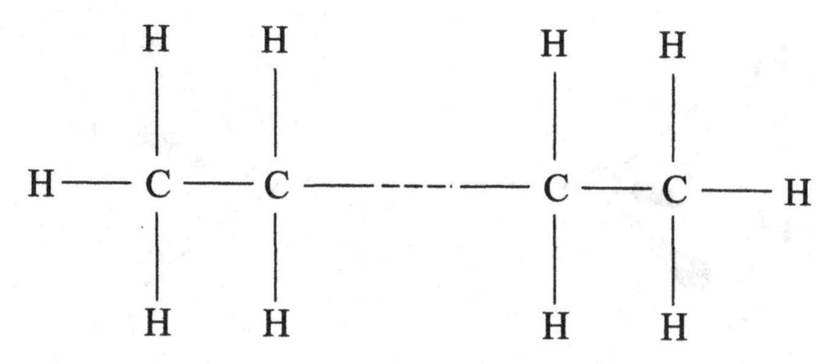

Figure 4.7: The chemical structure of an alkane molecule (C_nH_{2n+2}).

state $|k\rangle$ has the form:

$$\omega_k \langle k| z |0\rangle = \omega_k \sum_{am} \left(X^k_{ma} \langle m| z |a\rangle + Y^k_{ma} \langle a| z |m\rangle \right). \tag{4.60}$$

Starting from a single-particle basis, the RPA oscillator strengths and polarizabilities can again be readily calculated. Examples of such calculations on atoms, molecules, and clusters can be found in [337, 343, 356], and references therein.

4.8 Molecular Additivity Methods

4.8.1 Bond and hybrid polarizability methods

It is known that molecular polarizabilities cannot be simply computed by adding the polarizabilities of the constituent atoms of the molecule. However, reasonable values for average, *static* molecular polarizabilities can be calculated using simple semiempirical methods. There are several closely-related methods for using experimental data on polarizabilities to produce simple and accurate formulas for computing molecular polarizabilities.

The earliest of these methods, the *bond polarizability* method, used atomic refraction data [357, 358] or molecular refraction data [112] to deduce polarizability values that could be associated with specific bonds. This method was quite accurate at computing the refractive index of a large molecule, such as an alkane (C_nH_{2n+2}) — see Fig. 4.7, by adding the refractive indices associated with the different bonds present in the molecule. Somewhat later, Vogel [111] developed the *group polarizability* method that associated unique polarizabilities with specific molecular subgroups. For example, for the alkanes where $n > 1$ we have

$$\alpha_0(C_nH_{2n+2}) = (n-2)\alpha_0(CH_2) + 2\alpha_0(CH_3), \tag{4.61}$$

where the values of the polarizabilities for CH_2 and CH_3 are known from experimental data.

Another method is the atomic hybrid polarizability (*ahp*) method [359]. The *ahp* method assigns to each atom in a given state of hybridization an associated

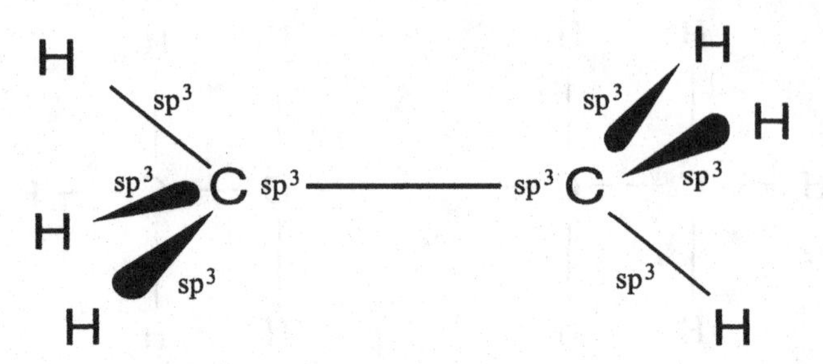

Figure 4.8: The bonding orbitals and structure of the ethane (C_2H_6).

polarizability that is unrelated to the specific atom at the other end of the bond. For example, for each atom A in a molecule in a particular state of hybridization we associate a hybrid polarizability $\alpha_A(\text{ahp})$. The total molecular polarizability is then the sum of the hybrid polarizabilities, i.e.

$$\alpha_0(\text{ahp}) = \sum_A \alpha_A(\text{ahp}). \tag{4.62}$$

The different states of hybridization of atomic orbitals to form molecular orbitals are denoted in the following way: σ, *te* (tetrahedral), *tr* (trigonal), *di* (digonal), and π. For example, ethane (C_2H_6) has six hydrogen atoms each with an associated hybrid polarizability of $\alpha_H(\text{ahp}) = 0.387$ Å^3 and two carbon atoms each with an associated tetrahedral hybrid polarizability of $\alpha_{CTE}(\text{ahp}) = 1.061$ Å^3 (see Fig. 4.8). Thus the hybrid polarizability of ethane can be estimated as

$$\alpha_{C_2H_6}(\text{ahp}) = 6 \times \alpha_H(\text{ahp}) + 2 \times \alpha_{CTE}(\text{ahp}) = 4.44 \text{ Å}^3. \tag{4.63}$$

This calculated result differs only slightly from the experimentally determined value [360, 361] of $\alpha_{C_2H_6}(\text{exp}) = 4.48$ Å^3.

Finally, the atomic hybrid component (*ahc*) method [115] can be used to estimate polarizabilities using the formula

$$\alpha_0(\text{ahc}) = \frac{4}{N}\left[\sum_A \tau_A(\text{ahc})\right]^2, \tag{4.64}$$

where τ_A is called the atomic hybrid component of atom A (in a given state of hybridization) and N is the total number of electrons in the molecule. This formula is based, though not derived, on an understanding of the results of variational perturbation theory and molecular orbital theory (see Ref. [115] for details). There is a one-to-one correspondence between the atomic hybrid components τ_A and the atomic hybrid polarizabilities $\alpha_A(ahp)$. A set of optimized values for both parameters can be found in Ref. [116]. In general, the two methods (*ahc* and *ahp*) tend to produce little difference. Both methods have been compared in detail for many different molecular groups and molecules [116]. The same reference [116] extensively compares both methods to experimental results.

A recent paper [73] has also extended the atomic hybrid component method to calculations of tensor polarizabilities of molecules.

4.8.2 Atom dipole interaction theory

Another semiempirical theory that associates a polarizability with each site in a molecule and attempts to incorporate the effect of neighboring sites is the atom dipole interaction model (see Ref. [362, 363] and references therein). In this picture, the atoms in a molecule are viewed as polarizable point particles, and the polarization of the molecule is calculated self-consistently. In other words, each point dipole is polarized by the external electric field as well as by the dipole field due to its neighbors. The problem is thus reduced to a matrix equation of dimension equal to the number of atomic dipoles. The atomic polarizability parameters can be fit for a particular compound, and then employed for other compounds in the same family.

An extension of the above technique is the atom monopole-dipole interaction model [364], which retains the view of atom as polarizable point particles but also assumes that charge transfer may occur between atoms in response to changes in local electric potentials. This formulation involves a second set of parameters to describe the degree of charge transfer, and leads to a correspondingly larger matrix problem. It was recently applied to the calculation of polarizability tensors of aromatic hydrocarbons and alkane molecules [362] and fullerenes C_{20} through C_{240} [365].

4.9 Inclusion of Vibrational and Rotational Degrees of Freedom

Throughout this chapter we have been concentrating on the electronic polarizability of atoms, molecules, and clusters. While in the great majority of cases this is indeed the dominant effect, the possibility of vibrational contributions to the total polarization should be kept in mind. Similarly, the effect of an external field on the rotational motion sometimes needs to be considered. Work along these lines has been much less developed than the treatment of the electronic polarizability. An extensive discussion and a list of references can be found in a recent review article [75].

There does not yet appear to exist a general qualitative understanding of the interplay of electronic and vibrational polarization. The majority of calculations proceed from the Born-Oppenheimer adiabatic approximation (i.e., the total molecular wavefunction is written as a product of separate electronic, vibrational, and rotational parts), add an external-field term to the Hamiltonian, and perform a numerical computation of the resulting dipole moment.

One approach is to consider the perturbation of the molecule as a whole. Using the complete molecular wavefunction one can extract energy corrections to various orders, and then try to dissect the result into the electronic, vibrational, and rotational components.

Alternatively, the calculation can proceed in two steps, following more closely the standard Born-Oppenheimer methodology. First, the electronic Schrödinger equation perturbed by the external field is solved, and a set of perturbed potential-energy surfaces for nuclear motion is obtained. Next, the rovibrational Schrödinger equation is used to obtain the rovibronic energies of the molecule in the presence of the electric field. From the combination of perturbed energies one extracts the total molecular polarizability. Ref. [75] refers to this approach as the *clamped-nucleus method.* The actual calculation may be carried out by one of the techniques discussed earlier, for example, by the finite-field method or by a set of perturbation-theory expressions.

An important aspect of the analysis is the relative orientation of the external field and the molecular axes. To relate the calculation to an observable quantity, usually both rotational (i.e., orientational) and thermal averaging over the populated rotational states is carried out. Similarly, the vibrational contribution may be specified either as that for a single vibrational state, or as a thermal average. Due to the evident complexity of the procedure, calculations have so far been restricted to diatomics and some other small molecules. A more complete description of these procedures, applied to diatomic molecules, was given in Section 3.3.2.

Interestingly, calculations have predicted that in certain systems the vibrational polarizability should greatly exceed the electronic contribution. For example, in certain hydrogen-bonded systems the former dominate by two orders of magnitude [76, 77]; the polarizability of endohedral fullerene complexes $[M@C_{60}]^+$ (M=metal atom) may be comparable with the electronic contribution or even exceed it by over an order of magnitude, depending on the temperature [80], see Section 3.5.3(c).

An interesting example of the interaction of an electric field with the rotational motion of a linear molecule is the creation of pendular states [366]. The interaction potential of a linear rotator with an external field E is

$$V_\alpha = -\frac{1}{2}\mathrm{E}^2(\alpha_{||}\cos^2\theta + \alpha_\perp \sin^2\theta), \tag{4.65}$$

which can be rewritten in the form $V_\alpha = A\cos^2\theta + B$. This is seen to be a double-well potential, whose energy levels and inter-well tunneling are of interest in rotational spectroscopy, alignment, and trapping experiments with both nonpolar and polar molecules.

4.10 Results from Specific Calculations

Here we compare a representative sample of calculations on different species to illustrate the success and applicability of the different techniques described above. The sample of species include the following: two atoms (He and Na), a small molecule (N_2), a large dielectric molecule or cluster (C_{60}), and sodium metal clusters (Na_n). Note that Table 3.3 in Section 3.4 compares theory and experiment for the polarizabilities of the nucleic acid bases adenine, cytosine, guanine, thymine, and uracil.

4.10.1 Helium atom

A vast literature exists on the calculation of the polarizability of helium. This is because helium is the simplest multielectron atom. Both its static and frequency-dependent polarizabilities have been accurately measured. Helium is frequently used as a benchmark for calculations because of the accuracy with which its polarizability is known and for this reason we include it here. The results for helium using a number of techniques are summarized in Table 4.1. The calculations are listed in chronological order. Some of the references [367–370] in the table also reported values of $\alpha_{He}(\omega)$ at various frequencies. Other recent calculations of $\alpha_{He}(\omega)$ at nonzero frequencies are given by Malykhanov [371], Dmitriev [372], and Bishop and Pippin [373].

A careful discussion of the theoretical and experimental situation as of 1976 is given by Glover and Weinhold [374]. They also presented rigorous lower and upper bounds on α_{He} that eliminate a significant number of calculated values reported in the literature. For this reason we only list those values calculated up to 1976 that fall within their bounds. For completeness, most values reported after 1976 are included. Results that do not fall within the two rigorous bounds are indicated by square bracket [ ]. One important experimental value quoted by Glover and Weinhold is in error. The refractive index measurement by Newell and Baird [375] was not corrected for virial coefficients (see Section 5.1.2) and we have entered the corrected value in our Table 4.1. This measurement (if you include error) is just outside the rigid bounds calculated by Glover and Weinhold. Thus, two of the six experimental measurements discussed in [374] fall within the rigorously-bounded range. Also, Langhoff and Karplus [376] used Pade approximants to extrapolate the refractive index measurements of Cuthbertson and Cuthbertson [377] to dc. The resulting value falls within the proper bounds and it is this number that is frequently cited as the experimental result (see Refs. [370, 378]).

There has been significant recent interest in the polarizability of helium [379–392]. An accurate experimental measurement of the helium polarizability was reported in 1980 by researchers working on dielectric constant gas thermometry (DCGT) [393] (see also [394, 395]). Using dielectric gas thermometry actually yields a value for the ratio A_ϵ/R, where

$$A_\epsilon = \frac{4\pi N_A \alpha_0}{3} \tag{4.66}$$

and R is the universal gas constant. In Eq. (4.66), N_A is Avogadro's number. We have provided an experimental value for the polarizability in Table 4.1 based on these DCGT measurements, where we assume $R = 8.31441$ J/mol. We should point out that a recent paper [391] uses the DCGT result for A_ϵ/R and a recent theoretical value for α_0 (see [389]) to obtain a more accurate value for R (actually for Boltzmann's constant k where $R = N_A k$). This most accurate experimental result differs from the recent, accurate theoretical values for α by 7-9 standard deviations. Presently, the most accurate numbers might be the theoretical results in Ref. [389], which includes a careful treatment of relativistic effects. We take special note of the fact that for 5 of the most recent calculations [385–387, 389, 392] the maximum difference in any pair is 0.015 %. According to Ref. [389], the relativistic correction is 0.0033 %. Finally, it is important to note that a variational-perturbation calculation by Buckingham and Hibbard in 1968 [396] differs from the recent accurate relativistic calculation by only 0.005%!

4.10.2 Sodium atom

Sodium is another popular atom for calculations. We include sodium because it is important as an atom, in molecules, and in clusters. From the experimental viewpoint, sodium is frequently the atom of choice for demonstrating a new technique or idea, e.g. atom interferometry. For the estimate of α_0 using the sum of weighted oscillator strengths (see Eq. (4.7)), we used the discrete oscillator strengths from the NBS compilation of Wiese, Smith, and Miles [409] and we used the continuum (photoionization) cross sections of Marr and Creek [410]. The contribution to the polarizability due to continuum transitions was 0.5%.

The measurement of sodium's polarizability by atom interferometry [411] is the most accurate measurement of the polarizability of a condensable material by nearly an order of magnitude. Two theoretical results [412, 413] agree with the experimental measurement.

4.10.3 Nitrogen molecule

Since nitrogen is naturally a diatomic molecule at room temperature, it is easy to measure its polarizability with standard techniques. For this reason the experimental measurements are quite accurate and it is a suitable benchmark for theoretical calculations. A comparison of the experimental and theoretical values are given in Table 4.3. Some of the theory papers evaluate the frequency-dependent polarizability [435, 436]. In many calculations, the parallel and perpendicular components ($\alpha_{\|}$ and $\alpha_{\perp}$) are given [435, 437–440]. For the static values presented in Table 4.3, the two most accurate experimental values lie within 1 standard deviation of each other. Only a single theoretical result [437] agrees with either of the most accurate experimental measurements. However, the theoretical calculations cluster around the experimental values fairly tightly. Except for the semiempirical additivity method, all the theoretical values in Table 4.3 are within 5 % of the experimental values.

The theoretical values quoted here are taken to be those calculated for N_2 in its ground state at its equilibrium separation. Many theoretical papers present results for α as a function of internuclear separation. This list is not exhaustive but it should be nearly complete for the years 1977-1996. Many early references can be found in the references presented here, e.g. [441].

An accurate value for the depolarization ratio $\kappa = (\alpha_{\|} - \alpha_{\perp})/3\alpha_0 = 0.131 \pm .003$ was measured at $\lambda = 632.8$ nm [81] (see Section 5.1.3).

4.10.4 Alkali clusters

(a) Overview. This section offers an overview of calculations on medium-size simple

Table 4.1: Helium polarizability results.

Technique	$\alpha_0(dc)$ (Å^3)	Ref.	Year
Experimental			
Refractive index (frequency of 47.7 GHz)	0.2045 ± 0.0003	[375]	1965
Dielectric constant	0.2057 ± 0.0004	[397]	1967
Pade approximation of refractive index	0.20506	[376]	1969
Dielectric constant	0.205053 ± 0.000010	[393]	1980
Theoretical			
Variational perturbation theory	0.204967	[396]	1968
Variational perturbation theory	0.2051	[367]	1968
Random-phase approximation	[0.1959]	[368]	1975
Rigorous upper bound	0.205103	[374]	1976
Rigorous lower bound	0.204878	[374]	1976
Variational perturbation theory	0.204964	[374]	1976
Variational perturbation theory	0.2043	[398]	1977
Variational perturbation theory	0.204956	[399]	1981
Variational perturbation theory	0.2049664	[400]	1981
Coupled Hartree-Fock	[0.1959]	[401]	1983
Multiconfigurational self-consistent field	0.2049	[369]	1985
Variational perturbation theory	0.204962	[402]	1987
Screened hydrogen model	0.204973	[403]	1988
Perturbation theory	0.204967	[404]	1989
Perturbation theory	0.20497	[405]	1989
Time-dependent coupled-cluster theory	[0.2058]	[370]	1991
Many-body perturbation theory	0.2051	[378]	1991
Distorted-wave Born approx. of e-He collisions	[0.2137]	[406]	1991
Hartree-Fock	[0.1960]	[407]	1992
Differential diffusion Monte Carlo	[0.204]	[379]	1993
Quantum Monte Carlo	0.20489	[380]	1993
Self-consistent field	[0.19593]	[381]	1993
Perturbation theory	0.20567	[408]	1993
Perturbation theory	0.2049679	[382]	1993
Coupled-cluster theory	0.2052	[383]	1994
Density-functional (correlated wavefns.)	0.20547	[384]	1994
Perturbation theory	0.20497521	[385]	1994
Variation-perturbation theory	0.2049679	[386]	1995
Variation-perturbation theory	0.204987	[387]	1995
Variation-perturbation theory	0.204979	[388]	1996
Relativistic no-pair theory	0.2049577	[389]	1996
Variational calculation	[0.1959]	[390]	1996
Variational calculation	0.204967973	[392]	1996

Table 4.2: Sodium polarizability results.

Technique	$\alpha_0(dc)$ (Å^3)	Reference	Year
Experimental			
Deflection	24.4± 1.7	[414]	1974
E-H balance	23.6± 0.5	[415]	1974
Atom interferometry[a]	24.11±0.06 ± 0.06	[411]	1995
Theoretical			
Oscillator strength	24.4	[279]	1959
Hartree-Fock	27.1	[416]	1964
Hartree-Fock	27.2	[417]	1965
Hartree-Fock	24.12	[412]	1967
Variational perturbation theory	28.74	[418]	1969
Variational perturbation theory	24.93	[419]	1971
Effective quantum numbers	23.8	[420]	1973
Semiempirical	23.9	[421]	1976
Multi-configuration perturbation theory	24.45	[422, 423]	1976
Pseudopotential	22.4	[424]	1977
Pseudopotential	24.35	[425]	1979
Semiempirical pseudopotential	24.3	[426]	1982
Hartree-Fock	23.63	[427]	1984
Self-consistent field	24.07	[413]	1984
Time-dependent coupled Hartree-Fock	28.34	[428]	1986
Oscillator strength	24.4	[429]	1989
Hartree-Fock	28.3	[430]	1991
Pseudopotential	24.0	[431]	1993
Perturbation theory	24.258	[408]	1993
Local Density Approximation	23.4	[432]	1995
Oscillator strength calculation	23.80	[433]	1997
Variation-perturbation method	24.599	[434]	1997
Oscillator strength	24.5	this work	1997

[a]The error here is quoted as ± statistical ± systematic.

Table 4.3: Nitrogen (N_2) polarizability results. These values represent the spherically averaged static dipole polarizability of N_2.

Technique	$\alpha_0(dc)$ (Å^3)	Reference	Year
Experimental			
Refractive index (frequency of 47.7 GHz.)	1.7396 ± 0.0004	[375]	1965
Dielectric constant	1.7403 ± 0.0008	[397]	1967
Dispersion of depolarization ratio	1.74 ± 0.03	[442]	1975
Evalulation of dispersion data	1.740	[443]	1977
Theoretical			
Hartree-Fock	1.694	[441]	1977
Antisymmetrized products of strongly orthogonal geminals	1.651	[444]	1978
Multiconfiguration self-consistent field	1.643	[445]	1979
Time-dependent Hartree-Fock	1.682	[446]	1979
Configuration interaction	1.720	[447]	1980
Second-order polarization propagator approximation	1.673	[448]	1982
Configuration interaction	1.7416	[437]	1983
Time-dependent coupled Hartree-Fock	1.689	[401]	1983
Hartree-Fock	1.686	[449]	1985
Self-consistent field	1.6912	[438]	1986
Self-consistent field	1.6989	[450]	1986
Self-consistent field	1.623	[451]	1986
Many-body perturbation theory	1.7059	[439]	1988
Many-body perturbation theory	1.750	[440]	1988
Multiconfiguration linear response	1.713	[436]	1989
Constrained dipole oscillator strength	1.740	[452]	1990
Semiempirical additivity method (ahc)	1.94	[116]	1990
Multiconfiguration self-consistent field	1.713	[435]	1993
Coupled-cluster theory	1.795	[383]	1994
Perturbation theory	1.713	[453]	1994
Variational calculation	1.749	[453]	1994
Time-dependent local density approx	1.7666	[454]	1996

metal clusters (number of atoms $n \gtrsim 8$), i.e., those which are beyond the scale of small molecules and in which the electronic shell structure is already present.[6] This is also the size range for which some experimental data (see Fig. 3.13) and a large number of theoretical results are available. Information on closed-shell alkali clusters is presented in Tables 4.4 and 4.5. As was described in Sec. 3.5.1, the polarizabilities of free cluster particles are enhanced with respect to the classical value thanks to the effect of electron spill-out.

The high degree of electron delocalization in metal clusters means that the calculation of polarizabilities is a true many-body problem; atomic orbital or bond additivity methods will not provide an adequate description of the mobile electron system.[7] Indeed, one sees from the tables that calculations rely on such techniques as the Random-Phase Approximation, density-functional theory, and various refinements.

Polarizability calculations with full inclusion of cluster geometries have been performed only for the smallest species; all other approaches start with the jellium model in which the ionic cores are approximated by a spread-out charge distribution [Sec. 3.5.1(c)]. More recently, ionic pseudopotentials have been used to augment the jellium approximation. To keep the models computationally tractable, the pseudopotentials are usually averaged into the jellium, that is, convoluted with the uniform positive charge distribution.

(b) Analytical RPA. Most of the calculations represented in the tables are numerical. However, an accurate analytical treatment of the response properties of finite clusters can also be developed [185]. It is based on the RPA linear responsemetal clusters formalism Eqs. (4.56), (4.57). As described in Sec. 4.7.3, the static polarizability can be found from these equations either directly by setting $\omega = 0$, or via a calculation of the dipole excitation spectrum. The latter calculation is greatly simplified by the fact that the dominant dipole transitions in clusters are the surface plasma resonances [Sec. 3.5.1(e)] which lie at relatively high frequencies ω_{cluster}. This allows one to expand the RPA polarization operator Eq. (4.57) in powers of $(\varepsilon_\lambda - \varepsilon_\lambda)/\omega_{\text{cluster}}$[193], which makes the integral equation (4.56) tractable.

It turns out (see [185] for details) that the static polarizabilities (as well as the dipole resonance frequencies) of metallic clusters are fully determined by the ground-state density distribution of the delocalized valence electrons. Specifically, the following relation is obtained for spherical clusters (cf. Table 1.1):

$$\alpha_0 = R^3 f(R), \quad f(R) = \frac{g(R) + q}{g(R) - 2q + 6q^2}. \tag{4.67}$$

Here R is the cluster radius, $g(r) \equiv n(r)/\rho_+$ is the density of the valence electrons expressed in terms of that of the positive cluster background ρ_+ (thus $g(R)$ is the

[6] A discussion of polarizability calculations for smaller Na_n clusters ($n \leq 6$) and a detailed list of references can be found in [432].

[7] The strongly correlated character of valence electron response is illustrated by an estimate cited in Ref. [472]. The RPA ground state comprises contributions of all $2np - 2nh$ excitations of even order, with approximately 10% of the electrons occupying orbitals above the Fermi level. This implies that the number of excited configurations grows dramatically with cluster size; in particular, no finite-order CI calculation would be realistic for any but the smallest clusters.

Table 4.4: Polarizabilities of closed-shell sodium clusters. Polarizability values are given in terms of the limiting value for a classical sphere: $\alpha = R^3$, where R is the sphere radius. For a metallic cluster of N atoms, the reference radius is taken to be $R = r_s N^{1/3}$ a.u., where r_s is the bulk Wigner-Seitz density parameter (r_s=3.96 for Na). The quoted experimental error includes systematic uncertainties as well as variance in values measured at different cluster beam conditions.

Method	Cluster Polarizability			Refs.
	Na_8	Na_{20}	Na_{40}	
Classical sphere (R^3)	74 Å^3	184 Å^3	368 Å^3	
Experiment	1.77 ± 0.03	1.68 ± 0.10	1.61 ± 0.03	[198, 199]
Jellium background models				
Minimization of extended	1.4		1.2	[209]
Thomas-Fermi functional			1.40	[455]
Modified Sternheimer sol'n	1.39	1.32	1.30	[456]
of KS eqns.	1.45	1.37	1.32	[457]
Exchange-only LDA	1.48-1.50	1.38-1.41	1.33-1.38	[458]
Weighted-density corrections	1.49	1.42	1.37	[457]
to V_{xc} in KS eqns.	1.81	1.63	1.53	[457]
Analytical RPA response	1.54	1.41	1.33	[185, 459]
Matrix RPA (LDA basis)	1.34	1.33	1.29	[460]
		1.37	1.33	[461]
	1.45	1.39	1.35	[337]
Matrix RPA (HF basis)	1.52	1.46	1.42	[337]
"Local RPA" (LDA basis)	1.39	1.31		[462]
TDLDA	1.41	1.34	1.30	[463, 464]
	1.48	1.44	1.43	[465]
	1.45	1.39	1.34	[337]
Self-interaction corrected	1.52	1.46	1.41	[464]
TDLDA	1.70	1.61	1.51	[464]
	1.66	1.59	1.56	[465]
Jellium with corrections for structural effects				
Elastically deformable jellium	1.45-1.50	1.30-1.32	1.23-1.25	[466]
Finite jellium surface thickness	1.71	1.61	1.56	[467]
Spherically-averaged V_{pseudo}		1.92	1.75	[461]
Optimized ionic structures + pseudopotentials (average values calculated from the polarizability tensor)				
D_{2d} symmetry	1.51			[468]
	1.31			[469]
	1.61			[470]
T_d symmetry	1.63			[468]
	1.43			[469]
Langevin molecular dynamics	1.43-1.55	1.27-1.47		[471]

Table 4.5: Polarizabilities of closed-shell potassium clusters. Polarizability values are given in terms of the limiting value for a classical sphere: $\alpha = R^3$, where R is the sphere radius. For a metallic cluster of N atoms, the reference radius is taken to be $R = r_s N^{1/3}$ a.u., where r_s is the bulk Wigner-Seitz density parameter (r_s=4.87 for K). The quoted experimental error includes systematic uncertainties as well as variance in values measured at different cluster beam conditions.

Method	Cluster Polarizability			Refs.
	K_8	K_{20}	K_{40}	
Classical sphere (R^3)	137 Å^3	342 Å^3	684 Å^3	
Experiment	1.79 ± 0.09	1.66 ± 0.13		[198, 199]
Jellium background models				
Modified Sternheimer sol'n of Kohn-Sham equations	1.32	1.28		[457]
Weighted-density corrections to V_{xc} in Kohn-Sham eqns.	1.38 1.68	1.33 1.52		[457] [457]
Analytical RPA response	1.49	1.37	1.30	[185, 459]
Matrix RPA (LDA basis)		1.28		[461]
Jellium with corrections for structural effects				
Spherically-averaged V_{pseudo}		1.92		[461]

electron density at the cluster edge), and the parameter q is a measure of the valence electron spill-out (already encountered in Sections 3.5.1(e) and 4.2.2):

$$q = \frac{1}{R}\int_R^\infty g(r)dr. \tag{4.68}$$

We see once again that the enhanced polarizability of small clusters is directly related to the spill-out effect. Thus Eq. (4.67) reflects the same physics as the sum rule estimate Eq. (3.92) but is more precise.

The density $g(r)$ is to be taken from a ground-state calculation, e.g., Hartree-Fock, LDA, etc. In fact, the predictions of Eq. (4.67) turn out to be rather insensitive to the details of $g(r)$, since only the overall amount of the spill-out enters [185, 473]. Thus even a semiclassical Thomas-Fermi density distribution has been shown to provide accurate results. The analytical results for polarizabilities and dipole resonances compare very well with other jellium-based models, and make it possible to understand the evolution of cluster properties over a wide size range. For example, an application to aluminum clusters is shown in Fig. 3.15.

(c) Discrepancies between theories and experiments. The tables reveal the striking fact that the majority of calculations are in a consistent disagreement with the experimental data on alkali cluster polarizabilities. Indeed, most theoretical entries fall below the measured values by as much as 20%.[8] The magnitude of the discrepancy is actually more serious, since one is most interested in understanding the deviation of cluster response from the classical limit: $\alpha = R^3$. (All-electron molecular calculations on smaller alkali clusters [432] also display this problem.)

Several suggestions for modifying the basic theory have been proposed; some are listed in the tables. They include introducing approximate non-local and self-interaction corrections into the density functionals; relaxing the sharp jellium surface by allowing it to acquire a smoother profile; and convoluting ionic core pseudopotentials into the positive jellium. The possible influence of thermal molecular motion [432, 474] and low-energy electronic excitations [185] also has been pointed out. All of these suggestions are reasonable, and succeed in raising the calculated polarizability values towards the experimental data. On the other hand, it is obvious that including *all* of them at once would predict values considerably in excess of the measurement. Thus the theoretical picture remains quite unsettled at the moment. It should be pointed out that the metal cluster polarizability issue is not a narrow parochial problem, since it clearly involves fundamental questions in the theory of many-particle systems.

Note that the experimental data on the polarizability of medium-sized clusters is scarce, limited to one series of measurements on a supersonic beam of neutral alkali clusters [198, 199]. (Dimer polarizabilities measured in this work are in good agreement with another recent measurement [475].) Clearly, further experimental information is needed.

[8] Interestingly, the same theoretical procedures do an adequate job of reproducing the values of giant collective resonance frequencies which are intimately related to the static polarizabilities (see Section 3.5.1(e)).

4.10.5 Fullerene: C_{60}

In Section 3.5.3 we commented on the keen interest in the polarizability of C_{60}. This high degree of interest is due to the novelty and beauty of these molecules, and to the fact that they lie at the crossroads between molecular and nanostructure science. As a result, their response properties have been analyzed by a variety of approaches, ranging from condensed-matter techniques to molecular structure calculations to nuclear physics-inspired approaches. Table 4.6 contains a listing of representative experimental and theoretical results for $\alpha(C_{60})$. It is intended to provide an impression of the variety of approaches used and the range of values derived for this quantity. (Refs. [1, 260, 268, 476] also give an overview of C_{60} polarizability results.) The table shows that the experimental results mostly cluster around $\alpha(C_{60}) \simeq 85 - 90$ Å^3, while theoretical calculations form a broader distribution centered at approximately the same value and extending up and down by about a factor of two.

As we noted earlier, no direct polarizability measurements of isolated fullerenes are yet available. The experimental values have been derived from the dielectric constant ε_0 of fullerite crystals and films by means of the Clausius-Mossotti relation Eq. (3.95), see Section 3.5.3 and Section 5.1.1. It is worth noting that this relation breaks down as the ratio of α to unit-cell volume approaches $3/4\pi$. The C_{60} polarizability is indeed high enough so that deviations from the Clausius-Mossotti value can be expected [477].

A lattice vibrational contribution of ≈ 2 Å^3 has been identified experimentally [266] (see Fig. 3.25); however, the theoretical calculations concentrate on the electronic polarizability.

Table 4.6: Examples of experimental and theoretical results for the polarizability of C_{60}. The experimental values are based on the dielectric properties of solid fullerites or thin films. Theoretical listings label the treatment of C_{60} structure used in the calculation.

Method	$\alpha_0(\text{Å}^3)$	Refs.
Experiment		
Microwave absorption	36	[478]
Transmission and Ellipsometry	84-87	[159, 266]
Near-infrared absorption	85	[479]
Capacitance	91	[480]
Electron energy loss	93	[481]
Theory		
Perturbation theory (tight binding)	36	[257]
Linear response (tight binding)	43	[482]
Pariser-Parr-Popple (extended Hubbard) Hamiltonian	48	[268]
Dielectric shell	48-55	[483]
RPA (tight binding)	57	[484]
Atom-monopole-dipole interaction	61	[252]
Coupled Hartree-Fock (Gaussian basis sets)	65 (lower limit)	[485]
Discrete interacting atomic dipoles	77	[253]
LDA (Gaussian basis sets)	78	[486]
LDA (LCAO)	78-101	[260]
Time-dependent Hartree-Fock	79	[487]
Conducting sphere (Sec. 3.5.3)	80	$\alpha = R^3$
Additivity (atomic hybrid components)	80	[488]
Time-dependent Hartree-Fock (NDO)	82	[489]
LDA+RPA (jellium shell)	82	[288]
LDA (Gaussian basis sets)	83	[477, 490]
RPA (atomic basis set)	86	[476]
LDA + RPA (non-local pseudopotentials)	89	[491]
TDLDA (jellium shell)	91	[492]
Linear response (spherical well)	116	[493]
TDLDA (jellium shell)	138	[494]
Valence-effective hamiltonian	154	[495]

Chapter 5

Experiment

5.1 Traditional Techniques

5.1.1 Dielectric constant measurements

The average polarizability can be accurately determined by measuring the dielectric constant ε. To make this measurement, a capacitor is constructed such that the volume between the plates can be evacuated or filled with the material to be measured. The dielectric constant is determined by taking the ratio of the vacuum capacitance C_0 to εC_0, the capacitance when the volume is filled. The dielectric constant can be measured by making the capacitor part of a resonant circuit and observing the change in resonance frequency when material is introduced between the capacitor plates [496]. A more common way of measuring the dielectric constant is by using an ac bridge circuit to compare the unknown capacitance to a capacitance standard [497–500]. This type of circuit is illustrated in Fig. 5.1. In these methods an ac frequency is used, but the frequency is small enough (typically 1–20 kHz) that the polarizability is little different from its static value.

The polarizability is related to the dielectric constant by the Debye equation

$$\alpha = \frac{3}{4\pi n}\left(\frac{\varepsilon - 1}{\varepsilon + 2}\right) - \frac{\mu_0^2}{3kT}, \tag{5.1}$$

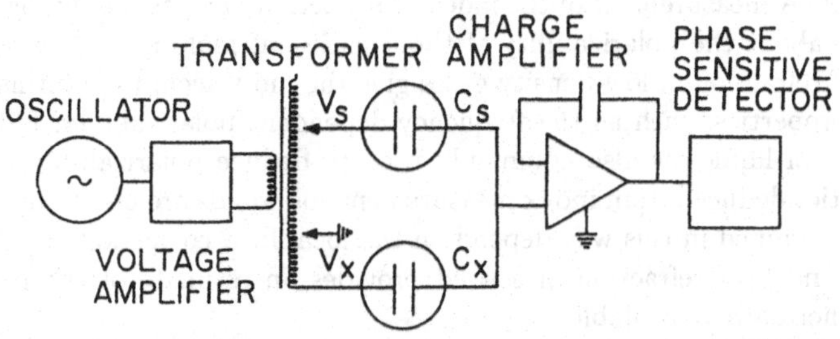

Figure 5.1: Bridge circuit used for comparison of a standard capacitance C_s and unknown C_x. The balance condition of no detector current is $C_x = C_s(V_s/V_x)$. (Reprinted with permission from Ref. [497].)

where n is the number density, μ_0 is the permanent electric dipole moment of the molecule, k is the Boltzmann constant and T is the absolute temperature [32]. The special case when $\mu_0 = 0$ is called the Clausius-Mossotti relation. The dielectric constant is nearly unity for a dilute gas so that $\varepsilon \approx 1 + 4\pi n[\alpha + \mu_0^2/(3kT)]$.

The Clausius-Mossotti equation is a good approximation for gases, but the dielectric constant can be measured precisely enough that corrections to Eq. (5.1) can be observed [497, 498]. These corrections are expressed by making a virial expansion of the Clausius-Mossotti function [501]:

$$\left(\frac{\varepsilon - 1}{\varepsilon + 2}\right)\frac{1}{d} = A_\varepsilon + B_\varepsilon d + C_\varepsilon d^2 + \cdots, \tag{5.2}$$

where $A_\varepsilon, B_\varepsilon, C_\varepsilon$ are the first, second, and third dielectric virial coefficients, $d = n/N_A$ is the molar density (mol/cm^3 — also called amount-of-substance density), and N_A is Avogadro's number. The first dielectric virial coefficient is

$$A_\varepsilon = \frac{4\pi N_A}{3}\left(\alpha + \frac{\mu_0^2}{3kT}\right). \tag{5.3}$$

The second and third dielectric virial coefficients depend on the intermolecular potential and on higher-order permanent moments and nonlinear polarizabilities of the molecules. Measurements are still not quite precise enough to accurately deduce the interatomic potential and higher-order moments from measurements of $B_\varepsilon, C_\varepsilon$. Further discussion of these points can be found in Ref. [502].

Measurements of the dielectric constant provide the most accurate measurements of the static or near-static polarizability for species which are a gas (or solid) at room temperature. The technique cannot be directly applied to deduce the polarizability of single particles if the species cannot be isolated in a gas form. For species which must be heated to form a gas, one must contend with complications such as dissociation (for molecules and clusters), excitation, ionization, and temperature gradients.

5.1.2 Refractive index

The index of refraction is the ratio of the speed of light in vacuum to the speed of light in a material. A measurement of the index of refraction of a gas, liquid, or solid yields information about the polarizability of the constituent molecules. Measurements of the index of refraction of low-density gases give the most accurate information about molecular properties, such as the frequency-dependent polarizability, but the index of refraction of liquids is also commonly used to deduce polarizabilities. However, polarizabilities deduced from index measurements of liquids are often imprecise, since values of α obtained in this way depend on the local field correction used — see Ref. [503]. The index of refraction of a solid provides an estimate of the polarizability when no other data is available.

Only a single index of refraction η can be measured for a gas, liquid, or amorphous solid if the molecules are not optically active[1], and if the sample is not in an applied

[1]A gas or liquid composed of optically active molecules will have different indices of refraction for

electric or magnetic field. (The Kerr effect, which involves the measurement of the index of refraction of a material in an electric field, is discussed in Section 5.1.4.) The index of refraction is related to the average polarizability α_0 by the Lorentz-Lorenz relation

$$\alpha_0 = \frac{3}{4\pi n}\left(\frac{\eta^2 - 1}{\eta^2 + 2}\right) \tag{5.4}$$

where n is the number density of the gas or liquid [32]. This equation is valid for non-polar molecules or at frequencies high enough that the permanent dipole moments cannot follow the electric field. A dipole term must be added if the frequency of the radiation is less than or comparable to rotational frequencies of the molecule. The Lorentz-Lorenz and Clausius-Mossotti equations are related by the Maxwell relation $\varepsilon = \eta^2$. Measurements of the index of refraction determine the polarizability at the frequency ν of the light used to make the measurement. The dc polarizability can be determined using long-wavelength radiation, or by extrapolating short-wavelength measurements to dc. In a dilute gas, the index of refraction is nearly unity so that $\eta \approx 1 + 2\pi n\alpha_0$.

The index of refraction can be determined by observing the angle of refraction at the interface between a sample and a reference material of index η_r, and applying Snell's law:

$$\eta \sin\theta = \eta_r \sin\theta_r, \tag{5.5}$$

where θ and θ_r are the angles of the light with respect to the normal at the interface, in the sample and in the reference material, respectively. For example, this method was used to determine the liquid and gas phase index of refraction of several refrigerants being considered to replace chlorofluorocarbon (CFC) compounds [499, 500, 504]. Such a measurement is necessary to estimate the temperature-dependent dipole moments of these molecules.

Accurate measurements of the index of refraction of a gas can be made interferometrically. One such technique is described briefly below. An evacuated cell is placed in one arm of a Michelson interferometer. This cell is filled as the fringe shift N^* is measured[2]. The index of refraction is related to the fringe shift by

$$2l(\eta - 1) = N^*\lambda \tag{5.6}$$

where l is the length of the cell, and λ is the wavelength of the light. An interferometric technique was used to investigate the temperature dependence of the polarizability of several molecules different wavelengths in the range between 633 nm and 325 nm [505] — see also the results obtained with a twin Michelson interferometer [506, 507]. Interferometric measurements have been made for substances for which a capacitance measurement of the dielectric constant is impractical. For example, Alpher and White

light which is right- or left-circularly polarized. If a molecule has a reflection symmetry, it will not be optically active. Many sugars and other biologically produced substances are optically active.

[2]Actually, the fringe shift N^* is measured during evaporation of the sample, i.e. while the cell is filling, for substances with low vapor pressures at room temperature. However, for substances which are gases at room temperature, N^* is measured during evacuation of the cell.

measured the polarizability of atomic nitrogen and atomic oxygen, using a shock tube to dissociate N_2 and O_2 [508].

Newell and Baird measured the refractive indices of several gases (air, Ar, CO_2, He, H_2, N_2, and O_2) at microwave frequencies (47.7 GHz) by observing the change in resonant frequency as gas was added to a cylindrical microwave cavity and to a Fabry-Perot resonator [375]. With the exception of oxygen (which has a resonance at 60 GHz), these measurements should be identical to the index of refraction at zero frequency.

To determine the polarizability from their stated value of $\eta - 1$, the density of the gas must be calculated from the measured pressure and temperature of the gas. The ideal-gas number density is

$$n_0 = \frac{P}{kT} \tag{5.7}$$

where P is the pressure. The index of refraction measurements are accurate enough that we must consider the non-ideal behavior to determine the number density. An adequate approximation is

$$n = n_0(1 - Bn_0/N_A) \tag{5.8}$$

where B is the second virial coefficient (cm^3/mol). For N_2 the fractional correction to ideal gas behavior is 5×10^{-4}, which is twice the quoted error in $\eta - 1$.

In a more exact description of the refractive index as a function of gas density, an expression similar to that for the dielectric constant Eq. (5.2) is used. We can write this relation [509, 510]

$$\frac{\eta^2(\lambda, n_0, T) - 1}{\eta^2(\lambda, n_0, T) + 2} = A_\eta(\lambda, T)d_0 + [B_\eta(\lambda, T) - A_\eta(\lambda, T)B(T)]d_0^2 + ... \tag{5.9}$$

where

$$d_0 = \frac{n_0}{N_A} = \frac{P}{RT}, \tag{5.10}$$

and A_η and B_η are the first and second refractive virial coefficients. The first virial coefficient is directly proportional to the mean polarizability and we can write

$$A_\eta(\lambda, T) = \bar{\alpha}(\lambda, T)\frac{4\pi N_A}{3}, \tag{5.11}$$

where we have converted the expression found in Ref. [509] to cgs units. The second virial coefficient $B(T)$ is defined by

$$\frac{P}{RT} = d + B(T)d^2 + C(T)d^3 + ... \tag{5.12}$$

where

$$d = \frac{n}{N_A} \tag{5.13}$$

is the density in moles (amount-of-substance density).

Modern techniques involving Michelson interferometers allow measurement of $\eta(\lambda, T)$ over broad wavelength, temperature, and density ranges (see the references of Section 5.2.2). Two advantages provided by such an approach are

1. Since η is measured continuously as a function of d (or n), then the polarizability $\alpha(\omega)$ can be obtained from the Lorentz-Lorenz relation Eq. (5.4) in the limit of small density, i.e. as $d \to 0$.

2. The same technique allows an accurate measurement of the second virial coefficient $B(T)$.

The index of refraction of a thin film of C_{60} was determined by measuring the thickness l of the film with a step profiler and observing the intensity of transmitted light as a function of wavelength [479]. The thin film acts like an etalon with a free spectral range (FSR) of

$$\Delta\nu = c/(2\eta l). \tag{5.14}$$

Measurement of the FSR and of the thickness yield $\eta = 2.0$. The measured density of C_{60} is 1.68 g/cm^3 (Ref. [511]) and the molecular weight is 720 amu, so that the number density is $n = 1.41 \times 10^{21}$ cm^{-3}. Using the Lorentz-Lorenz relation we find that $\alpha_0 = 85$ Å^3. See Section 4.10.5 for a review of other results for C_{60}.

Refractive indices of thin films can also be measured by the m-lines method, which is treated separately in Section 5.2.1.

5.1.3 Rayleigh scattering

Rayleigh scattering is the non-resonant scattering of light by a polarizable particle much smaller than a wavelength of light. The oscillating electric field of the incident light induces an oscillating dipole moment which re-radiates light in a dipole pattern. For a spherically symmetric molecule and linearly-polarized light, the induced dipole moment is parallel to the incident polarization and is proportional to the polarizability of the particle. If the molecule is not spherically symmetric then the induced dipole moment depends on the orientation of the molecule, and Rayleigh scattering causes depolarization of the incident light.

The amount of Rayleigh scattering depends on both the average polarizability α_0 and the anisotropy κ of the polarizability tensor:

$$\kappa^2 = [(\alpha_{xx} - \alpha_0)^2 + (\alpha_{yy} - \alpha_0)^2 + (\alpha_{zz} - \alpha_0)^2]/(6\alpha_0^2), \tag{5.15}$$

where $\alpha_{xx}, \alpha_{yy}, \alpha_{zz}$ are the principal polarizabilities of the molecule. For a highly symmetric molecule such as C_{60}, $\kappa = 0$. For linear or symmetric-top molecules $\alpha_{zz} = \alpha_{\parallel}, \alpha_{xx} = \alpha_{yy} = \alpha_{\perp}$ so that

$$\kappa = (\alpha_{\parallel} - \alpha_{\perp})/(3\alpha_0). \tag{5.16}$$

The most common setup for observing Rayleigh scattering is shown in Fig. 5.2. A laser linearly polarized along $\mathbf{e}$ is used to illuminate a gaseous sample. Scattered radiation of polarization $\mathbf{q}$ is detected after it passes through a polarizer. The mean radiation rate per unit solid angle (the steradiance) for a single particle, averaged over molecular orientations, is

$$\langle S_0 \rangle = I(2\pi\nu/c)^4 \frac{1}{5}\alpha_0^2[3\kappa^2 + \cos^2\theta(5 + \kappa^2)] \tag{5.17}$$

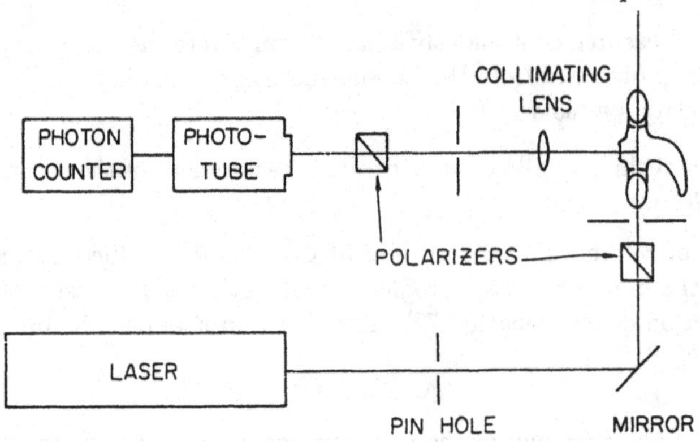

Figure 5.2: Experimental apparatus for measuring Rayleigh scattering. (Reprinted with permission from Ref. [442].)

where I is the incident intensity and θ is the angle between $\mathbf{q}$ and $\mathbf{e}$. For a dilute gas, the scattered light from individual particles is incoherent and the steradiance from each particle can be added together. If the number of illuminated particles is known, it is possible to measure both α_0^2 and κ^2. The limitations of this classical expression are discussed in detail by Bridge and Buckingham [512].

The depolarization ρ_0 is defined as the ratio of the minimum to maximum steradiances for the direction of scattering in question. If the light is observed in a direction which is perpendicular to both the propagation direction and the polarization of the incident light then the depolarization is

$$\rho_0 = 3\kappa^2/(5 + 4\kappa^2). \tag{5.18}$$

The depolarization is independent of number density, and determines κ^2 independent of α_0.

Rayleigh scattering is most often used to determine the depolarization (and hence κ^2) of a gas in a cell [442, 512, 513]. However, Rayleigh scattering from a beam of particles was used to measure the polarizability of C_{60} [514]. The measured polarizability for C_{60} was an order of magnitude larger than calculations and other data, perhaps indicating the difficulty in making a scalar polarizability measurement using Rayleigh scattering.

5.1.4 Electro-optic Kerr effect

The electro-optic Kerr effect is the creation of optical birefringence in an isotropic fluid by a strong electric field. This field is called the orienting field, since its effect is to orient the dipoles in the fluid. If the electron distribution of an oriented dipole is asymmetric, either because the molecular polarizability is anisotropic or because of non-linear distortions of the molecule, then the response of the fluid will be different for light polarized parallel ($\|$) and perpendicular ($\perp$) to the orienting field. In the

Kerr effect, the difference of indices of refraction $\eta_{\parallel} - \eta_{\perp}$ is proportional to the square of the orienting electric field E. (The Pockels effect is a linear electro-optic effect which exists only for certain crystals that lack a center of symmetry.) The orienting field may be produced between parallel plates with a dc or oscillating voltage. The oscillation frequency is then low enough that the permanent dipole moments of the molecules follow the field. This is the situation which will be assumed below. However, optical fields can be used as the orienting field, in which case the permanent dipole moments cannot align themselves with the field. It is assumed that the orienting field is much larger than the field of the light used to observe the Kerr effect.

The Kerr constant B is defined as

$$B = \frac{\eta_{\parallel} - \eta_{\perp}}{\lambda_0 \mathrm{E}^2}, \tag{5.19}$$

where λ_0 is the wavelength of the probe light and E is the magnitude of the static field. The molar Kerr constant ${}_mK$ is also commonly reported in the literature, since it is an additive quantity in the limit that interactions between molecular species can be ignored in a mixture:

$${}_mK = \frac{6\lambda_0 \eta B V_m}{(\eta^2+2)^2(\varepsilon+2)^2} = \frac{6\eta V_m}{(\eta^2+2)^2(\varepsilon+2)^2}\frac{\eta_{\parallel} - \eta_{\perp}}{\mathrm{E}^2}, \tag{5.20}$$

where η is the average index of refraction, ε is the *dc* dielectric constant, and V_m is the molar volume ($\mathrm{cm^3\ mol^{-1}}$). A general expression for ${}_mK$ is derived in Refs. [32, 515]. For symmetric-top molecules:

$${}_mK = \frac{4\pi N_A}{81}\left[\frac{9\alpha_0^e\kappa^e}{5kT}\left(\alpha_0\kappa + \frac{\mu_0^2}{3kT}\right) + \frac{2\mu_0\beta^{0e}}{3kT} + \gamma^{0e}\right], \tag{5.21}$$

where β^{0e} is the mean first hyperpolarizability and γ^{0e} is the mean second hyperpolarizability. The superscript 'e' denotes quantities measured at the frequency of the light. The values of α_0 and κ without superscripts refers to their values for static fields. The superscript '0' on β^{0e} and γ^{0e} refers to the static field which induces the polarization of the fluid. These hyperpolarizabilities differ from the static hyperpolarizabilities and from those responsible for non-linear harmonic generation. It is often assumed that the hyperpolarizabilities are nearly equal to their dc values. Equation (5.21) is a classical expression. The quantum theory of the Kerr effect has been developed by Buckingham and co-workers, and is needed when considering light molecules (H_2, D_2, see Ref. [516]) or when the probe is near a resonance of the molecule [517–519]. To determine the anisotropy κ of the polarizability from Kerr effect measurements, the average polarizability must be measured in some other way (typically index of refraction or dielectric constant measurements) since κ appears only in the product $\alpha_0\kappa$.

For a spherical-top molecule μ_0, κ, and β vanish, and the Kerr effect determines γ^{0e}. For polar molecules, only one term depends on T^{-2} so that $\alpha_0^e\kappa^e$ can be separated from the other terms if the temperature dependence of ${}_mK$ is measured. For non-polar molecules Eq. (5.21) becomes

$$ {}_mK = \frac{4\pi N_A}{81}\left[\frac{9\alpha_0^e\kappa^e\alpha_0\kappa}{5kT} + \gamma^{0e}\right]. \tag{5.22} $$

Measuring ${}_mK$ as a function of T enables the independent determination of γ^{0e} and the product $\alpha_0^e\kappa^e\alpha_0\kappa$. Often Kerr effect measurements are combined with measurements of Rayleigh scattering (which measures $|\kappa^e|$) to separate the terms in Eq. (5.21).

A typical setup for measuring the Kerr effect is shown in Fig. 5.3. Polarizer P1 polarizes the light at 45° with respect to the electric field between two parallel plates. The light passes through the material between the two plates and then through P2, a polarizer which is crossed with respect to the first polarizer. The Kerr effect causes the light to become elliptically polarized as it passes through the material with the applied electric field. Light transmitted by the crossed polarizers oscillates at twice the frequency of the applied field, since the Kerr effect is proportional to E^2.

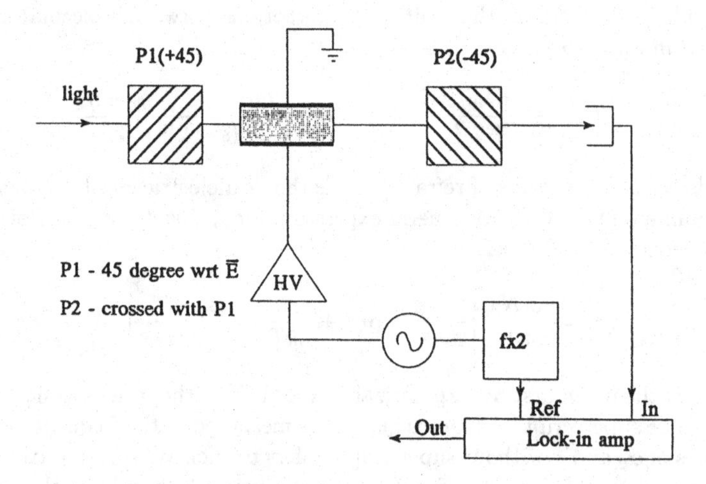

Figure 5.3: Experimental setup for measuring the Kerr effect.

Böttcher and Bordewijk [32] have an excellent discussion of the Kerr effect, including measurements before 1978. For a more recent application of the Kerr effect see Ref. [520].

5.1.5 Beam deflection

A straightforward way to determine the polarizability of a molecule is to observe the deflection of a collimated molecular beam which is passed through a static, inhomogeneous electric field. In modern-day experiments, the electric field is produced using a "2-wire" deflecting field. The result is a relatively large region over which the product of the electric field and the field gradient is constant. Defining the field to be along

$\hat{z}$, the force on an atom in this field is

$$\mathbf{F}_{\mathrm{E}} = \alpha_{zz} \mathrm{E} \frac{d\mathrm{E}}{dz} \hat{z} \tag{5.23}$$

where $\mathbf{F}_{\mathrm{E}}$ is the electric force on a molecule, α_{zz} is the zz component of the polarizability tensor, and E is the magnitude of the electric field. The spatial deflection of a molecule is approximately proportional to $1/v^2$, where v is the magnitude of the velocity of the molecule: a slower molecule will spend more time in the field region, and therefore receive a larger change in momentum; and a slower molecule takes longer to reach the detector, so that the displacement is larger for the same change in momentum.

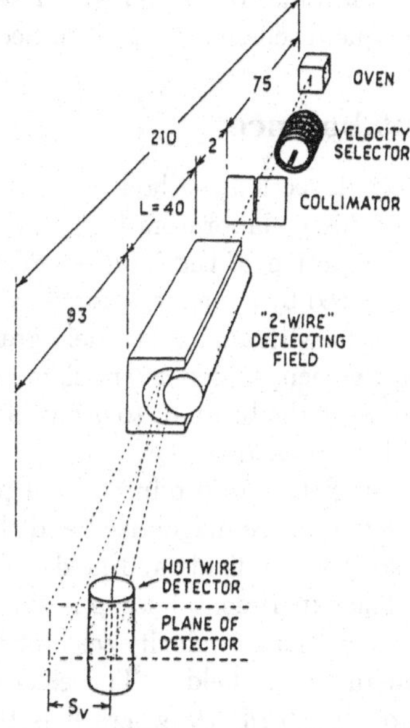

Figure 5.4: Schematic diagram of the apparatus used by Hall and Zorn for the measurement of alkali-metal polarizabilities. (Reprinted with permission from Ref. [414].)

A typical setup is shown in Fig. 5.4. The source in this case is an oven with a narrow rectangular orifice. The particles pass through a rotating-disk velocity selector in order to minimize errors associated with an imperfect knowledge of the velocity distribution. A slit placed before the deflecting field, in conjunction with the opening of the oven, limits the transverse velocity and position distribution of the beam. Deflection is measured by a hot wire detector. The oven and hot-wire detector are useful for measurements involving alkali metals, but other materials might require a different source and/or detector.

Scheffers and Stark [21] in 1934 first used electric deflection to measure the polarizabilities of alkali atoms (Li, K, Cs) to accuracies of 20% to 50%. In 1936 Scheffers and Stark [521, 522] used this method to measure the polarizability of atomic hydrogen—one of the few experimental measurements of the polarizability of hydrogen. Many of the early electric deflection measurements are inaccurate, with errors often attributed to a lack of knowledge of the velocity distribution. In 1974 Hall and Zorn [414] used a velocity selector in their electric deflection measurements of the polarizabilities of alkali metals, and demonstrated that the assumption of a Maxwellian velocity distribution leads to erroneous polarizability measurements. Electric deflection measurements made since 1974 have either employed a velocity selector, or have taken care to ensure that the distribution is known. Recent deflection experiments have measured the polarizabilities of alkali halide dimers [523], homonuclear and heteronuclear alkali dimer molecules [475], and alkali clusters Na_n and K_n (see Fig. 3.13) The latter measurement took advantage of the very narrow velocity distribution of supersonic beams. The polarizability of Al and Al clusters (Fig. 3.15) was measured by deflection using a technique discussed in detail in Section 5.2.3.

5.1.6 E-H gradient balance

Closely related to the direct deflection method is the E-H gradient balance method developed by Bederson and his collaborators [415, 524–527]. In this technique the electric force, caused by interaction of the induced electric dipole moment with an electric field gradient, is balanced by a magnetic force, caused by interaction of the effective magnetic dipole moment with a magnetic field gradient. Balancing the electric and magnetic forces yields a velocity-independent signal so that it is not necessary to know the velocity distribution of the beam or to use velocity selection. The technique has been applied to several atoms and to H_2.

The electric and magnetic fields are produced by a pair of electrodes in the two-wire configuration. The electrodes are magnetized and there is an electrical potential difference between them, so that the fields and field gradients are all approximately in the same direction, $\hat{z}$. The experimental setup is similar to the electric deflection setup shown in Fig. 5.4, except that no velocity selector is needed and the pole pieces produce both electric and magnetic fields. The electric force on an atom in the magnetic substate m_F is $\alpha_{zz}(m_F)\mathrm{E}\partial\mathrm{E}/\partial z$ where E is the magnitude of the electric field and α_{zz} is the zz component of the polarizability. The magnetic force on an atom is $\mu_{\text{eff}}(m_F)\partial\mathrm{H}/\partial z$ where μ_{eff} is the effective *magnetic* moment of the atom in the magnetic field H. The forces are balanced when

$$\alpha_{zz}(m_F)\mathrm{E}\frac{\partial \mathrm{E}}{\partial z} = -\mu_{\text{eff}}(m_F)\frac{\partial \mathrm{H}}{\partial z}, \tag{5.24}$$

independent of the velocity of the atom. The electric and magnetic fields are assumed to be congruent, i.e. $(\partial\mathrm{E}/\partial z)/\mathrm{E} = (\partial\mathrm{H}/\partial z)/\mathrm{H}$, so that the balance condition can be rewritten as

$$\alpha_{zz}(m_F)\mathrm{E}^2 = -\mu_{\text{eff}}(m_F)\mathrm{H}. \tag{5.25}$$

In practice the polarizability is measured at a fixed magnetic field, scanning the electric field and observing the undeflected atoms as shown in Fig. 5.5. If $\mu_{\text{eff}}(m_F)$ is known, then $\alpha(m_F)$ can be determined provided that E, and H are known. The most accurate measurements using this technique use a nearly simultaneous measurement of an atom with a known polarizability (for example, 2 $^3S_1$ metastable helium). If the same magnetic field is used when balancing the reference atom then

$$\alpha_{zz}(m_F) = \alpha_r \left[\frac{\mu_{\text{eff}}(m_F)}{\mu_r}\right] (\frac{V_r}{V})^2, \tag{5.26}$$

where V is the potential applied to the pole pieces and the quantities with an 'r' subscript refer to the reference atom. The tensor polarizability of the atom is completely determined by measuring $\alpha_{zz}(m_F)$ for two or more m_F.

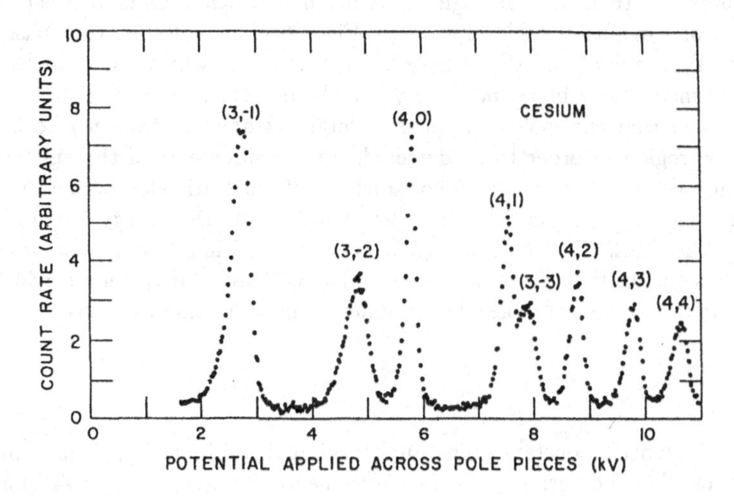

Figure 5.5: Scan of the electric potential applied across the pole pieces at a fixed magnetic field strength. The peaks result from a balancing of the electric and magnetic forces acting on cesium atoms in different (F, m_F) sublevels. The separation of the peaks is due entirely to the dependence of μ_{eff} on (F, m_F), since the polarizability of cesium is a scalar. (Reprinted with permission from Ref. [415].)

A disadvantage of the technique is that not all atoms or molecules have an appropriate ratio of polarizability to magnetic moment to make it practical to balance the electric and magnetic forces. For example, the alkali metal dimers were measured by deflection [475], since their magnetic moments are on the order of nuclear magnetons. In this measurement the magnetic field was used to remove atoms from the beam of particles emitted from an oven. A measurement of the polarizability of metastable $^3D_3$ mercury could not be made using the E-H balance technique because its polarizability was too small. The polarizability of this state was instead measured using electric deflection [526]. Atoms with a large nuclear spin also present difficulties, since the large number of magnetic sublevels means that there is a low intensity in any one peak. This complicated the measurement of the polarizability of indium, such that the E-H gradient balance measurement was supplemented by pure deflection measurements [527]. The E-H gradient balance technique has been used to measure the polarizability of Li, Na, K, Rb, Cs, and $^3P_2$ metastable Ne, Ar, Kr, Xe (Ref. [415]),

H_2 (Ref. [528]), $^3P_2$ metastable Hg (Ref. [526]), Tl (see Ref. [26]), and most recently In (Ref. [527]).

5.1.7 Atomic beam resonance

Molecular-beam resonance experiments have been used to measure α_2, the anisotropic part of the polarizability tensor. However, they are insensitive to the scalar polarizability. In a typical molecular-beam resonance experiment [529] molecules are emitted by a source and then pass through the A magnet, which is used to select molecules of a particular magnetic sublevel m_F—see Fig. 5.6. The molecules then pass through an interaction region, and then through the B magnet, which allows molecules of a particular magnetic sublevel m'_F to pass to the detector. In a "flop-in" experiment $m'_F = m_F$ so that the molecules must undergo a transition from m_F to m'_F in the interaction region in order to be detected. In measurements of the anisotropic polarizability, the interaction region contains a uniform static electric field $\mathrm{E}\hat{z}$ and an oscillating electromagnetic field. The static field splits the energy levels of the m_F states by the Stark effect (which depends on $|m_F|$). The oscillating electromagnetic field induces transitions from m_F to m'_F. The oscillation frequency ν which causes the maximum number of molecules to undergo the transition from m_F to m'_F is

$$\nu = \frac{1}{2h}|\alpha_{zz}(m_F) - \alpha_{zz}(m'_F)|\mathrm{E}^2, \tag{5.27}$$

where h is Planck's constant. The anisotropic polarizability α_2, defined in Section 3.2.1, is completely determined by a measurement of $|\alpha_{zz}(m_F) - \alpha_{zz}(m'_F)|$ for a single pair of magnetic sublevels.

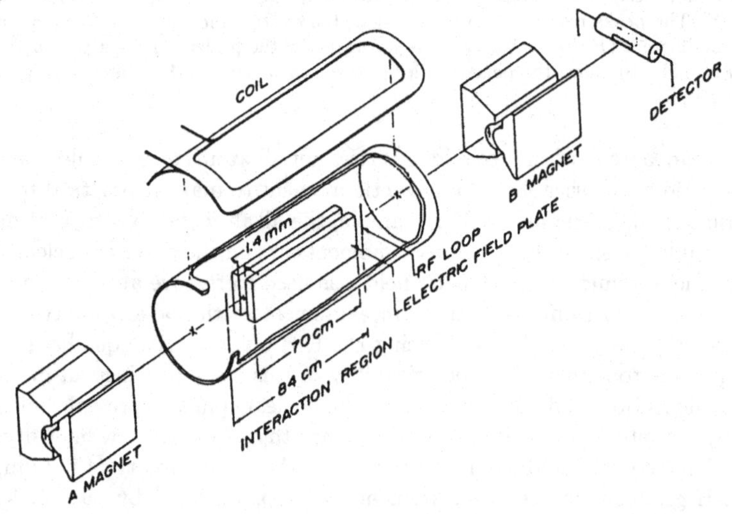

Figure 5.6: Sketch of the apparatus used in the beam-resonance technique (with one of four magnetic-field coils shown).(Reprinted with permission from Ref. [530].)

Very precise measurements of α_2 have been made for several atoms. For example, Gould *et al.* have measured α_2 of ^{85}Rb to be $\alpha_2 = (-2.72 \pm 0.14) \times 10^{-6}$ Å^3 [530, 531]. This is more than seven orders of magnitude smaller than the scalar (isotropic) part of the polarizability tensor. The polarizability of rubidium is nearly scalar because $J = 1/2$. For an atom with angular momentum J, physical properties will have spherical tensor components of maximum rank $L = 2J$ (see Section 2.2). The presence of a rank $L = 2$ (tensor) polarizability is due to hyperfine structure in the excited states.

English and MacAdam [532] and MacAdam and Ramsey [533] were the first to apply the beam resonance technique to molecules, making accurate measurements of the polarizability anisotropy of H_2 and D_2. The results of this pioneering work are an order of magnitude more precise than other measurements of these molecules using Rayleigh scattering [512] or beam deflection [528].

5.2 New Techniques

5.2.1 M-lines

The m-lines method was first used to measure the refractive index and thickness of thin film waveguides [534, 535]. Initially, the technique relied on the measurement at a single frequency, relying on the use of a TEM_{00} mode laser beam. Further developments extended the technique to polychromatic sources, allowing dispersion curves for thin films to be measured [536, 537]. In 1983, Ding and Garmire [538] significantly improved the method by extending it to leaky quasi-waveguide thin films. This allowed measurement of materials whose refractive index is *less* than that of the substrates that support them and it removed the need to apply pressure to the thin film (which can change the properties of the film). Also, the difficult task of prism coupling, present in the lossless waveguide version, is eliminated since the coupling prism is the substrate in the quasi-waveguide method. The quasi-waveguide method has allowed researchers to measure the polarizability of large polymers formed as thin films [98] or dissolved in a thin-film matrix of polymethylmethacrylate (PMMA) [100] — see Sec. (3.4).

The m-lines method relies on the observation of several lines that appear simultaneously when light is optimally coupled from a prism into a thin film — see Fig. 5.7.

The different lines that are observed on the screen only occur at particular angles of incidence $\theta = \theta_m (m = 0, 1, 2, ...)$ for the light on the prism. Here the index m corresponds to the mode number of the wave (leaky or lossless) that propagates in the thin film. Each waveguide mode m will have a particular angle of incidence θ_m that will optimize the coupling of the incident light into that mode. However, when the condition is optimal for a particular mode, there will be a few modes on either side that will also experience significant coupling and hence many lines are visible on the screen. The angle of incidence on the prism base θ determines which mode is most efficiently coupled, because this angle determines the phase velocity v^i of the

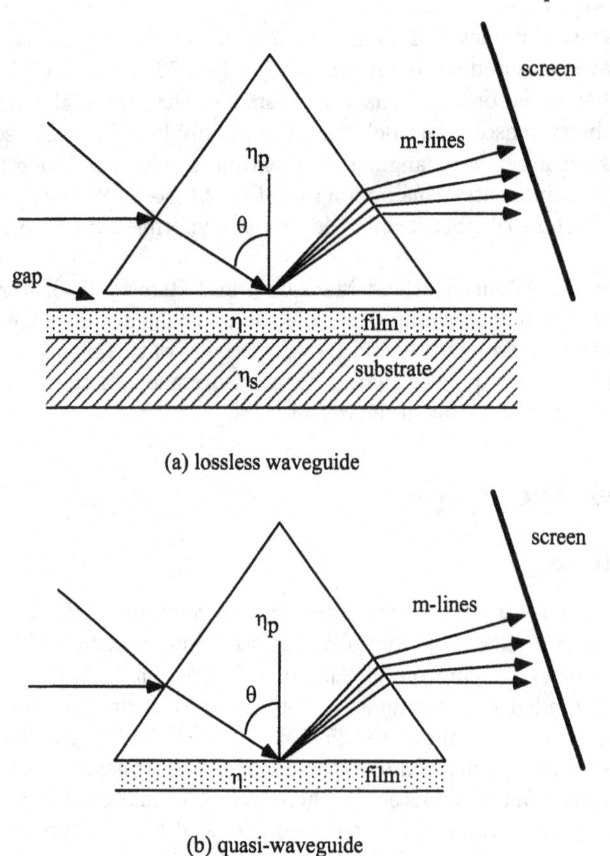

Figure 5.7: A diagram showing the m-lines method when light is prism-coupled into a lossless guided mode in the thin film (a) or into a leaky mode in a thin film (b).

incident wave along the x-direction in the prism and in the gap, where

$$v^i = \frac{c}{\eta_p \sin\theta}, \tag{5.28}$$

where c is the speed of light in vacuum, and η_p is the refractive index of the prism. When the incident phase velocity equals the phase velocity of the mth characteristic mode in the thin film ($v^i = v_m$), strong coupling into that mode occurs. By experimentally measuring the angles θ_m at which the coupling is optimized, the effective refractive index for the mth mode, N_m, can be determined from

$$N_m = \frac{c}{v_m} = \eta_p \sin\theta_m. \tag{5.29}$$

For a thin film of index η and thickness d on a substrate of index η_s, the refractive index of the mth mode, N_m, can also be calculated [535]. These values depend on the frequency of light ω and on its polarization ρ, where $\rho = 0$ for TE polarization and $\rho = 1$ for TM polarization. Thus we can write $N_m = N_m(\omega, \rho, \eta_s, d, \eta)$. The parameters ω, ρ, η_s can all be measured by independent means. Hence by comparing calculated values of N_m for different values of d and η to measurements of N_m, we can determine d and η. The mode index N_m must be measured for at least two modes m to determine the thin-film parameters d, η. If more than two modes can be measured, then the results can be checked for consistency and the accuracy of the results can be improved. The calculations of N_m for a lossless waveguide must done numerically with a computer; the calculations are long but relatively simple [535]. For a lossless waveguide the refractive index of the film must exceed the refractive indices of the materials located on either side of it.

For a quasi-waveguide, where coupling is to leaky modes, the index of refraction of the thin film is less than the substrate and the substrate acts as the optical coupler — see Fig. 5.7(b). Here an analytic expression for the refractive index of the thin film η can be written that is accurate to 1 part in 10^6. To eliminate the thickness as a parameter, we need to use the refractive indices of two different modes l, m and we have [538]

$$\eta = N_l^2 + \frac{(l+1)^2}{(m+1)^2 - (l+1)^2}(N_l^2 - N_m^2). \tag{5.30}$$

Thus, once the mode refractive indices N_m, N_l are found from the measured angles θ_m, θ_l, a hand held calculator can be used to determine η from Eq. (5.30).

In both cases, the procedure is to rotate the prism-film setup until you reach the angle θ_m that optimizes the m-line pattern on the screen for the mth mode. Using Eq. (5.30), this allows a value of the mode index N_m to be found. From any two modes values for d and η are found. In the end, to improve accuracy, both the lossless waveguide and quasi-waveguide techniques vary d and η to minimize the expression

$$\sigma(d, \eta) = \sum_m [\bar{N}_m - N_m(d, \eta)]^2, \tag{5.31}$$

where $\bar{N}_m$ is the effective refractive index of the mth mode determined from the measurements and $N_m(d, \eta)$ is the calculated effective index. This last step improves the accuracy of the results in both cases and the final accuracy for each case is about the same: for the lossless waveguide $\delta\eta = 16 \times 10^{-5}$ and for the quasi-waveguide $\delta\eta = 4.1 \times 10^{-5}$. A major difference in the two techniques is that the lossless waveguide measures refractive indices for high index materials, while the quasi-waveguide method is for lower index materials. For more details about the lossless waveguide method, see Ref. [535]; for more details about the quasi-waveguide method, see Ref. [538].

5.2.2 Dispersive Fourier-transform spectroscopy

Dispersive Fourier-transform spectroscopy (DFTS) is an interferometric technique used to measure the complex refractive index of gases and liquids, which can be related to the mean polarizability via the Lorentz-Lorenz equation given in Sec. 5.1.2.

The ability of DFTS to simultaneously determine both the real and imaginary parts of the refractive index (or polarizability) is its key feature. It avoids the use of the Kramers-Krönig relations that are frequently used to find the imaginary part of the polarizability, α'', from a knowledge of its real part over a large wavelength range — see Sec. 2.5. DFTS in the infrared and far infrared wavelength ranges is a common method for studying the refractive index of liquids [539–541]. Recently, this capability of dispersive Fourier-transform spectroscopy for measuring α in the visible wavelength range was demonstrated for the first time by Hohm and coworkers [295]. However, since dispersive Fourier-transform spectroscopy is a relative technique, it must be accompanied by at least one absolute measurement. The absolute measurement, usually done at the HeNe laser wavelength of 632.8 nm, can be carried out by the same apparatus. This technique has been successfully developed and applied to many species including Cd [542], Hg [296], Zn [24], Ar, H_2, O_2 [507], NO_2/NO_4 mixtures [295, 543], As_4 [544], I_2 [545], and $CBrF_3$ [546].

To account for the complex refractive index and polarizability, the Lorentz-Lorenz relation Eq. (5.4) must be slightly modified. A complex refractive index $\tilde{\eta}$ can be written

$$\tilde{\eta}(\bar{\nu}) = \eta(\bar{\nu}) + i\frac{a(\bar{\nu})}{4\pi\bar{\nu}}, \tag{5.32}$$

where $a(\bar{\nu})$ is the absorption coefficient of the particles at wavenumber $\bar{\nu} = 1/\lambda$, where λ is the wavelength of the light. The Lorentz-Lorenz relation becomes

$$\tilde{\alpha} = \frac{3}{4\pi n}\frac{\tilde{\eta}^2(\bar{\nu}) - 1}{\tilde{\eta}^2(\bar{\nu}) + 2}. \tag{5.33}$$

Rewriting Eq. (5.32) as $\tilde{\eta}(\bar{\nu}) = \eta_R(\bar{\nu}) + i\eta_I(\bar{\nu})$, and noting that $\tilde{\alpha} = \alpha' + i\alpha''$, the real and imaginary parts of the complex polarizability are

$$\begin{aligned} \alpha' &= \frac{1}{A}[(\eta_R^2 + \eta_I^2)^2 + \eta_R^2 - \eta_I^2 - 2] \\ \alpha'' &= \frac{6\eta_R\eta_I}{A}, \end{aligned} \tag{5.34}$$

where

$$A = \frac{4}{3}\pi n[(\eta_R^2 + \eta_I^2)^2 + 4(\eta_R^2 - \eta_I^2 + 1)]. \tag{5.35}$$

Here n is the number density.

Like the traditional method for measuring the refractive index using a Michelson interferometer (see Sec. 5.1.2), DFTS allows determination of the complex refractive index over a broad range of frequencies, e.g. 11,000-22,000 cm^{-1} [295]. The basic experimental setup is to record the intensity that results from the interference of white light from the two arms of a Michelson interferometer while the path length in one arm is modulated [547]. A highly simplified sketch of the apparatus is given in Fig. 5.8. A white light source is collimated and split, then the two beams traverse the two arms of the interferometer, one arm containing the sample cell and the other a reference cell. Very careful control over the temperature and pressure of the gases in the interferometer is important because the refractive index depends on both of these parameters — see the discussion at the end of Sec.5.1.2. The result of linearly changing the path length in one arm is to produce an optical path difference Δw between the two arms of the interferometer

$$\Delta w = [\eta_R(\bar{\nu}) - 1]l - \delta + \delta_I(\bar{\nu}), \tag{5.36}$$

where η_R is the real part of refractive index of the gas in the sample cell, l is twice the length of the sample cell, $\delta/2 = l_1 - l_2$ is the path difference between the two arms with both cells empty, and $\delta_l(\bar{\nu})$ is a small correction term for the residual dispersion of the apparatus without gas. In Ref. [295], the path difference δ was determined by illuminating a second interferometer, called the reference interferometer and coupled to the first interferometer, with monochromatic light from a HeNe laser. A detailed discussion of such a twin Michelson interferometer is given by Hohm and Kerl [548] and by Goebel and Hohm [544].

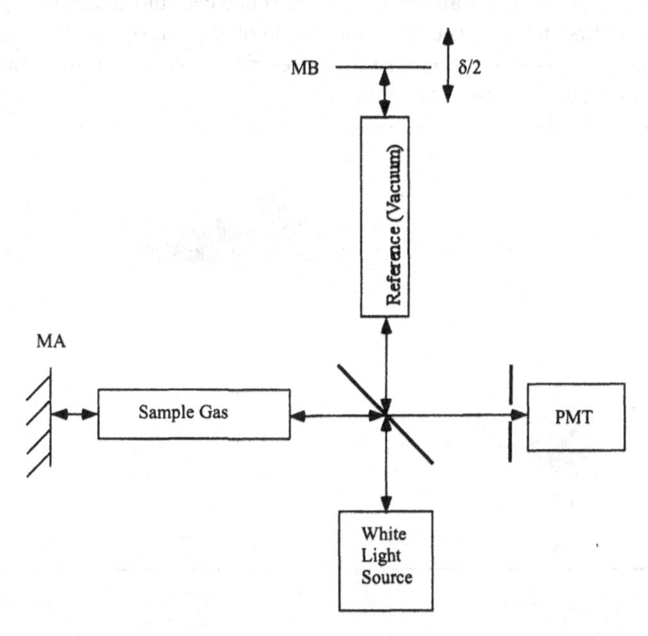

Figure 5.8: A diagram showing the basic elements used in a twin Michelson interferometer for using dispersive Fourier transform spectroscopy to measure the complex polarizability $\tilde{\alpha}$.

The main experimental measurement consists of a pair of dispersive Fourier interferograms, one without gas and with gas in the sample cell. An example of interferograms from a recent measurement are given in Fig. 5.9. The interferograms are simply recordings of the intensity of light from the output of the interferometer that has passed through a pinhole and was detected by a photomultiplier tube. To understand how the refractive index is determined from the interferograms, it is important to describe the dependence of the intensity recorded in an interferogram as a function of changing path length, δ, of the interferometer. Note that the difference in phase of the light in one arm relative to the other is

$$\Delta\phi = 2\pi\Delta w\bar{\nu}. \tag{5.37}$$

For monochromatic light, the intensity of light at the detector (point of observation, P) is

$$I(P) = (I_1 + I_{20}) + 2(I_1 I_{20})^{1/2}\cos\Delta\phi, \tag{5.38}$$

where

$$I_1 = I_{10}e^{-al}. \tag{5.39}$$

Here we have used the Beer-Lambert law to express the absorption experienced by light passing through the first arm of the interferometer, which contains the sample cell and gas. The quantity I_{10} corresponds to the intensity in arm 1 when the sample cell is evacuated and no absorption occurs, i.e. $a = 0$. The quantity I_{20} is the intensity in arm 2, containing an evacuated reference cell and hence it experiences no absorption. The first term of the right-hand-side of Eq. (5.38), in parentheses, is the background intensity and the second term gives rise to the interferogram pattern as $\Delta\phi$ is varied by changing the path length δ.

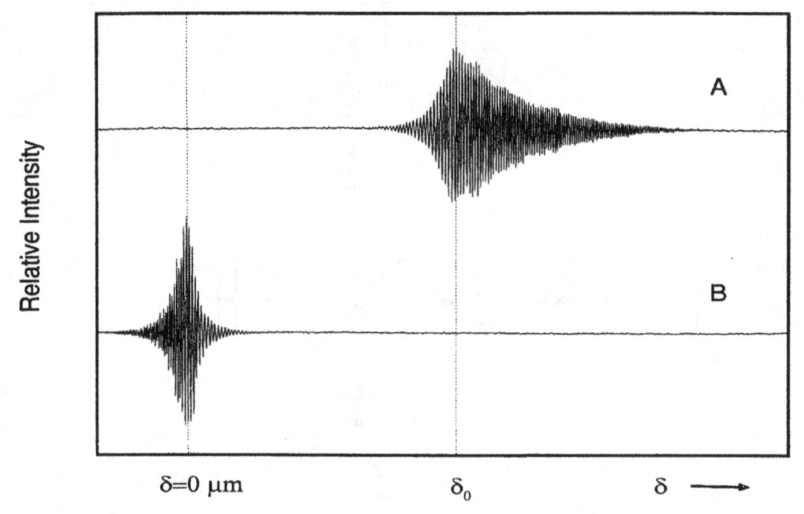

Figure 5.9: Sample interferogram (A) of As_4-vapor at $T = 900$ K, $d_{As_4} = 4.504$ mol/m^3 and corresponding apparatus interferogram (B).(Reprinted with permission of U. Hohm)

In the case of white light, Eq. (5.38) becomes an integral over frequencies (in wavenumbers) detectable by the photomultiplier tube

$$I(P) = \int_{\bar{\nu}_l}^{\bar{\nu}_u} [I_1(\bar{\nu}) + I_{20}(\bar{\nu})]d\bar{\nu} + \int_{\bar{\nu}_l}^{\bar{\nu}_u} 2[I_1(\bar{\nu})I_{20}(\bar{\nu})]^{1/2}\cos\Delta\phi\, d\bar{\nu}. \tag{5.40}$$

In practice, the spectral response of the photomultiplier tube limits the wavenumber range over which the intensity is detectable. However, its frequency response is broad enough that, to good approximation, the substitutions $\bar{\nu}_l \to 0$ and $\bar{\nu}_u \to \infty$ are appropriate for Eq. (5.40). The resulting expression for the recorded interferogram is

$$I(\delta) - I(\infty) = \int_0^\infty 2[I_1(\bar{\nu})I_{20}(\bar{\nu})]^{1/2} \cos \Delta\phi(\bar{\nu}, \delta) d\bar{\nu}, \tag{5.41}$$

where

$$I(\infty) = \lim_{\delta \to \infty} I(\delta) = \int_0^\infty [I_1(\bar{\nu}) + I_{20}(\bar{\nu})] d\bar{\nu}. \tag{5.42}$$

From relations given above, the interferogram intensity difference can be written [547]

$$I(\delta) - I(\infty) = \int_{-\infty}^\infty A(\bar{\nu}) e^{i2\pi\bar{\nu}\delta} d\bar{\nu}, \tag{5.43}$$

where

$$A(\bar{\nu}) = B(\bar{\nu}) e^{-la(\bar{\nu})/2} e^{-i2\pi\bar{\nu}l[n(\bar{\nu})-1]}, \tag{5.44}$$

and

$$B(\bar{\nu}) = 2[I_{10}(\bar{\nu})I_{20}(\bar{\nu})]^{1/2} e^{-i2\pi\bar{\nu}\delta_l(\bar{\nu})}. \tag{5.45}$$

Thus, the interferogram can be expressed as a Fourier transform of a function $A(\bar{\nu})$, where $A(\bar{\nu})$ is simply the Fourier amplitude of the intensity interferogram difference at wavenumber $\bar{\nu}$. In contrast to the Fourier spectrum generated by the system in the presence of a gas, the factor $B(\bar{\nu})$ is simply the Fourier spectrum generated by the system in the absence of gas, where the complex refractive index is unity, i.e. $\tilde{\eta}(\bar{\nu}) = 1$.

By taking the Fourier transform of a pair of interferograms, the two spectra $A(\bar{\nu})$ and $B(\bar{\nu})$ are generated with appropriate computer programs. Once these spectra are calculated from the measurements, they can be related to the absorption coefficient $a(\bar{\nu})$ and the refractive index $\eta(\bar{\nu})$ using the relations given by Kerl and Häusler [547]. For the absorption coefficient we have

$$a(\bar{\nu}) = -\frac{1}{l} \ln \left[\frac{|A(\bar{\nu})|^2}{|B(\bar{\nu})|^2} \right]. \tag{5.46}$$

Here the modulus $|z|$ of a complex number $z = x + iy$ (x, y real) is given by $|z| = \sqrt{x^2 + y^2}$. The expression for the refractive index is slightly longer

$$\eta(\bar{\nu}) - 1 = \frac{\delta_0}{l} + \frac{[\Phi_A(\bar{\nu}) - \Phi_B(\bar{\nu})]}{2\pi\bar{\nu}l}. \tag{5.47}$$

The absolute phases $\phi_S, S\epsilon\{A, B\}$ are defined by

$$\Phi_S(\bar{\nu}) = \phi_S(\bar{\nu}) + 2\pi m_S, \tag{5.48}$$

where ϕ_S is the principal value of the phase and m_S is an integer. The phase ϕ_S is given by

$$\phi_S = \tan^{-1}\left(\frac{\text{Im}S(\bar{\nu})}{\text{Re}S(\bar{\nu})}\right) \begin{cases} +\pi & \text{if Re}S < 0 \text{ and Im}S \geq 0 \\ -\pi & \text{if Re}S < 0 \text{ and Im}S < 0. \end{cases} \tag{5.49}$$

Here δ_0 is the difference between the origins of the sample A and the background B interferograms. For further discussion of the conversion of interferograms to Fourier spectra and the connection to the refractive index and absorption, consult the work of Kerl and Häusler [547] and references therein. We remind the reader that in addition to the interferograms either one refractive index and δ_0, or alternatively two refractive indices measured at two different wavenumbers $\bar{\nu}_1$ and $\bar{\nu}_2$ must be known to calibrate the scale for the relative values measured by DFTS.

The accuracy of the polarizabilities that result from measurements of η and a by dispersive Fourier transform spectroscopy depends on species. Some numbers for $|\Delta\alpha_0|/\alpha_0$ from recent measurements are: NO_2 uncertainty 0.1 % [295], N_2O_4 uncertainty 2% [295], Cd uncertainty 3% [542]. The accuracy of polarizabilities measured using this technique on nonabsorbing, room-temperature gas species is more typically 0.05% [546]. Advantages of this technique are its high accuracy, its wide wavelength range, and the fact that it is capable of a simultaneous measurement of both the real and imaginary parts of the polarizability via the refractive index and absorption of the light by the gas (see especially References [295, 545]). At present an important disadvantage of the technique is that it can only be applied to measure the refractive index η and the absorption coefficient a of macroscopic quantities. Determinations with single molecules or with molecular beams are not possible.

The same group that has extended the capability of DFTS to the visible range has also determined polarizabilities from refractive index measurements made using a different Fourier transform technique that also uses a Michelson interferometer [507, 509, 549–552]. In these measurements, the interferometer is illuminated simultaneously by several different *monochromatic* light sources (lasers) and the pressure of the gaseous sample is lowered to zero, keeping the end mirror fixed ($\delta = 0$). This produces an intensity which depends on the density d, that is $I(d)$. The result is a very fine superposition of sinusoidal functions. Taking the Fourier transform *with respect to the amount-of-substance density* d yields a peak spectrum from which precise polarizabilities at the different wavelengths can be obtained. The accuracy of such polarizabilities are typically on the order of 0.05%. If the gas does not have significant absorption at the wavelengths used, a reliable dispersion curve for α can be generated from the data.

5.2.3 Position sensitive time-of-flight

A clever technique for measuring dc polarizabilities was recently developed and applied to Al_n (n=1,2, and > 15) — see Fig. 3.15 [210, 211, 553]. The method is based on the traditional approach of deflecting a neutral beam using a nonuniform electric field. The novel aspect of the technique is its detection scheme using a position-sensitive time-of-flight spectrometer and laser photoionization — see Fig. 5.10.

Here a cluster source based on laser vaporization is used to generate an Al cluster beam with clusters of up to several thousand aluminum atoms. The clusters traverse a long region (12.5 cm) with an inhomogeneous, dc electric field — see Fig. 5.10. This aspect is no different than the 'original' deflection experiments of Scheffers and Stark

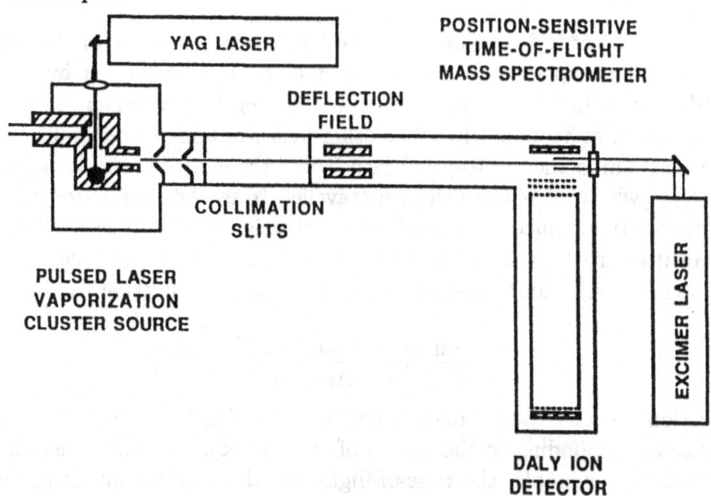

Figure 5.10: The experimental setup for the time-of-flight detection of the polarizability of aluminum clusters. (Reprinted with permission from Ref. [211].)

— see Section 5.1.5. Each cluster is deflected by an amount that is directly proportional to its polarizability and inversely proportional to its mass. After deflection the clusters enter a long drift tube (100 cm) that allows the deflected clusters to further spread out in the transverse direction. At the end of this tube the clusters enter a region where they are ionized by absorbing photons of energy 6.41 eV ($\lambda = 1973$ Å) produced by an excimer laser. Particular attention is paid to adjusting the ionization volume to generate the maximal single-shot signal. After ionization occurs an electric field is pulsed on to accelerate the ionized clusters orthogonal to the original drift direction. This electric field and two others are used to form a time-of-flight mass spectrometer. After a 100 cm drift path the cluster ions are detected by a Daly ion detector.

To gain a physical understanding of how the position-sensitive time-of-flight spectrometer affects the cluster beam, refer to Fig. 5.11. The transverse spatial profile of the deflected beam in the ionization region is converted into a time profile by the time-of-flight spectrometer. In addition the time-of-flight spectrometer plays its traditional role of separating out different clusters in time according to their charge-to-mass ratios. The actual measurement consists of finding the average shift in the signal due to a given cluster mass when the deflecting field is on compared to when the deflecting field is off. For each beam pulse this time shift is determined for all the different cluster masses. The shift in time Δt for a particular cluster mass can then be related to the transverse deflection d of that cluster when the field is on. The deflection d is the transverse displacement at the ionization region when the deflection field is on versus off. The deflection of a given cluster mass is given by

$$d = \left(\alpha_0 \mathrm{E}\frac{d\mathrm{E}}{dz}\right)\frac{K}{mv^2}, \tag{5.50}$$

where K is a geometric factor and v is the longitudinal velocity of the cluster. The quantity d is the displacement of an object of mass m acted on by a force $F = \alpha_0 \mathrm{E}(d\mathrm{E}/dz)$ for a time $t = L/v$, where L is the length over which the force acts — see the discussion in Section 5.1.5. The other factor of v in the denominator is due to the time the beam is allowed to propagate before the effects of the force are observed ($t_{obs} = D/v$), where D is the distance travelled from the interaction region to the point of observation. Since an accurate determination of K and $\mathrm{E}(d\mathrm{E}/dz)$ is difficult, these quantities are calibrated by deflecting lithium. Then the polarizability of an aluminum cluster α_{Al} can be deduced by comparing its deflection to that of lithium, i.e.

$$\frac{\alpha_{Al}}{\alpha_{Li}} = \frac{d_{Al}}{d_{Li}} \frac{m_{Al}}{m_{Li}} \left(\frac{v_{Al}}{v_{Li}}\right)^2 . \tag{5.51}$$

The deflection of each cluster is determined by measuring the time shift in the centroid of the pulse corresponding to the arrival of clusters of a particular mass at the Daly ion detector. For example, the time-of-flight signal for aluminum atoms is given in Fig. 5.12. The shift in the centroid of the peak corresponds to the deflection of the Al atoms by the deflecting field. The centroid of each peak is calculated by finding the mean time-of-arrival of the cluster ions of a given mass through numerical integration. The *time shift* Δt in the centroids corresponding to undeflected and deflected cluster ions is simply the difference in the mean arrival times in the two cases, i.e.

$$\Delta t = \int t I_d(t) dt - \int t I_u(t) dt. \tag{5.52}$$

The quantities I_d and I_u correspond to *normalized* cluster ion intensities in the deflected and undeflected beams respectively, that is

$$\int I_k(t) dt = 1, \qquad k = u, d. \tag{5.53}$$

The accuracy of the technique is competitive with other methods. For example, the measurements on atomic aluminum yield a scalar polarizability of $\alpha_0 = 6.8 \pm 0.3\text{Å}^3$, while the measurements of the higher clusters ($n > 15$) are accurate to 15%.

The technique offers several advantages that should be noted. It uses a universal detector that can be applied to any particle species that remains stable upon ionization. It is a position-sensitive measurement that uses differences in timing to deduce differences in position. In addition, the deflections of all the different cluster types in the beam are measured for each pulse. This has obvious advantages in minimizing systematic errors such as shot-to-shot beam variations, long-term changes over the course of data-taking in cluster densities and distributions, etc. Finally, the entire deflection profile for the beam is recorded each shot, in contrast with deflection experiments where a slit is slowly scanned to map out the beam profile.

As a final note, we point out that to measure the polarizabilities of semiconductor clusters Becker and coworkers (see Section 3.5.2) scan a narrow laser beam through the time-of-flight ionization region. This narrow laser beam acts like a translating slit.

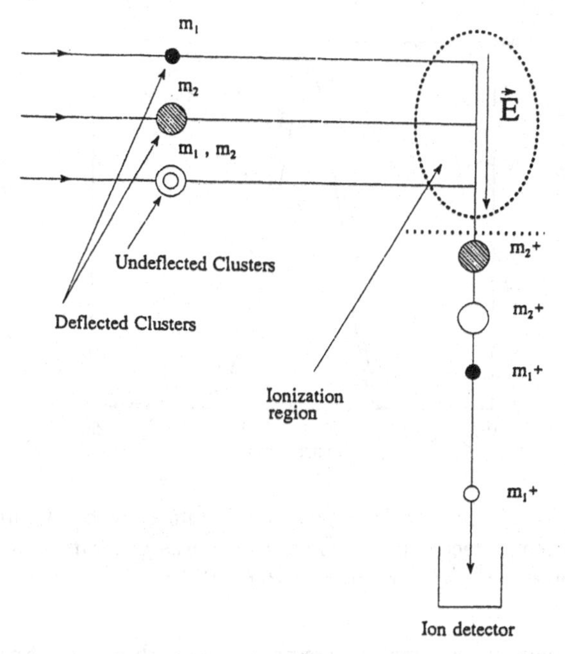

Figure 5.11: A sketch showing how the time-of-flight spectrometer distributes the deflected and ionized clusters in time.

5.2.4 Light force

A novel method for measuring the ac polarizabilities of atoms, molecules and clusters is based on applying large light forces using pulsed lasers with good longitudinal coherence [1, 22, 280, 554]. To understand the light-force technique, consider the simple picture of a molecule with a scalar polarizability, α_0, located in a *static* electric field **E**. If the electric field is a function of position, then a molecule will experience a conservative force $\mathbf{F}_C$ given by [as in Eq. (5.23)]

$$\mathbf{F}_C = \alpha_0 \mathrm{E} \boldsymbol{\nabla} \mathrm{E}. \tag{5.54}$$

Thus molecules in regions with large fields and large field gradients experience large forces. The technique takes advantage of the large fields and large field gradients that modern pulsed lasers are capable of producing. In the light-force technique a beam of molecules is sent through a standing wave formed by an intense, pulsed laser — see Fig. 5.13. Here, two of the three molecules traverse regions of large $\mathrm{E}\boldsymbol{\nabla}\mathrm{E}$ and receive significant deflections. The other molecule traverses an antinode of the standing-wave field and is not deflected.

A molecule in an *oscillating field* that is detuned far from any resonances also experiences a time-averaged force that is conservative. The conservative force only affects the motion along one dimension (the propagation direction of the laser, $\hat{x}$).

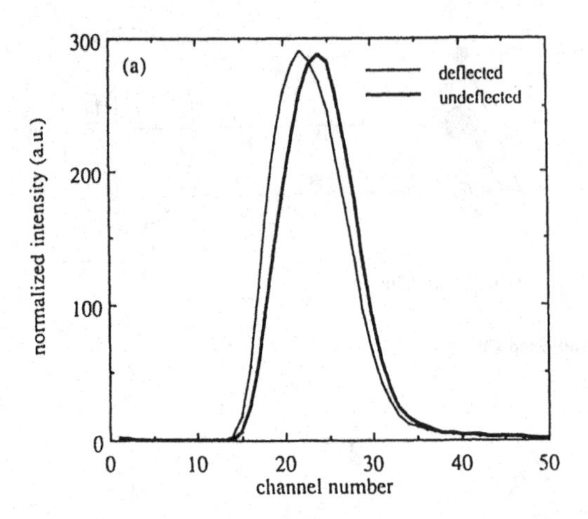

Figure 5.12: A plot of the data for atomic aluminum showing the undeflected peak and the time-shifted deflected peak. Similar data was generated for the other mass clusters. (Reprinted with permission from Ref. [211].)

The equation of motion for a molecule experiencing such a force can be written [1]

$$\frac{d^2x}{dt^2} = \frac{dv}{dt} = V(y,z)f(t)\sin(2kx), \tag{5.55}$$

where t is the time, $k = 2\pi/\lambda$ is the wavevector of the light-force beam, and the characteristic velocity V is

$$V(y,z) = \frac{16\pi^2}{M\lambda c}\mathcal{F}(y,z)\alpha_{zz}, \tag{5.56}$$

where λ is the wavelength of the standing wave laser, M is the mass of the molecule, and

$$\mathcal{F}(y,z) = \frac{c}{8\pi}\int_{-\infty}^{\infty} \mathrm{E}^2(y,z,t)dt \tag{5.57}$$

is the fluence (energy/area) of the laser pulse. In Eq. (5.55) $f(t)$ is the slowly-varying envelope of the laser pulse. Physically, the characteristic velocity corresponds to the maximum kick that can be given to a molecule by the light force. Here, the light is assumed to be linearly-polarized along the z-axis. The polarizability in Eq. (5.56) for an atom of angular momentum J in sublevel m is given in Eq. (3.19).

To detect an effect due to a conservative force, the molecules must be prepared in a nonuniform distribution in phase space. This is a direct result of Liouville's theorem. Molecules are prepared by phase-space filtering the molecular beam with a pair of slits. After passing through a pair of slits, the molecules traverse the standing-wave region, where the polarization force redistributes the molecules in phase space. The effects of the force can be detected by measuring the redistribution of molecules in

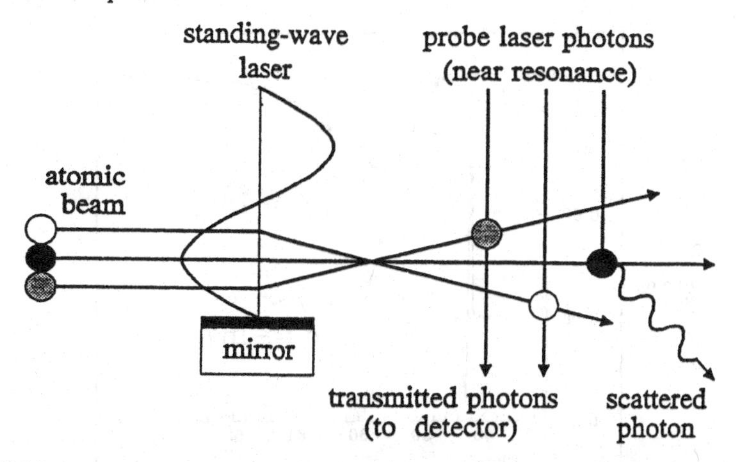

Figure 5.13: The change in the velocity distribution due to the standing-wave laser may be measured by observing the absorption of resonant light just downstream of the standing-wave region. Atoms that receive significant velocity kicks are Doppler detuned from resonance and scatter few photons. Atoms that experience no net force remain tuned on resonance and continue to scatter photons at a high rate.

configuration space and/or by measuring the redistribution of molecules in velocity space.

To date, the redistribution in velocity space has been determined by measuring the change in absorption of resonant light that occurs due to the Doppler shift — see Fig. 5.13. The atoms are probed slightly downstream from the standing wave with resonant cw light. The absorption coefficient is $a = -\ln T$, where T is the transmittance of the atomic beam. The absorption coefficient is a convolution of the x-velocity distribution $p(v)$ and the absorption cross section $\sigma(v)$. The absorption coefficient can be written

$$a(\bar{v}_i, V_{\alpha_0,\alpha_2}) = nl \int_{-\infty}^{+\infty} p(v - \bar{v}_i; V_{\alpha_0,\alpha_2})\sigma(v)dv, \tag{5.58}$$

where nl is the column density of molecules, $\sigma(v)$ is the absorption cross section of the molecules. The quantity $\bar{v}_i$ is the mean initial transverse velocity of the atoms that pass through the second slit. The absorption coefficient is measured for atoms unaffected by the light force (a_i) and for atoms that experienced the peak light force (a_f) — see Fig. 5.14. This pair of measurements is made for a series of values for the mean velocity $\bar{v}_i$. The mean velocity is changed by moving the second slit relative to the first. The results for uranium are given in Fig. 5.15. The data of Fig. 5.15 is fit by assuming the initial $p(v)$ is the triangular distribution formed by a pair of matched slits and by assuming $\sigma(v)$ is a Lorentzian function of v. The best fit of Eq. (5.58) to the data (Fig. 5.15) is obtained by adjusting the scalar and tensor polarizability values. In fact, under reasonable assumptions [22], the final absorption coefficient a_f is simply a function of α_0, α_2 and the initial coefficient a_i, independent of the assumed shapes of $p(v)$ and $\sigma(v)$.

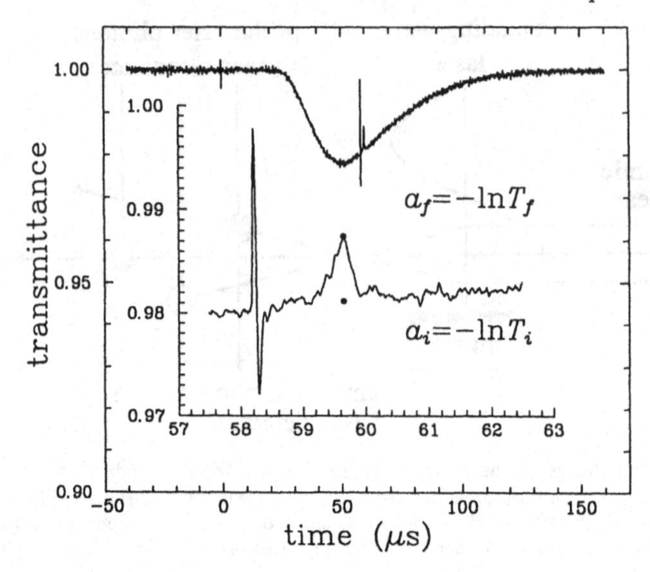

Figure 5.14: Raw data for uranium showing the transmittance of the probe laser versus time. The expanded inset shows the firing of the standing-wave laser (58.2 μs) and increased transmittance due to atoms affected by the light force (59.6 μs).

Recently, this technique was used to measure polarizability values for uranium at $\lambda = 1.06\mu$m [22]. The results are $\alpha_0(\omega) = 24.9 \pm 1.5$ Å^3 and $\alpha_2(\omega) = 0.7 \pm 1.7$ Å^3. These ac values can be converted to dc values using a ratio calculated using the oscillator strength sum rule. The resulting dc values are $\alpha_0(dc) = 20.3 \pm 1.4$ Å^3 and $\alpha_2(dc) = 0.7 \pm 1.8$ Å^3. The advantages of this technique are that it is an absolute measurement, it can be used to measure tensor components, and it can measure ac polarizabilities. It can also be applied to ions, since the fundamental interaction between the bare charge of the ion and the field averages to zero with an ac field that changes sign every femtosecond or so. Of course, static field techniques are incapable of directly measuring induced polarizabilities of ions.

5.2.5 Atom interferometry

An elegant atom-interferometry technique for measuring dc polarizabilities was recently demonstrated using sodium [411]. It is one of the first applications of the emerging field of atom interferometry. The interferometer used in these experiments consists of three mechanical gratings, each with slit separations of 200 nm and open fractions of 39.5%. The physical layout of the gratings that form the interferometer is given in Fig. 5.16. This layout is equivalent to a Mach-Zender interferometer, where for zero path length difference "white-light" interference occurs [555]. Under this condition for the atom interferometer, all velocities (or "wavelengths") interfere constructively to provide an intensity maximum. Each grating acts like a beamsplitter.

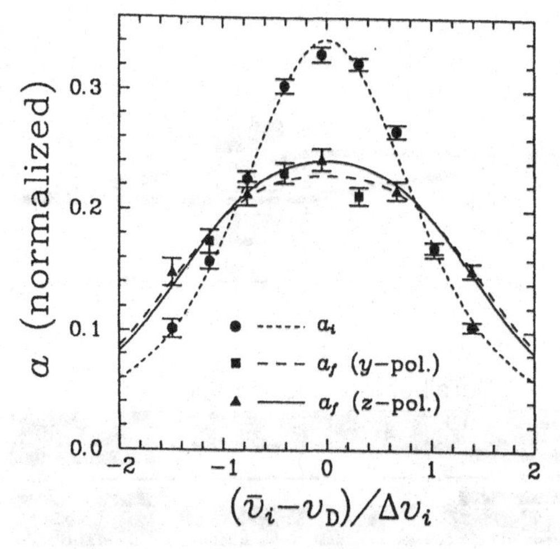

Figure 5.15: Typical set of absorption data for uranium. The standing-wave polarization alternates between y and z along the abscissa. The fit curve is drawn for a fluence typical of the fluence measured for each point.

The path of the two separated beams is shown in Fig. 5.16. Each beam is diffracted three times: once as zeroth order and twice as the first-order.

The essence of the technique is to apply an electric field to one of the separated beams thereby shifting the phase of its quantum-mechanical wavefunction with respect to the other beam. The phase difference is detected in the interference pattern. The beam that traverses the uniform electric field is shifted in energy by the potential

$$U(x) = \frac{1}{2}\alpha_0 \mathrm{E}^2, \tag{5.59}$$

where E is the applied electric field. The JWKB approximation gives an expression for the evolution of the phase of the wavefunction as a function of position

$$\psi(x) \propto e^{i\int_{x_0}^{x} k(x')dx'}\psi(x_0), \tag{5.60}$$

where the wavevector as a function of the classical path is given by

$$k(x) = \frac{1}{\hbar}\sqrt{2m(W - U(x))} \simeq k_0 + \frac{U(x)}{\hbar v}. \tag{5.61}$$

Here k_0 is the wavevector in the absence of a field, W is the total energy, and $v = \hbar k_0/m$.

The difference in phase between the two beams is simply an integral of the potential energy corresponding to the Stark shift Eq. (5.59). The Stark shift in energy causes a shift in the phase of the wavefunction of the atoms in the field compared to atoms in field-free regions. The phase shift is

$$\Delta\phi = \frac{1}{\hbar v}\int_{x_0}^{x_f} U(x)dx. \tag{5.62}$$

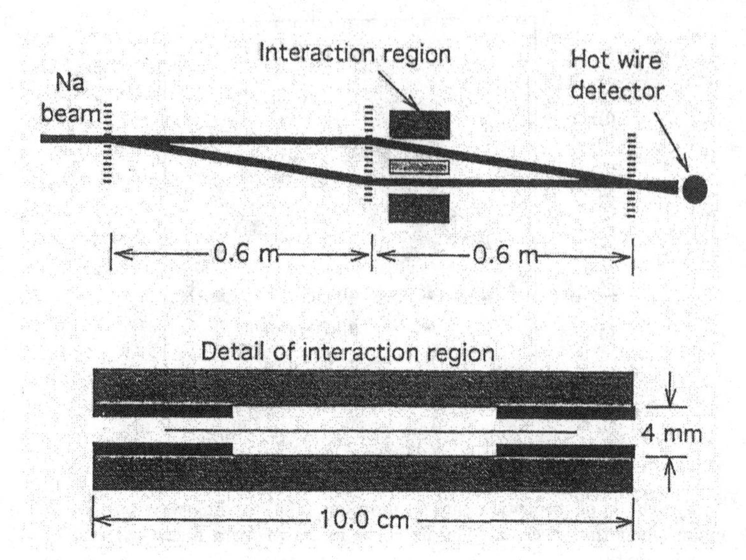

Figure 5.16: A sketch of the atom interferometer showing the gratings use to produce interference fringes. The field is applied in the upper half of the interaction region (or the lower half, but not both). (Reprinted with permission from Ref. [411].)

where $x_f - x_0$ is the length of the electric-field region. Substituting Eq. (5.59) into Eq. (5.62) gives

$$\alpha_0 = \frac{\Delta\phi_{\text{Stark}}}{\mathrm{E}^2 L_{\text{eff}}} 2\hbar v. \tag{5.63}$$

Here L_{eff} is the effective length of the electric field region which includes corrections for edge-field effects.

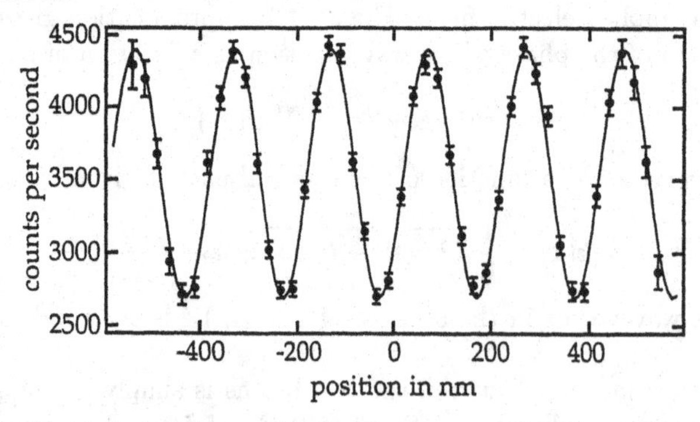

Figure 5.17: A plot of the interference fringes for sodium under the conditions described in the text. Also shown is a fit to the fringe data. (Reprinted with permission from Ref. [411].)

The actual measurement consisted of scanning the third grating, which changed the interference signal at the location of a hot wire detector (see Fig. 5.16), and counting the number of atoms striking the detector at each grating position. The result was an interference fringe pattern as a function of position — see Fig. 5.17. In the measurement of the polarizability of sodium, the fringe visibility, defined as the ratio $(N_{max} - N_{min})/N_{max}$ where N is the number of counts, was 35%. The shift in phase for a given pattern was deduced to within 10 milliradians. By fitting the phase shift $\Delta\phi_{\text{Stark}}$ as a function of field (applied voltage) as shown in Fig. 5.18, the polarizability was determined with an accuracy of 0.3%. This is a factor of 6 better than other best measurements of sodium (see Table 4.2) [414, 415]. It is within a factor of 8 of the best measurements of any species [497]. The advantages of this technique are its accuracy, its use of a simple, uniform electric field, and that it is an absolute measurement that relies only on a knowledge of the electric field.

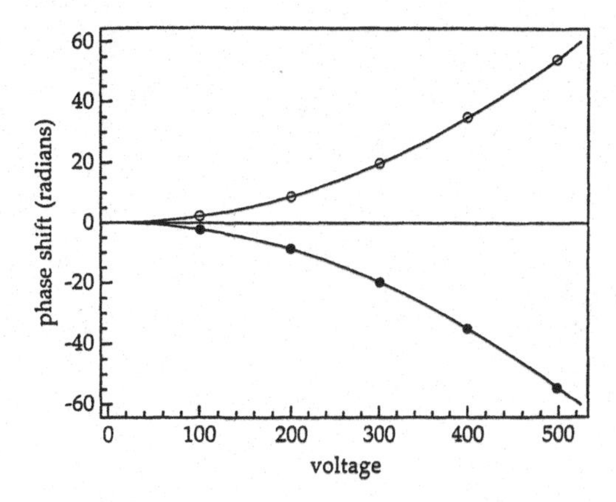

Figure 5.18: A plot of the phase shift versus voltage on the deflection plates. The quadratic dependence of $\Delta\phi$ versus V is quite clear. The open circles correspond to having the electric field in the upper half of the interaction region. The solid circles correspond to a field applied in the lower half. (Reprinted with permission from Ref. [411].)

Chapter 6

Manifestations of Polarization Properties

6.1 Optical Properties

6.1.1 Light scattering: refractive index and optical absorption

The polarizability is the fundamental, electronic property of a single particle that determines the common optical properties (e.g. index of refraction and absorption coefficient) of a system of particles. The macroscopic equivalent to the electric polarizability is the electric susceptibility χ. The susceptibility describes the scattering of light by a collection of particles. We begin our description of light scattering by considering a collection of particles (a gas cell of atoms for instance) with an incident light beam whose electric field is given by

$$\mathbf{E}(\mathbf{r},t) = \frac{\mathcal{E}}{2}\left[\hat{\varepsilon}e^{i(\mathbf{k}\cdot\mathbf{r}-\omega t)} + c.c.\right]. \tag{6.1}$$

Here we have assumed a plane wave is incident on the cell with electric field polarization along the direction indicated by the unit vector $\hat{\varepsilon}$. The Maxwell equations describe the changing electric field propagating in a region of space containing a medium of particles. From the Maxwell equations a wave equation can be derived (see, for example, the text by Jackson [56])

$$\nabla^2\mathbf{E}(\mathbf{r},t) - \frac{1}{c^2}\frac{\partial^2}{\partial t^2}\mathbf{E}(\mathbf{r},t) = \frac{4\pi}{c^2}\frac{\partial^2}{\partial t^2}\mathbf{P}(\mathbf{r},t), \tag{6.2}$$

where $\mathbf{P}(\mathbf{r},t)$ is the linear polarization of the medium at position $\mathbf{r}$ and time t. The linear polarization is the net dipole moment per unit volume of material and it is related to the applied electric field through the electric susceptibility χ

$$\mathbf{P}(\mathbf{r},t) = \chi\mathbf{E}(\mathbf{r},t). \tag{6.3}$$

For the ith component of the linear polarization oscillating at a frequency ω we can write

$$\begin{aligned} \mathrm{P}_i(t) &= \frac{\mathrm{P}_i(\omega)}{2} e^{-i\omega t} + c.c. \\ &= \frac{1}{2}\sum_j \chi_{ij}(\omega)\mathrm{E}_j(\omega)e^{-i\omega t} + c.c. \end{aligned} \quad (6.4)$$

The ijth component of the frequency-dependent electric susceptibility tensor $\chi_{ij}(\omega)$ can be derived from first principles in quantum mechanics, similar to the derivation for the polarizability tensor given in Appendix B.2. For an explicit quantum-mechanical derivation for the susceptibility tensor see, for example, the book by Weissbluth [556]. The result is

$$\chi_{ij}(\omega) = \frac{N}{\hbar}\sum_{l\neq k} \rho_{kk}\left[\frac{(\mathrm{p}_i)_{kl}(\mathrm{p}_j)_{lk}}{(\omega_{lk} - \omega - i\Gamma_{lk}/2)} + \frac{(\mathrm{p}_j)_{kl}(\mathrm{p}_i)_{lk}}{(\omega_{lk} + \omega + i\Gamma_{lk}/2)}\right]. \quad (6.5)$$

Here N is the number of particles in a unit volume and ρ_{kk} is the kth diagonal element of the density operator. Comparing this to our expression for the polarizability tensor [see Eq. (2.58) in Section 2.4], we find that

$$\chi_{ij}(\omega) = N\sum_k \rho_{kk}[\alpha_{ij}(\omega)]_{kk} = N\mathrm{Tr}[\hat{\rho}\hat{\alpha}_{ij}(\omega)]. \quad (6.6)$$

To conveniently understand the role of the susceptibility in scattering, we simplify to the case where the electronic response of the medium is isotropic, so the susceptibility tensor χ_{ij} simplifies to a scalar χ. By substituting Eqs. (6.1) and (6.3) into the wave equation Eq. (6.2), we can find

$$k^2 - \frac{\omega^2}{c^2} = \frac{4\pi\omega^2}{c^2}\chi(\omega). \quad (6.7)$$

This can be rewritten as

$$\frac{k^2c^2}{\omega^2} = 1 + 4\pi\chi(\omega). \quad (6.8)$$

The susceptibility χ (like α) is in general complex, and analogous to Eq. (2.87) for α, we can write for χ

$$\chi = \chi' + i\chi''. \quad (6.9)$$

If we define

$$\frac{kc}{\omega} = \eta + i\kappa, \quad (6.10)$$

then from Eq. (6.8) and Eq. (6.10) we can write

$$1 + \chi' = \eta^2 - \kappa^2 \quad (6.11)$$

$$\chi'' = 2\eta\kappa. \quad (6.12)$$

Here η is the refractive index and κ is the extinction coefficient. The scattering of light by particles entails forward scattering (which depends on the refractive index, η)

and absorption (which depends on the extinction coefficient, κ), in which the light is ultimately scattered in all directions. A very lucid and detailed quantum-mechanical treatment of the relation of the refractive index and the extinction coefficient to the quantum-mechanical scattering amplitude at frequency ω, $f(\omega)$, is given by Sakurai [557]. Here we will continue to discuss light scattering from a classical viewpoint.

The electric field of a plane wave in the medium Eq. (6.1) has its phase modified in a way that is summarized by Eq. (6.10). Using this expression, we can rewrite the phase of the electric field associated with light propagating along the z axis as

$$\begin{aligned} i(kz - \omega t) &= i\left[\frac{\omega}{c}(\eta + i\kappa)z - \omega t\right] \\ &= i\omega(\eta\frac{z}{c} - t) - \frac{\omega}{c}\kappa z. \end{aligned} \tag{6.13}$$

Thus the phase of the electric field has been modified by the introduction of the factor η into the space dependent part. The refractive index modifies the phase of the propagating electric field. The electric field amplitude has been modified by the factor $e^{-\omega\kappa z/c}$, and hence the transmitted intensity will increase (amplification) or decrease (absorption) with distance in the medium depending on whether the extinction coefficient κ is negative or positive. The intensity of light transmitted through a medium can be found by taking the cycle-averaged value of the Poynting vector (see, for example, [56]). The result is that the intensity at a position z in the medium is given by

$$I(z) = I(0)e^{-Kz}, \tag{6.14}$$

where the attenuation or absorption coefficient is

$$K = \frac{2\omega\kappa}{c}. \tag{6.15}$$

Here $I(0)$ is the intensity of the light beam when it enters the region of the medium. The effect of scattering on the electric field of the light beam is sketched in Fig. 6.1. Here we have shown the common case where the extinction coefficient is positive and the field is attenuated.

The physical significance of the different real and imaginary parts of the electric susceptibility (and polarizability) has now been made explicitly clear. If there is no absorption then $\kappa = 0$, which implies, from Eq. (6.12) that the imaginary part of the susceptibility is zero, i.e. $\chi'' = 0$. Then, from Eq. (6.11), we see that $\chi' = \eta^2 - 1$. Thus the real part of the susceptibility (polarizability) is associated with the refractive index of the medium, while the imaginary part of the susceptibility (polarizability) is associated with the absorption of the medium. If the frequency of light is chosen close to that of a resonance for the particles, then the maximum absorption occurs at the resonant frequency where $\eta = 1$ and so $\chi' = 0$. Likewise for frequencies far from any resonance the absorption will be quite small ($\kappa \simeq 0$ and $\chi'' \simeq 0$) and the refractive index will depend on the value of χ', i.e. $\chi' = \eta^2 - 1$. The behavior of the refractive index and the absorption coefficient near a resonance mimics the behavior of the real and imaginary parts of the polarizability, whose shapes near a resonance have been sketched previously (see Fig. 2.5).

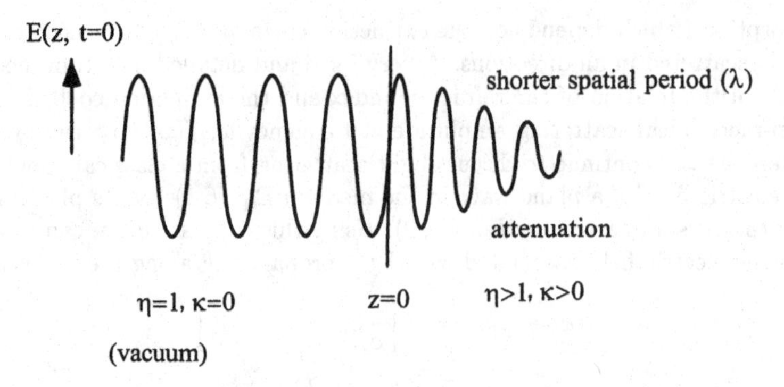

Figure 6.1: A sketch showing the effect of scattering by the medium on a light beam incident from left-to-right. The beam, propagating along z, encounters a medium at $z = 0$ with index $\eta > 1$ and extinction coefficient $\kappa > 0$. A snapshot in time shows the electric field at different points in space.

We again point out that the dielectric constant is, like the electric susceptibility, a macroscopic property of a collection of particles that is described microscopically by the electric polarizability. The dielectric constant is given by the relation [see Eq. (6.8)]

$$\mu\epsilon = 1 + 4\pi\chi, \tag{6.16}$$

where μ is the magnetic permeability of the medium.

Many important aspects of scattering require a different or modified picture from the one presented here. For example, when a particle is large enough so that its size is on the order of the wavelength ($r \simeq \lambda$) new, interesting resonances appear. This Mie scattering is important in aerosol physics and in understanding light scattering in the Earth's atmosphere. For a more complete description of light scattering, the reader should consult the literature. The literature on light scattering is extensive; as a starting point we suggest the reader begin by consulting one of the many texts that treat light scattering, some of which cover specialized aspects, such as small-particle scattering, optical rotation, etc. (see, for example, [72, 158, 558–562])

We have discussed the impact of the light-particle interaction on the light; in the next section we consider one consequence of this interaction on the particle: light forces.

6.1.2 Polarization forces

Use of the mechanical properties of light, i.e. the use of light forces to manipulate material particles, has blossomed in the last decade. The optical Stern-Gerlach method [563], the recent application of light forces to polarizability measurements (see Section 5.2.4), the use of optical tweezers to manipulate small biological structures [564], and the use of light to trap and cool atoms [565] are all excellent examples of the use of polarization forces. The recent success at cooling alkali atoms to form a Bose-Einstein condensate highlights the importance and great future of light force applications. The light force is actually used to precool the atoms to prepare them for the time-orbiting potential (TOP) trap [566].

Here we discuss polarization forces from a classical viewpoint with an emphasis on the central role the particle's polarizability plays in the interaction. Only forces arising from induced dipole moments will be considered. For the simplest case, a particle in a static field, there is an induced dipole moment $\mathbf{p}$ proportional to the inducing field $\mathbf{E}$, i.e.

$$\mathbf{p} = \boldsymbol{\alpha} \cdot \mathbf{E}. \tag{6.17}$$

The particle will experience a force only if the field is inhomogeneous. For a non-uniform electric field the force on the particle is

$$\mathbf{F} = \mathbf{p} \cdot \nabla \mathbf{E}. \tag{6.18}$$

In the static field case the force is conservative, that is it can be expressed as the negative gradient of a potential energy U given by

$$U = -\frac{1}{2}\mathbf{E} \cdot \boldsymbol{\alpha} \cdot \mathbf{E}. \tag{6.19}$$

Next, consider a particle interacting with a harmonic field of the form

$$\mathbf{E}(\mathbf{x}, t) = \frac{1}{2}\left[\mathbf{E}_0(\mathbf{x})e^{-i\omega t} + c.c.\right], \tag{6.20}$$

where ω is the angular frequency of the applied field. Eventually the particle will acquire a steady-state, oscillating dipole moment $\mathbf{p}$ that also oscillates at the angular frequency ω:

$$\mathbf{p}(\mathbf{x}, t) = \frac{1}{2}\left[\mathbf{p}_0(\mathbf{x})e^{-i\omega t} + c.c.\right]. \tag{6.21}$$

From the discussion in Section 2.3, the amplitude of the induced dipole moment $\mathbf{p}_0$ will be proportional to the polarizability at frequency ω, i.e. $\alpha(\omega)$:

$$\mathbf{p}_0 = \boldsymbol{\alpha} \cdot \mathbf{E}_0(\mathbf{x}). \tag{6.22}$$

Recall that any second-rank tensor like α can be written $\alpha = \alpha' + i\alpha''$, where α' is Hermitian and $i\alpha''$ is anti-Hermitian (both α' and α'' are Hermitian). Work is done on the particle by the field at a rate

$$P(\mathbf{x}, t) = \mathbf{E}(\mathbf{x}, t) \cdot \frac{\partial \mathbf{p}(\mathbf{x}, t)}{\partial t}. \tag{6.23}$$

This expression only considers the work done on the particle due to its induced dipole.

For a harmonic field of the form given above, Eq. (6.20), the time-averaged rate at which work is done by the field is

$$< P > = \frac{\omega}{2}\mathbf{E}_0^* \cdot \boldsymbol{\alpha}'' \cdot \mathbf{E}_0. \tag{6.24}$$

Thus the dissipation of energy by the particle is given by the anti-Hermitian part of the polarizability. This is simply due to absorption and agrees with the results of the previous section.

The time-averaged force $<\mathbf{F}>$ experienced by a particle in a harmonic field is the sum of a conservative and a dissipative force, i.e.

$$<\mathbf{F}> = <\mathbf{F}_c> + <\mathbf{F}_d>. \tag{6.25}$$

The conservative force, also known as the dipole or gradient-field force, is the negative gradient of a time-averaged potential energy $<U>$ that is similar to the form of Eq. (6.19) and we can write

$$<\mathbf{F}_c> = -\nabla <U> = -\nabla\left(-\frac{1}{4}\mathbf{E}_0^* \cdot \alpha \cdot \mathbf{E}_0\right). \tag{6.26}$$

The conservative force depends only on the Hermitian part of the polarizability.

In contrast, the dissipative force, also called the scattering or spontaneous force, depends solely on the anti-Hermitian part of the polarizability

$$<\mathbf{F}_d> = \frac{1}{4i}\mathrm{E}_{0,k}^* \alpha_{kl}'' \nabla \mathrm{E}_{0,l} + c.c., \tag{6.27}$$

where repeated indices are summed over. For the special case of a plane wave where the electric-field amplitude can be written

$$\mathbf{E}_0(\mathbf{x}) = \mathcal{E} e^{i\mathbf{k}\cdot\mathbf{x}}, \tag{6.28}$$

the time-averaged conservative force is zero and the dissipative force can be written

$$<\mathbf{F}_d> = \frac{\mathbf{k}}{2}\mathcal{E}^* \cdot \alpha'' \cdot \mathcal{E} = \frac{\mathbf{k}}{\omega} <P>. \tag{6.29}$$

Hence the dissipative force can be written as the product of the photon absorption rate $<P>/\hbar\omega$ and the momentum per absorbed photon $\hbar\mathbf{k}$.

In Section 2.5 there are plots of both the Hermitian and anti-Hermitian parts of the polarizability versus frequency near a resonance (see Fig.2.5). From the plot of α'' versus frequency, ω, it is clear that the work done on an induced dipole by a harmonic field peaks on resonance. This is the basis for optical molasses [567] and for the cooling of atoms in optical traps [565]. In both these cases light from a laser is tuned near an atomic resonance. The dipole force dominates when the light frequency is far from resonance since then $\alpha' \gg \alpha''$. The classical expressions for these time-averaged forces on an induced dipole in a harmonic field are derived in detail (from the basic Lorentz force expression) in the thesis of Kadar-Kallen [554].

The next two sections briefly describe optical traps and the role polarizabilities play in their behavior. Presently, optical traps are of two main types:

Magneto-optic trap — based on the absorptive or non-Hermitian part of the polarizability, it uses the dissipative light force to trap [see Eq. (6.29)]. This trap has been successfully used to condense atoms into a quantum-degenerate state (although the magneto-optic trap must be turned off and a standard magnetic quadrupole trap must be used to cool and trap atoms to temperatures below about 10 μK to form a condensate).

Dipole-force trap — based on the refractive index or Hermitian part of the polarizability, it uses the conservative light force to trap. This trap has been used as an optical tweezers to manipulate and study phenomena of biological importance.

6.1.3 Optical cooling and trapping of atoms

Here we briefly discuss optical cooling and trapping and we relate these phenomena to the polarizability. An extensive literature on optical cooling and trapping exists (see the review in Ref. [568] and for a more technical discussion see Refs. [565, 569] and especially Ref. [570]).

(a) Atom cooling. In 1933, Otto Frisch [571] used light from a sodium lamp to deflect a beam of sodium atoms from their light-free path. Maximum absorption of the lamp light occurs due to its resonant band of wavelengths. The photons are absorbed from one direction, yet the scattered photons are emitted in all directions. Hence momentum is absorbed by the atoms along the axis of the light (see Fig. 6.2). This is an example of a mechanical effect on particles due to the scattering force (dissipative force). In 1985, twenty-five years after the advent of the laser, it was demonstrated that the scattering force is powerful enough to slow down and stop a beam of atoms [572]. The atoms were cooled from 300 K to about 100 mK. The cooling laser was fixed in frequency and the atomic levels were Zeeman-shifted so that the *atomic resonant frequency was tuned* as the atoms moved through a gradient magnetic field — see Fig 6.3. This motion of the atoms through a magnetic field gradient allows the tuning of the resonant frequency to compensate for the Doppler shift in resonant frequency caused by the cooling and slowing of atoms.

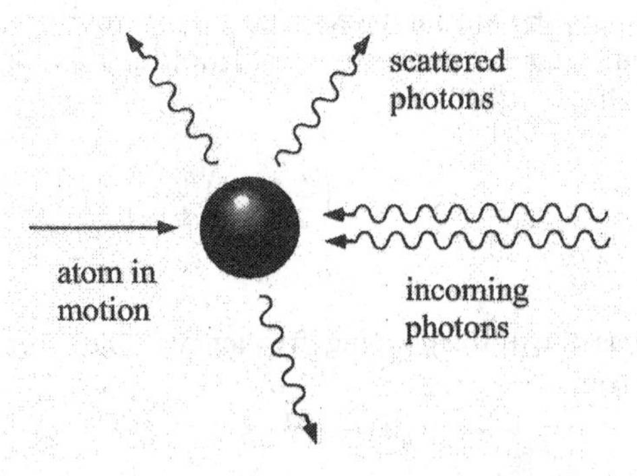

Figure 6.2: Sketch showing an atom absorbing photons from one direction and scattering them in random directions due to spontaneous emission.

Optical cooling relies on the dissipative force — here a laser is continuously tuned to be resonant with a subset of atoms moving toward it. Each atom that is on resonance absorbs photons from the laser beam and reradiates the absorbed photons over all space through spontaneous emission (see Fig. 6.2). The force on each atom

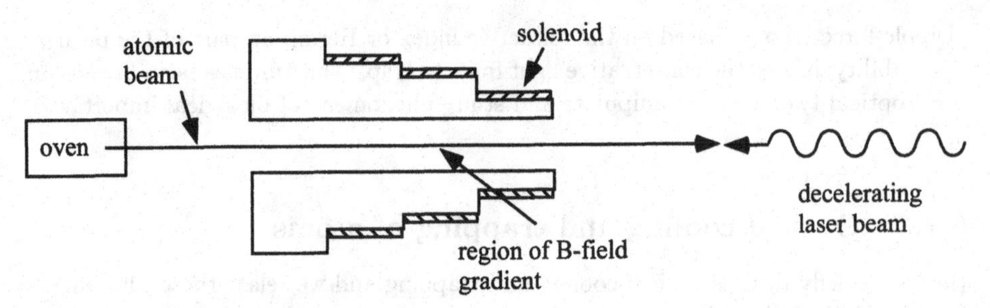

Figure 6.3: A simplified sketch of the experimental setup used by Phillips et al. to slow and cool atoms in a magnetic field gradient. For details see Refs. [572, 573].

in this case is given by [568]

$$\begin{aligned} \mathrm{F}_d &= \hbar k \frac{< P_{scatt} >}{\hbar\omega} \\ &= \hbar k \frac{1}{2\tau}\left(\frac{2I/I_s}{1 + 2I/I_s + 4(\omega_L - \omega_0 + \boldsymbol{k}\cdot\boldsymbol{v})^2\tau^2}\right), \end{aligned} \tag{6.30}$$

where $\boldsymbol{k}$ is the wavevector of the light, $\boldsymbol{v}$ is the atomic velocity, τ is the natural lifetime of the excited state, I is the intensity of the light, and I_s is the saturation intensity, given by

$$I_s = \frac{\hbar\omega_0}{\sigma\tau}, \tag{6.31}$$

where σ is the peak cross section for scattering a photon of resonant frequency ω_0. From the optical theorem, Section 2.5, the relation between the photon scattering cross section and the non-Hermitian part of the polarizability is $\sigma = 4\pi\omega\alpha''/c$. Using this relation between σ and α'' and using the definition $I = c\mathrm{E}_0^2/8\pi$, we can rewrite the ratio I/I_s found in Eq. (6.30) as

$$\begin{aligned} \frac{I}{I_s} &= \left(\frac{1}{2}\alpha''\mathrm{E}_0^2\right)\frac{\tau}{\hbar} \\ &= (\omega_R\tau)^2, \end{aligned} \tag{6.32}$$

where $\omega_R \equiv \mathrm{pE}_0/\hbar$ is the Rabi flopping frequency. Thus, for unsaturated beams, where $I \ll I_s$, we have

$$\mathrm{F}_d = \hbar k \omega_R, \tag{6.33}$$

which is directly proportional to the polarizability α''. This expression is identical to the earlier result Eq. (6.29).

In the Doppler cooling case, the force is proportional to the polarizability. The lowest temperature range to which alkali atoms can be cooled is limited by the spontaneous nature of the emission of photons from excited state atoms in the trap. This temperature range is 100-300 μK for the alkalis. The atoms essentially undergo diffusion as a result of the random direction of spontaneously-emitted photons and due to fluctuations in the cooling laser intensity. To cool further, a force independent of laser intensity must be present. A careful analysis of these cooling mechanisms, which become important as the atoms cool, is given in the lucid paper by Dalibard and Cohen-Tannoudji [574]. In the case of standing waves generated by counter-propagating laser beams of $\hat{\sigma}_+$ and $\hat{\sigma}_-$ polarization[1], the new cooling force can be written

$$F = \beta v, \tag{6.34}$$

where v is the atomic velocity and the damping coefficient is

$$\beta = \frac{120}{17} \frac{-\delta\Gamma}{5\Gamma^2 + 4\delta^2} \hbar k^2. \tag{6.35}$$

The expression Eq. (6.35) for β is the result of an analysis on an atom with the ground state angular momentum $J_g = 1$ and excited state angular momentum $J_e = 2$. The natural linewidth of the excited state is given by Γ, whereas δ corresponds to the detuning between the cooling laser frequency ω_L and the atomic resonance frequency ω_0, i.e. $\delta = \omega_L - \omega_0$. Although the force does not appear to depend directly on the polarizability, the motional orientational cooling can be related to a vector polarizability [575]. Note that in Eq. (6.35) the linewidth Γ depends on the same matrix elements of the dipole operator (squared) that occur in the expressions for the polarizability. This cooling mechanism can achieve temperatures of about 10 μK.

Atomic Bose-Einstein condensates are produced using evaporative cooling whereby an rf magnetic field pumps the hotter atoms in the Boltzmann tail from trapped states to unstable states (nontrapped states). At this point the laser fields have all been turned off and the cooling and trapping solely rely on the configuration of static and time-varying magnetic fields. This last stage of cooling does not depend on the polarizability — for details see Refs. [566, 576, 577].

(b) Atom trapping. In 1983 Ashkin and Gordon [578] derived the Optical Earnshaw Theorem: "it is impossible to trap a small dielectric particle at a point of stable equilibrium in free space using only the scattering force of radiation pressure". Simply stated, it implies that the Poynting vector of any configuration of static laser beams must have zero divergence. If the force is proportional to the intensity, then the force field must also have zero divergence (i.e. there is no confining force). For a short time (3 years) it was thought that the Optical Earnshaw Theorem precluded a trap based on the scattering force of radiation pressure. Hence the first optical trap, which relied on the ability to optically cool sodium atoms, was based on the dipole force (conservative force) [567]. The first atom trap relied on the scattering force Eq. (6.30) produced by two counterpropagating beams to slow and cool atoms. The cooling laser was tuned a little to the red of a transition (longer wavelength) so that, when an atom moving toward one of the beams would see the light as being Doppler-shifted to the

[1] In this polarization case, the cooling mechanism is called motional orientational cooling. The case of two counterpropagating beams with perpendicular linear polarization gives rise to cooling by the 'Sisyphus effect' (or Sisyphus cooling). For details see Ref. [574]. Actually, more recently it has become common to refer to both polarization configurations for cooling as Sisyphus cooling.

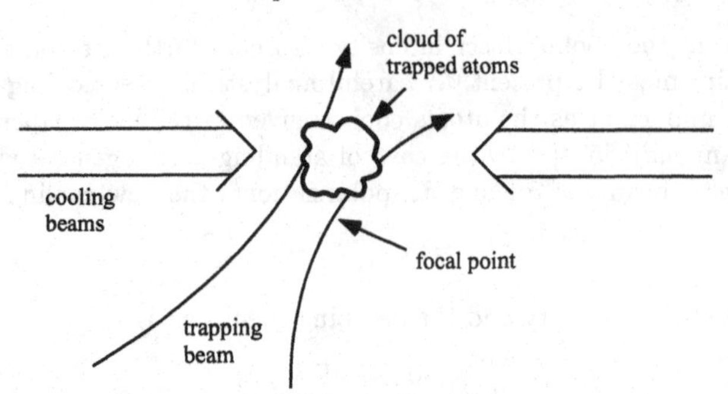

Figure 6.4: Sketch of the laser beams used in the first optical atom trap. This trap requires two different forces, the scattering force to cool and the gradient force to trap. The cooling and trapping was alternated at frequencies of about 0.1-1 MHz; the trapping lifetime was ~ 1 s. The physical size of the trapping region was small, with a confinement diameter of about 10 μm.

blue in its rest frame, the frequency of light would appear closer to resonance than if the atom were at rest. The increased scattering of the light slowed the atoms down. To actually trap the cooled atoms, a conservative (gradient-field) force was applied to the atoms (see Fig. 6.4). By alternating cooling and trapping, sodium atoms were confined to a small region of diameter $\simeq 10\mu$m. Hence the first atom trap required both parts of the polarizability to succeed (the scattering force uses α'' to cool and the gradient force uses α' to trap).

Current *atom* traps are nearly all based on the magneto-optic trap, which utilizes an idea that circumvents the Optical Earnshaw Theorem and allows the scattering force to be used to trap atoms. In 1986 Pritchard and others [579] showed that the Optical Earnshaw Theorem did not necessarily apply to atoms. It is easy to produce conditions whereby the internal degrees of freedom of atoms allow a scattering force that is not proportional to the intensity. Shortly thereafter, a trap using scattering forces was demonstrated; in fact, the trapping potential was much deeper than for traps using the gradient force [580]. This trap made use of a magnetic field that varied with position and has come to be known as a magneto-optic trap (MOT).

The basic geometry of a magneto-optic trap is shown in Fig. 6.5. Atoms are trapped at the intersection of six laser beams, each pair of which are counterpropagating. We note that four intersecting laser beams, in the proper geometry, are the minimum number necessary for a trap. This intersection produces a potential well in space that traps atoms in a small volume at the intersection. To understand the physics behind the magneto-optic trap, consider a single pair of oppositely-directed laser beams. For simplicity, consider the case of trapping an atom whose ground state has no electronic angular momentum, i.e. $J = 0$. The trapping laser couples the ground state to an excited state with $J = 1$. In the center of the trap there is a significant magnetic field gradient produced by a pair of anti-Helmholtz coils. The magnetic field itself is zero at the center. If the light is properly polarized, e.g. $\hat{\sigma}_+$ polarization for propagation along $\hat{z}$ and $\hat{\sigma}_-$ polarization for propagation along $-\hat{z}$ (in the atom's frame), then the space-dependent Zeeman shift in the energy levels of

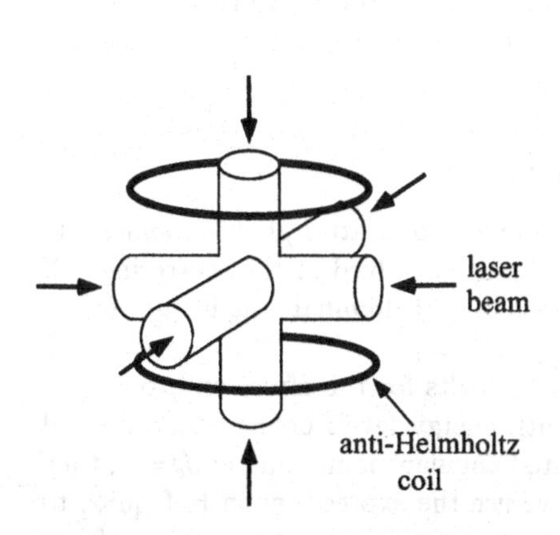

Figure 6.5: A magneto-optic trap (MOT) usually consists of an evacuated chamber or cell with at least six optical ports in pairs along three orthogonal axes. A laser beam is split into three separate beams and these three beams traverse the cell arms and intersect at the center. Each beam is reflected back onto itself to from three standing wave axes. The magnetic field gradient is provided by a pair of coils run with counter circulating currents: an anti-Helmholtz configuration.

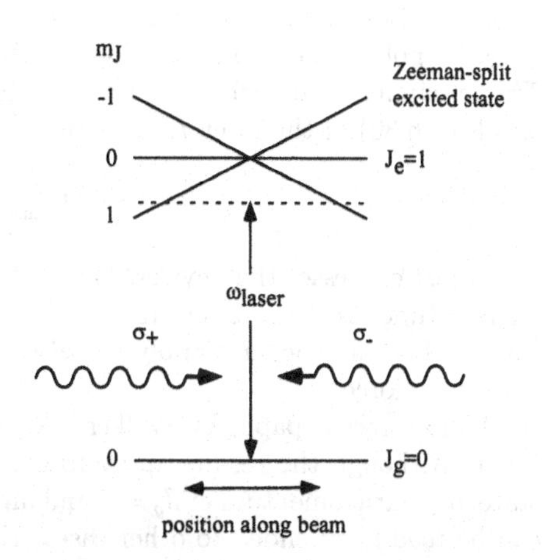

Figure 6.6: The Zeeman shift in levels as a function of position in a MOT for a ground state $J_g = 0$ and an excited state $J_e = 1$. Depending on the position along the axis shown, photons from only one beam (direction) will be absorbed due to selection rules and the polarization of the two counterpropagating beams. The diagram corresponds to the $\hat{\sigma}_+ - \hat{\sigma}_-$ polarization configuration for each pair of counterpropagating laser beams.

the excited state will create conditions whereby an atom moving along $\hat{z}$ will reach a position where the $\hat{\sigma}_-$ photons are resonant with it and will push the atom back toward the center of the trap. The same result applies to an atom moving along $-\hat{z}$ and $\hat{\sigma}_+$ photons (see Fig. 6.6).

The force on an atom in a MOT consists of a combination of a restoring force, due to the magnetic field gradient, and a damping force, corresponding to Doppler and molecular orientational cooling. Assuming small displacements and small velocities, the force can be expressed as a linear function of displacement and velocity, i.e.

$$\mathbf{F} = -\kappa \mathbf{x} - \beta \mathbf{v}, \tag{6.36}$$

where the spring constant of the trap depends on the magnetic field gradient. For example, from induced-orientation theory on an atom with transitions from $J_g = 1 \rightarrow J_e = 2$ [581], the force component along a single dimension $F = -\kappa z - \beta v$ has a spring constant

$$\kappa = \frac{g_{J_g} \mu_B}{\hbar k} \frac{dB}{dz} \beta, \tag{6.37}$$

where the damping constant β is given by Eq. (6.35). Here g_{J_g} is the magnetic sublevel degeneracy of the ground state, which has angular momentum J_g. The quantity

μ_B is the Bohr magneton, and B is the magnetic field. This expression assumes a $\hat{\sigma}_+ - \hat{\sigma}_-$ polarization configuration for each pair of counterpropagating laser beams. The maximum force, which will occur when the light is on resonance and is intense [see Eq. (6.30) in the limit $I/I_s \gg 1$], is

$$\mathrm{F}_{max} = \frac{\hbar k}{2\tau}, \tag{6.38}$$

which just expresses the physical fact that an atom cannot scatter photon momentum $\hbar k$ in a time less than 2 lifetimes of the excited state involved in the scattering. Of course, this rule does not apply if real transitions do not dominate, as in the case of the dipole force.

A more recent paper by Walker [582] derives results for the forces on atoms in a MOT. Although the results apply to atoms with energy levels defined by a ground state angular momentum of $J_g = 0$ and an excited angular momentum of $J_e = 1$, they can be readily extended to other cases. Here we use the expressions in Ref. [582] to show an explicit dependence of the MOT force on polarizability for the case of low intensity and with the laser beams in the $\hat{\sigma}_+ - \hat{\sigma}_-$ polarization configuration. The dipole moment can be written

$$\mathbf{p} = \boldsymbol{\alpha}_v \times \mathbf{E}, \tag{6.39}$$

where $\boldsymbol{\alpha}_v$ is the vector polarizability or gyrotopic polarizability associated with atoms in a magnetic field. Here the vector polarizability can be written

$$\boldsymbol{\alpha}_v = \mathrm{K}\boldsymbol{\Omega}, \tag{6.40}$$

where $\boldsymbol{\Omega} = g_J \mu_B \mathbf{B}/\hbar$ is the Larmor frequency and μ_B is the Bohr magneton. The parameter K can be expressed as a function of the laser detuning from resonance δ, the Rabi frequency ω_R [which depends on the intensity — see Eq. (6.32)], the linewidth Γ, and the wavelength λ

$$\mathrm{K} = \frac{3\lambda^3\Gamma}{32\pi^3} \frac{-\delta\Gamma - i(\delta^2 - \Gamma^2/4 + \omega_R^4/\Gamma^2)}{(\delta^2 + \Gamma^2/4 + 2\omega_R^2)(\delta^2 + (\Gamma/2 + \omega_R^2/\Gamma)^2)}. \tag{6.41}$$

Substitution of the dipole expression Eq. (6.39) in the force expressions given by Walker (specifically Eq.(21) in Ref. [582]) will, after some vector algebra, yield the result

$$\mathbf{F}_{\mathrm{MOT}} = \frac{8\pi}{\lambda c} I (\boldsymbol{\alpha}_v)_z \hat{z}. \tag{6.42}$$

Here the z-axis is defined as the axis along which the two laser beams are counterpropagating. For further details of this interesting approach to trapping forces in a MOT consult Ref. [582].

Finally, it is noteworthy that a recently constructed cesium atom trap was based solely on the gradient light force [583, 584]. In these experiments, cesium atoms were first cooled and trapped in a MOT and then trapped in a potential well of depth 115 μK by a focused CO_2 laser beam ($\lambda = 1.06\mu$m). The size of the focal region was $\sim 10^{-4}$ cm^3. Due to the long wavelength used and due to the dominance of the gradient force, such a trap has been labeled a QUEST (quasi-electrostatic trap).

6.1.4 Optical tweezers and light forces

Atom traps use the scattering force to optically confine atoms to a small region of space. In contrast, optical traps — also called optical tweezers — rely on the gradient force to optically confine particles in a small region of space. Optical tweezers are used to move and manipulate objects of microscopic size such as small metal or dielectric beads, single cells, organelles, and bacteria. An optical tweezers consists of a focused beam of laser light that traps objects just beyond the focal plane of the light (see Section 6.1.2). Optical trapping of small latex spheres of about 1 μm diameter was first demonstrated in the early 1970's by Ashkin [585]. These first traps used a *pair* of weakly-focused counterpropagating laser beams. In the mid-1980's Ashkin and collaborators [586] demonstrated the first single-beam gradient force optical trap, or optical tweezers. The development of *single-beam* optical traps has caused the technique to blossom, particularly in its application to biological systems (see, for example, Refs. [564, 587–589]). Here we will review the physics behind optical tweezers because it is based on the polarizability of the particle being trapped.

The basic experimental setup for a single-beam gradient force optical trap is shown in Fig. 6.7. The light for the optical trap is usually fed in through a microscope objective onto a water-immersed sample containing the particle or biological specimen of interest. For good trapping, the light beam should have a TEM_{00} spatial mode, which is Gaussian in profile. The trapping zone is the region on the light-beam axis just past the focal plane. The reason for this will be discussed shortly. Particles can be trapped over a fairly broad size range: diameter $\sim 0.05 - 200\mu$m. Typical laser powers used are 1-100 mW; at power levels close to or exceeding 100 mW many biological specimens will explosively heat or overheat enough to die. Typical forces that can be achieved are in the range of 1-40 pN. The lasers that are used to form optical tweezers are usually in the visible green to infrared portion of the spectrum (wavelengths roughly in the range 0.5-2.0 μm).

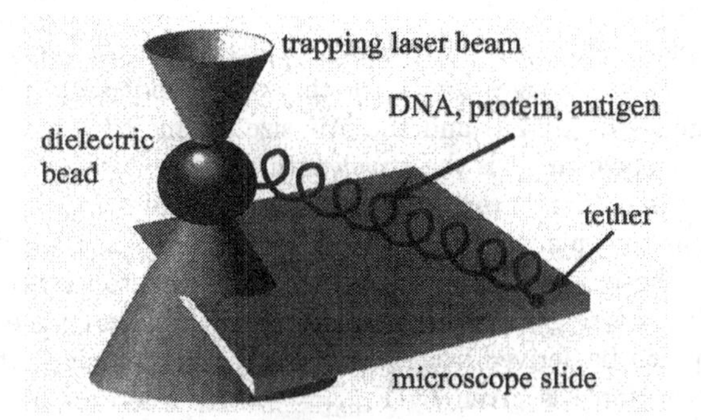

Figure 6.7: Optical tweezers are used in biological systems to micromanipulate small biological particles such as organelles, viruses, etc. Illustrated here is the trapping of a dielectric sphere that has been attached to a biological specimen. Examples of such systems include: latex microspheres attached to microtubules which were connected to kinesin enzymes [590]; polystyrene beads attached to DNA molecules [588]; polystyrene spheres attached to antigens which are bound to an antibody coated surface [589]. The sample is placed on a microscope slide for viewing and manipulation.

Optical tweezers are based on the gradient force, which was discussed in Section 6.1.2 where it was labelled the conservative force F_c. Simplifying the notation in Eq. (6.26), the gradient force can be written

$$\mathbf{F}_{\text{grad}} = -\frac{1}{4}\alpha\nabla|E_0|^2, \tag{6.43}$$

where E_0 is the electric field amplitude of the wave associated with the laser beam — see Section 6.1.2. Here $\alpha \equiv \alpha'$ is the real part of the polarizability. By considering the potential energy of a particle in the light field, it is easy to see that a particle will seek regions where the light intensity is highest. To understand this, note that the expression for the potential energy of a particle located in a light field is

$$U = -\frac{1}{4}\alpha|E_0|^2. \tag{6.44}$$

Since $\alpha > 0$ in most cases, the potential energy U is minimized by moving to regions of maximum electric field (or intensity).

Another force is present whenever particles are illuminated by light — it is called radiation pressure. Consider a large optical sail that totally reflects all light that hitting it. If only illuminated on one side, the sail will be forced to move in the same direction as the incoming light. The photons, in reversing direction upon reflection, push the sail. This force is also conservative and depends upon the real part of the polarizability α'. Radiation pressure that arises from scattering of light via the imaginary part of the polarizability α'' is perhaps more accurately labelled the spontaneous force [579]. The spontaneous force is the dominant force component for magneto-optic atom traps. For a collection of particles, like those that would form an optical sail, the refractive index is used to describe the light scattering process. Recall that the refractive index is related to the microscopic polarizability — see Section 6.1.1. Since radiation pressure tends to push a particle in the same direction as the laser beam, this force must be smaller than the z-component of the gradient force in order for the optical trap to work.

Our physical picture for optical tweezers depends on the size range of the particles being trapped. The two ranges for which there are good physical models are the Rayleigh size regime ($d \lesssim 0.2\lambda$) and the Mie size regime ($d \gtrsim \lambda$) , where d is the 'diameter' of the particle and λ is the wavelength of light. Physical insight is clearest in the Mie regime since simple ray optics suffices to explain trapping in this case. A set of ray diagrams to explain trapping in the Mie size regime is given in Fig. 6.8.

For a particle whose size (diameter d) is small compared to the wavelength of light ($d \lesssim \lambda$), Rayleigh scattering occurs. In this case, an expression for the scattering force in an optical trap can be derived from some very basic Rayleigh scattering formulas. First note that the force is given by $F_{\text{scat}} = \hbar k R$, where $\hbar k$ is the momentum carried by a single photon in the laser beam and R is the scattering rate (number of scattered photons/time). The scattering rate is simply the incident flux of photons $I/\hbar\omega$ (number of incident photons/time·area) times the scattering cross section σ (units of area), i.e.

$$R = \sigma\frac{I}{\hbar\omega}. \tag{6.45}$$

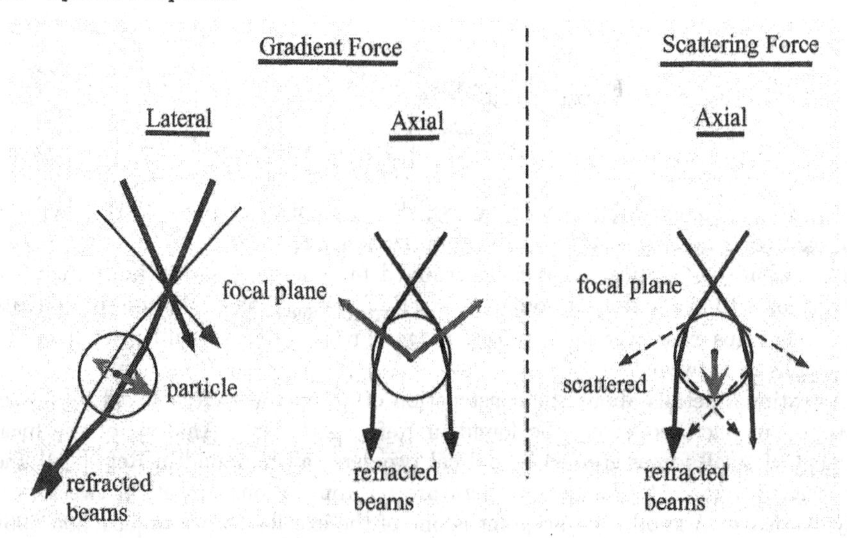

Figure 6.8: The diagrams above illustrate the forces that operate on a particle just beyond the focal plane of light in optical tweezers. Light rays are represented by darker arrows and the resulting momentum kicks (forces) on the particle are represented by lighter arrows at the particle center. Light ray intensity is indicated by arrow thickness. The left and middle diagrams illustrate the gradient force. For lateral motion, since the light is more intense near the central axis than off axis, the refracted light nearer the beam center provides a larger force toward the center than the force directed away from the beam center provided by the weaker light farther off axis. The net lateral force always points toward the central axis of the laser beam. Only the gradient force contributes to lateral confinement. In contrast, the axial force is a combination of the gradient and scattering forces. The middle diagram illustrates the gradient force along the axis, which always points toward the focal plane, regardless of whether the particle is above or below the focus. The scattering force (right diagram) always points in the propagation direction of the beam and so the only region that has counterbalancing forces along the axis is below the focus, where the gradient force and scattering force point in opposite directions.

Here I is the intensity of the light and ω is the angular frequency of the laser light which is related to wavelength by $\omega = 2\pi c/\lambda$ (c is the speed of light). An expression for the Rayleigh scattering cross section σ of a dielectric particle of radius r is derived in texts on electromagnetism [56]

$$\sigma = \frac{8\pi}{3} k^4 r^6 \left(\frac{\eta_{rel}^2 - 1}{\eta_{rel}^2 + 2} \right)^2 , \qquad (6.46)$$

where $\eta_{rel} = \eta_p/\eta b$ is the relative index of the particle η_p to the index of the background fluid η_b (e.g. water). Noting the definition of the wavevector $k = 2\pi\eta/\lambda$, we can finally write

$$F_{\mathrm{scat}} = \frac{I}{c} \frac{128\pi^5 r^6}{3\lambda^4} \left(\frac{\eta_{rel}^2 - 1}{\eta_{rel}^2 + 2} \right)^2 . \qquad (6.47)$$

The gradient force is given by the expression Eq. (6.26) in Section 6.1.2 which can be written

$$\begin{aligned} \mathbf{F}_{\mathrm{grad}} &= \frac{1}{4}\alpha\boldsymbol{\nabla}\mathrm{E}_0^2 \\ &= \frac{2\pi\eta_b^3 r^3}{c}\left(\frac{\eta_{rel}^2-1}{\eta_{rel}^2+2}\right)\boldsymbol{\nabla}I. \end{aligned} \tag{6.48}$$

To obtain the final expression in Eq. (6.48), the expression for the polarizability of a small particle is needed — see Eq. (2.10) in Section 2.1.

The stability of optical traps is determined by analyzing the ratio of the axial gradient force to the scattering force, i.e. $R=F_{\mathrm{grad}}/F_{\mathrm{scat}}$ [586]. We simply reiterate that stable traps exist over the complete range of particle sizes from Rayleigh to Mie and regions in between.

Operational details about the construction of an optical tweezers and its implementation in microscopes can be found in Refs. [591–593]. Analysis of the measurement of small forces applied by optical tweezers can be found in Ref. [594] and references therein. At present the literature on applications of optical tweezers is fairly extensive. A good starting point is one of the excellent reviews [591, 595, 596]. Interesting recent applications include cell sorting [597], measurement of elastic properties of DNA molecules [587, 588], the use of force measurements to detect very small concentrations (femtomolar) of free antigens in solutions [589], and the determination of binding forces of single proteins, like kinesin [590].

6.1.5 Nanostructure fabrication

Nanostructure fabrication using atom optics is interesting and has great potential. Nanostructures are patterns or structures placed on materials; they have a size in the range of $\lesssim$ 100 nm — see Fig. 6.9. Research and development of nanostructures is driven by the need for greater speed in electronic circuits and by the need for higher density archival storage. Although much nanostructure research centers around magnetic phenomena, such as the development of giant magnetoresistive devices [598], ideas from atom optics are being applied to the problems of nanostructure fabrication (see [599] and references therein). Here we briefly describe the recent success using laser-focused atom deposition to form a two-dimensional array of atomic 'dots'. The resulting array is very regular due to the accurate periodicity achievable with the optical standing-wave field used to form the pattern. This process relies on the enhanced polarizability of atoms at frequencies close to resonance, and we will concentrate on this aspect of the nanostructure fabrication process. Details on the resulting structures and the operational and technical advantages of this optical process over more traditional techniques for nanostructure fabrication, such as electron-beam lithography, can be found elsewhere [599]. Two important applications may emerge from the atom-optics technique: (1) use of the arrays as a calibration standard for lithographic processes and (2) production of a very regular array of small magnetic elements.

The basic experimental setup for producing the nanostructures shown in Fig. 6.9 is given in Fig. 6.10. Nanostructure fabrication relies on separate application of

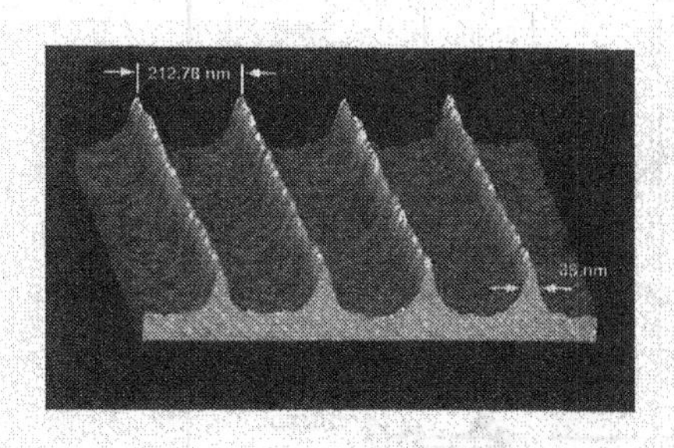

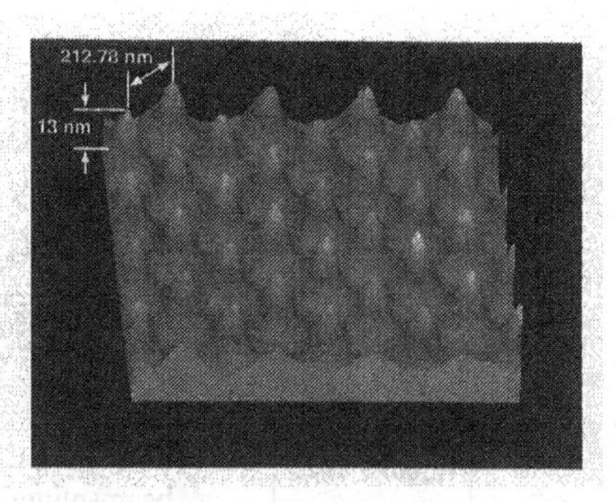

Figure 6.9: Two atomic force microscope images of nanostructures formed by laser-focused atom deposition of Cr. In (a) a regular pattern of lines of height 8 ± 1 nm have been formed by focusing atoms using a standing wave laser field in one dimension. In (b) a regular array of dots of height 13 ± 1 nm have been formed by focusing atoms using two standing wave laser fields in two dimensions. (Reprinted with permission from [599].)

two light forces: the scattering force to first collimate the atoms and the gradient force to focus the atoms. The light used to focus and deposit the atoms onto a substrate, typically silicon, is split into two parts. The atoms first encounter the top laser beam, which has traversed an acousto-optic modulator (AOM) to shift the laser beam frequency to the red side of the atomic resonance. This laser is retroreflected back on itself to form a standing wave field that has a lin $\perp$ lin configuration, which means that the two counterpropagating beams are linearly polarized along orthogonal directions. Producing an orthogonal linear polarization in the counterpropagating laser beam is achieved by placing a quarter-wave plate just in front of the mirror. It is the same standing wave configuration used in the Sisyphus cooling discussed in Section 6.1.3. The purpose is the same here as in an optical trap: to cool the atoms. This cooling of the atoms in the transverse direction achieves a necessary collimation of the atomic beam before the bottom laser beam uses the gradient force to focus the atoms onto the substrate.

The gradient force does the actual focusing of the atoms to form the nanostructure pattern — see Fig. 6.11. The light for this part is detuned *above* the resonant frequency of the atoms. A standing-wave field is also formed from this light and the effect is discussed in Section 5.2.4 — see especially Fig. 5.13. Note here that the two counterpropagating beams have the same polarization. The interaction between the particles and the standing-wave laser field is governed by the same relation for the potential energy U as in optical tweezers [see Eq. (6.44)]

$$U = -\frac{1}{4}\alpha|\mathrm{E}_0|^2, \tag{6.49}$$

where E_0 is the electric field amplitude of the wave associated with the laser beam — see Section 6.1.2. Here $\alpha \equiv \alpha'$ is the real part of the polarizability. From this expression we can see why the atoms are focused in the nodal regions of the standing-wave field. Since we are detuned to the blue side of the resonance, $\alpha < 0$ [see Eq. (2.5)

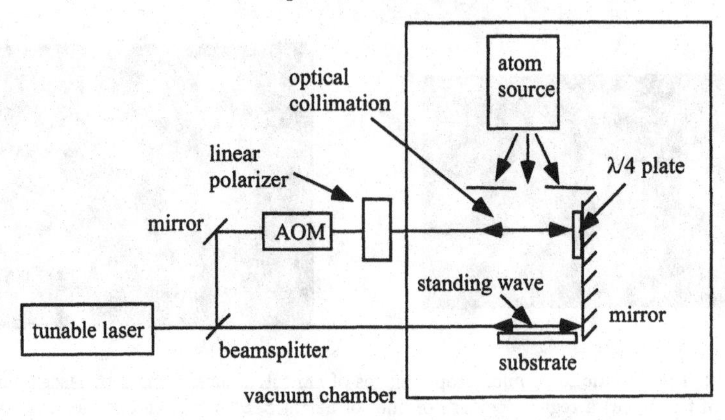

Figure 6.10: A simplified sketch of the experimental apparatus used to collimate and focus atoms into nanostructure patterns on material surfaces. The case shown is for the one-dimensional pattern produced in Fig. 6.9(a). In the two-dimensional case a pair of laser standing-wave fields orthogonal to each other are formed just above the substrate surface. The key here is to form the two standing waves from light that is orthogonally polarized — this keeps the two standing wave patterns from interfering with each other. For details see Ref. [599] and references therein.

in Section 2.1]. Note that with a negative polarizability, the potential is minimized at regions where the field is *weakest* — at the nodes.

A detailed discussion of the trajectories of atoms focused using a standing-wave laser field that produces the conservative potential given in Eq. (6.49) can be found in Ref. [600]. The trajectory analysis is from a particle optics point of view.

Although so far chromium and aluminum atoms have been used to fabricate nanostructures using this atom-optics technique, there has also been some work on applying atom optics to manipulate metastable rare gas atoms, which can expose lithographic resist materials [601]. Since laser technology and the field of atom optics are such active areas of development, we might expect rapid progress in their use for nanostructure fabrication.

6.2 Long-range Interactions

6.2.1 Interactions between charged and neutral particles

(a) Polarization potential. An atom or a molecule can be polarized by the Coulomb field of a nearby charged particle, for example an ion or an electron. The resulting attractive interaction has important consequences for charged-particle scattering, transport properties, and spectroscopy. Consider first an isotropic polarizable system [602]. At not-too-short distances, the leading contribution is that of the dipole moment induced by the Coulomb field of the incoming charge Ze: $\mathrm{p} = \alpha Ze/r^2$, where α is the polarizability. The force with which the dipole and the charge attract each other is given by $\mathrm{F}= (2p \cdot Ze/r^3)\cos\varphi$, where φ is the angle between the direction

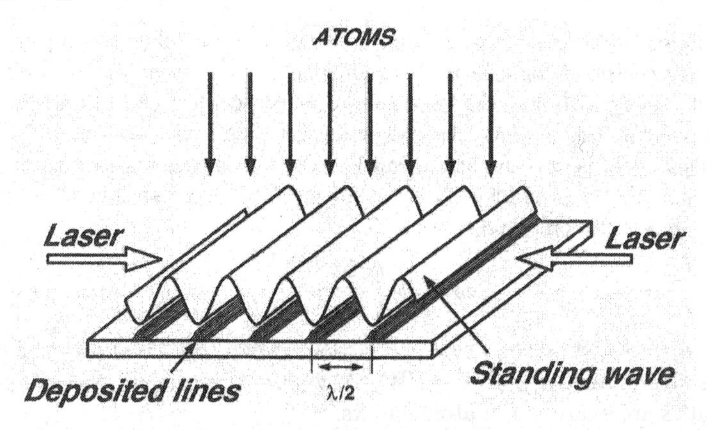

Figure 6.11: A diagram showing the one-dimensional standing-wave field pattern formed by the intersection of two counterpropagating laser beams. The atoms that traverse the standing-wave field are focused at the nodes of the field — see text for details. (Reprinted with permission from [599].)

of the induced dipole and the line joining it to the point charge. In the isotropic case considered here, $\varphi = 0$. Thus the force is F$= 2\alpha Ze^2/r^5$, corresponding to an attractive potential energy

$$V_{pol} = -\frac{\alpha(Ze)^2}{2r^4}. \tag{6.50}$$

For a perfectly conducting sphere $\alpha = R^3$; this of course is just the attraction of the charged particle to its own image charge [56].

In the case of an anisotropic molecule the angle φ between the induced moment and the incoming charge is no longer zero and more complicated expressions result. For axially-symmetric ("cylindrical") molecules one finds [603]

$$V_{pol} = -\frac{\alpha_0(Ze)^2}{2r^4}[1 + \kappa(3\cos^2\theta - 1)], \tag{6.51}$$

where α_0 and κ are the average dipole polarizability and the anisotropy, respectively:

$$\begin{aligned} \alpha_0 &= \tfrac{1}{3}(\alpha_\parallel + 2\alpha_\perp), \\ \kappa &= (\alpha_\parallel - \alpha_\perp)/(3\alpha_0), \end{aligned} \tag{6.52}$$

and θ is the angle between the molecular axis and the charge Ze [cf. Eqs. (3.71) and (5.16)]. Corresponding expressions for a point charge interacting with a tetrahedral molecule can be found in [28, 604], where other combinations (e.g., spherical-dipolar) and higher-order terms are also discussed. A concise table of common interaction potentials between atoms, molecules, and ions is assembled in [605].

(b) Low-energy electron attachment and photodetachment. The polarization potential V_{pol} may be strong enough to capture the incoming charge. The classical trajectories of a particle moving in an r^{-4} potential were determined by Langevin [606] (an English-language translation is given in [607]; see also [608–610]), who found

that for impact parameters greater than a certain critical value b_0 the projectile will swing by the center of force in a "dibrachoidal" orbit, never approaching it closer than a distance $b_0/\sqrt{2}$, and will be scattered off at some angle. However, if the impact parameter is less than b_0, the projectile will spiral into the center. Thus if no large repulsive core is present, and a capture or rearrangement reaction can proceed at close approach, one can identify $\sigma_0 = \pi b_0^2$ with the cross section for a reactive or attachment collision. One finds

$$\sigma_0 = 2\pi \left(\frac{(Ze)^2 \alpha}{\mu v^2} \right)^{1/2}, \tag{6.53}$$

where μ and v are the reduced mass and relative velocity for the collision; for electron projectiles this becomes $\sigma_0 = (2\pi^2 \alpha / W)^{1/2}$, where W represents the energy and all the quantities are expressed in atomic units.

A quantum-mechanical analysis of scattering in such a potential [608] leads to essentially the same result. It is found that the integral capture cross section oscillates sinusoidally about the classical value, with a deviation of no more than several percent. This holds down to the energy range characterized by

$$\left[8(Ze)^2 \alpha \mu^2 W / \hbar^4\right]^{1/4} \gtrsim 1, \tag{6.54}$$

which for electrons corresponds to $W \gtrsim (8\alpha)^{-1}$ with the collision energy and target polarizability again expressed in atomic units.

For a strongly polarizable system, e.g., a cluster, the condition is satisfied already at rather low energies. Indeed, with the polarizability of a metallic cluster of N atoms approximately given by $\alpha \sim R^3 = a_0^3 r_s^3 N$ (see Sections 3.5.1 and 4.10.4), we see that the electron capture cross section becomes applicable already at millivolt energies. This behavior has been observed experimentally [218] in collisions of sub-eV electrons with sodium clusters Na_N ($N = 8 - 40$). The inelastic scattering cross sections were found to rise with decreasing energy to values as high as several hundred Å^2, in agreement with Eq. (6.53) and the high metal cluster polarizability.

As the collision energy drops below the range given by Eq. (6.54) and approaches zero, one finds the limiting value [608]

$$\sigma^{capture}(W) \underset{W \to 0}{\rightarrow} 4\pi \left(\frac{e^2 \alpha}{2W} \right)^{1/2}, \tag{6.55}$$

(written for an electron projectile, $Z = 1$). This is exactly twice the classical capture cross section Eq. (6.53) above. The low-energy $1/v$ behavior is in agreement with the general theorem for inelastic s-wave scattering [332]

If the polarizability of the target is not high, the polarization capture cross section may become significant only at very low impact energies, in which case the limiting expression is appropriate. As the collision velocity increases, $\sigma^{capture}$ may be overcome by other contributions, e.g., by other inelastic channels or even by the geometrical cross section of the target. The interpolation formula [611]

$$\sigma^{capture}(W) \approx \frac{\pi}{2W} \left(1 - e^{-4\sqrt{2\alpha W}}\right) \tag{6.56}$$

(all quantities in atomic units) has been found to make an excellent fit to electron attachment cross sections of such molecules as SF_6 in the meV energy range (below the vibrationally inelastic threshold) [612].

The inverse problem to that considered above is also of interest: how does the long-range force affect the removal of a bound electron from a negative ion, as in a photodetachment experiment? This problem was considered theoretically by O'Malley [613], who found that near threshold the total cross section for spherically symmetric systems should have the form (in atomic units)

$$\sigma_l^{detachment} = const \cdot k^{2l+1}\left[1+\frac{4\alpha k^2 \ln k}{(2l+3)(2l+1)(2l-1)}+O(k^2)\right], \tag{6.57}$$

where k is the momentum of the outgoing electron and l is its angular momentum. The k^{2l+1} dependence is the well-known Wigner law [614]. The overall prefactor in Eq. (6.57) is governed by the dipole transition matrix element of the specific system, but the correction term in brackets depends on the long-range polarization force.

This effect was observed very clearly in the photoelectron spectra of a beam of cold C_{60}^- cluster ions [261]. Since the C_{60} molecule is both highly symmetrical and very polarizable (see Sections 3.5.3 and 4.10.5), it is an ideal system to observe a manifestation of Eq. (6.57). The first detachment channel in this case has $l = 0$, and the equation reduces to

$$\sigma_{l=0}^{detachment} = const \cdot k[1-(4\alpha/3)k^2 \ln k]. \tag{6.58}$$

Figure 6.12 shows that the experimental data are in excellent agreement with the predicted behavior. The steepness of the yield curve is dramatically modified by the polarization potential. Earlier, indications of similar behavior were found in the photodetachment spectrum of Au^- [615].

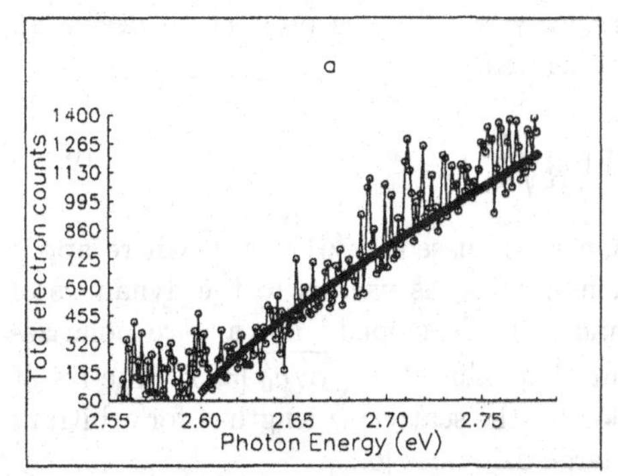

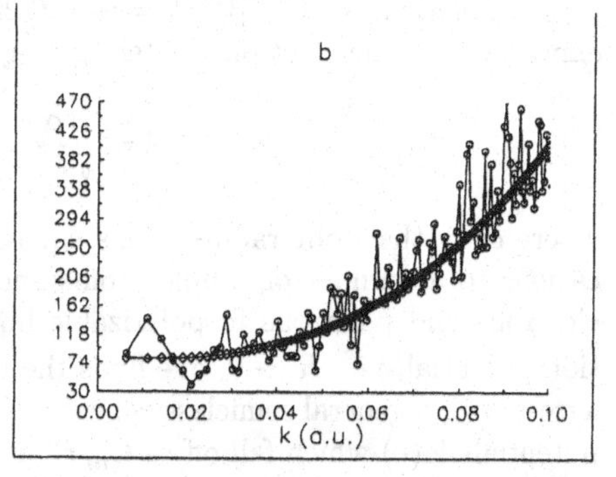

Figure 6.12: Threshold detachment spectra of cold C_{60}^-. Photoelectron yield is shown as a function of (a) photon energy, and (b) electron momentum. Polarization effects strongly modify the shape of the yield curve, which would be much sharper if these effects were absent. The open dots are the experimental data, and the diamond curve is the theoretical form, Eq. (6.58). The only fitting parameters are the electron affinity (determined to be 2.6 eV) and the overall proportionality constant. (Reprinted from Ref. [261] with the kind permission of Elsevier Science Publishers.)

(c) Other polarization effects in scattering. The calculation of attachment cross sections described above assumed that a capture or rearrangement reaction would proceed with 100% probability once the incoming projectile is captured in the long-range polarization field. In other words, the scattering center was assumed to act as sort of a perfect absorber. Of course, this picture is not necessarily valid, and in general one must also pay attention to the short-range part of the scattering interaction. Indeed, once the projectile spirals towards the scattering center it may turn out that the reaction channel is forbidden by a selection rule or is less than 100% efficient, or that the short-range field is strongly repulsive, or that its effective range exceeds the calculated Langevin value. These and other possibilities depend on the details of the particular problem and open a vast field of research.

In many cases, it turns out that the effect of the polarization interaction is to introduce a *correction* into the elastic or inelastic scattering cross section. For example, in low-energy elastic scattering the effective-range expansion for the s-wave cross section is modified to [616]

$$\sigma_0(k) = 4\pi \left(A + \alpha \frac{\pi}{3} \frac{e^2 \mu}{\hbar^2} k + ... \right)^2 , \tag{6.59}$$

where A is the scattering length. When A is negative, the cross section vanishes for

$$k' \approx -\frac{3}{\pi} \frac{A}{\alpha} \frac{\hbar^2}{e^2 \mu} \tag{6.60}$$

(up to correction terms of higher order in k). The actual observed elastic scattering cross section will not actually vanish, since it has additional contributions from higher partial waves, but it will exhibit a deep minimum at electron momentum k'. This is the famous *Ramsauer-Townsend effect.*

For the special model of a hard-core repulsive potential [$V(r) = \infty$ for $0 < r < R$] plus a polarization tail [$V(r) = -\alpha e^2/2r^4$ for $r > R$], it is possible to derive an expression for the electron scattering length as well:

$$A = \sqrt{\frac{\alpha}{a_0}} \cot \left(\frac{1}{R} \sqrt{\frac{\alpha}{a_0}} \right) , \tag{6.61}$$

where a_0 is the Bohr radius. This equation is discussed in [617–619], where applications to a number of simple atoms and molecules, as well as to the dynamics of electrons and positrons in polarizable liquids, are developed. For a pure polarization potential, i.e., $R \to 0$, one finds the limiting value $A = \sqrt{\alpha/a_0}$ [333]. The result agrees with a general semiclassical expression for the scattering length A for arbitrary potentials $V(r)$ which fall off as C_n/r^n at large distances [620].

Additional references to theoretical treatments of low-energy electron scattering in the presence of a polarization potential can be found in [621, 622]. Ref. [623] contains a set of lectures devoted to the behavior of slow electrons in gases and liquids.

The definition of an appropriate potential, valid for *all* values of r, for use in calculating elastic and inelastic electronic and ionic cross sections is actually a very non-trivial issue. Indeed, in general one has to account not only for the electrostatic polarization potential, but also for the exchange interaction (important when the incoming electron wave overlaps with the target electrons) and non-adiabatic effects. A discussion of these questions is well beyond the scope of this book, and we refer the reader to the extensive literature on electronic collisions (see, e.g., [602, 624–628]; Ref. [629] also contains a thorough summary of other review articles on the theory of electron scattering by molecules.) In passing, we mention that phenomenological potentials of the type

$$V = -\frac{\alpha e^2}{2r^4}\left(1 - e^{-(r-r_0)^6}\right) \quad \text{or} \quad V = -\frac{\alpha e^2}{2\left(r^2 + r_0^2\right)^2}, \tag{6.62}$$

where r_0 is an empirically chosen parameter, have been frequently used for the description of charged particle scattering by atoms and molecules [629, 630].

(d) Ion reactions and ionic mobility. The Langevin cross section is applicable not only to electron scattering but to ion-molecule collisions as well. It is found that a number of chemical reactions involving ions (e.g., $He^+ + CO \rightarrow C^+ + O + He$, $Ar^+ + D_2 \rightarrow ArD^+ + D$, etc.) can proceed with cross sections $\sim 10^1 - 10^2$ Å^2 for energies of typically up to a few eV. These values are significantly enhanced above what one would find if both structures were ground-state *neutrals,* and the observations are very well explained by the Langevin formula [609, 631]. Reactions rates are commonly expressed in terms of the *rate coefficient,* $k = v\sigma$; and for Langevin cross sections one finds from Eq. (6.53)

$$k_0 = v\sigma_0 = 2\pi(Ze)\sqrt{\frac{\alpha}{\mu}}, \tag{6.63}$$

i.e., a constant reaction rate coefficient.

At higher scattering energies the calculated orbiting cross sections become comparable to the gas-kinetic cross sections, and the original assumptions of the model no longer hold. In general, considerations similar to those enumerated in the preceding section apply, and it becomes necessary to refine the theory. These questions are discussed in a large number of sources on atomic and molecular scattering, see, e.g., [632, 633]. These references also consider the case of anisotropic molecules [cf. Eq. (6.51)].

Ion-neutral collisions are also the defining mechanism in the phenomena of transport and diffusion of ions in gases [603, 634]. The *mobility,* K, is defined as the coefficient relating the applied electric field to the ion drift velocity which develops as a result:

$$\mathbf{v}_d = K\mathbf{E}. \tag{6.64}$$

A calculation of mobility based on transport theory requires the knowledge of collision integrals of the type

$$Q(W) = \int (1 - \cos\theta)\sigma(\theta, W)\sin\theta d\theta, \tag{6.65}$$

where $\sigma(\theta, W)$ is the differential cross section for ion-gas collisions. Of course, the polarization potential Eq. (6.50) is critically important here. Langevin's treatment

[606], performed in the limit of pure polarization forces (plus hard-core repulsion), provides an expression valid for low temperatures and low electric field values:

$$K = \frac{36}{\sqrt{\alpha\mu}} \mathrm{cm}^2/(\mathrm{V\text{-}s}), \tag{6.66}$$

where the polarizability α is measured in atomic units and the reduced mass μ in units of the proton mass [634]. In fact, the orbiting cross sections form a good approximation to the diffusion cross section, because the final directions of the spiraling collisions are very nearly random, while those of the nonspiraling trajectories have small deflections and make a negligible contribution to the integral [Eq. (6.65)] [602]. Further discussions of the kinetic theory of ions in gases and its refinements can be found in the aforementioned references.

(e) Rydberg orbits. So far we have focused primarily on scattering phenomena, that is, on the translational motion of electrons and ions in a polarization field. In addition, the polarization force can influence, or even give rise to, bound states as well.

An important example of such influence is provided by the high Rydberg orbits of alkali atoms (see [635] for an extensive discussion of Rydberg atoms; the material below follows this book). For example, in a Rydberg Na atom the valence electron orbits around the Na^+ core, attracted to it by the Coulomb force and at the same time polarizing it. In particular, states of high angular momentum l are excluded from the core by the centrifugal barrier and do not penetrate it. As a result, the properties of the high-l states can be described purely in terms of the polarizability of the ion core, a model first proposed by J.Mayer and M.Goeppert-Mayer [636].

The binding energy of a Rydberg state is commonly represented in terms of the "quantum defect" δ_l as follows:

$$W_{nl} = -\frac{1}{2(n-\delta_l)^2} \cong -\frac{1}{2n^2} - \frac{\delta_l}{n^3} + \ldots \tag{6.67}$$

(in atomic units). A high-l Rydberg electron is slowly moving in a hydrogenic orbit around the ion core. In this adiabatic limit (when intervals between Rydberg states are much smaller than excitation energies of the core) one can write for the energy of this electron:

$$W_{nl} \cong -\frac{1}{2n^2} - \frac{\alpha}{2}\left\langle \frac{1}{r^4} \right\rangle, \tag{6.68}$$

where the first term is the hydrogenic energy and the second is the expectation value of the polarization potential Eq. (6.50). The expectation values of various powers of r are well known, and one finds that for $l \gg 1$ and $n \gg l$, $\langle r^{-4} \rangle \cong 3/(2n^3 l^5)$. Substituting this into Eq. (6.68) and comparing with Eq. (6.67), we derive for the quantum defect:

$$\delta_l \cong \frac{3\alpha}{4l^5}, \tag{6.69}$$

where α is the polarizability of the alkali *positive ions* in atomic units.

With their high values of polarizability, metal and fullerene clusters can be expected to support strongly bound Rydberg orbits. The polarization potential of a neutral cluster may in fact be sufficient to bind an electron all by itself. Spectra indicating that such "image-charge-bound" states exist in a ring-shaped gold cluster Au_6^- were reported in [217]. There have also been theoretical studies of dielectric clusters supporting surface electron states bound by the image forces, see [637, 638] and references therein.

(f) Long-range binding in ionic molecules. The attractive force of a polarized atom or molecule can bind not only electrons, as in the example of Rydberg systems above, but atomic ions as well. Such long-range ion-atom diatomics were studied in an experiment on ArM^+ complexes (M=Ni, Co, V, Zr) [639]. It was demonstrated that vibrational spectroscopy provides a novel way to extract the polarizability of argon.

Underpinning this technique is the fact that close to the dissociation limit, the vibrational energy level spacing of all such molecules will have a universal structure. Indeed, if the asymptotic potential energy curve is written in the form

$$U(r) = (\text{short-range repulsive core}) - \frac{C_n}{r^n}, \tag{6.70}$$

then semiclassical WKB quantization predicts that the spacing of vibrational energy levels $G(v)$ (v=vibrational quantum number) will satisfy the following equation:

$$\begin{aligned} \Delta G(v) &\equiv \frac{1}{2}\left[G(v+1) - G(v-1)\right] \\ &= -K_n \left[G(v) - D\right]^{(n+2)/2n}. \end{aligned} \tag{6.71}$$

Here $\Delta G(v)$ is an approximation to the derivative of the energy with respect to the vibrational quantum number: $\Delta G(v) \approx dG(v)/dv$. The quantity D is the energy of the dissociation limit for the particular electronic state being probed, so $[G(v) - D]$ is the binding energy of the v-th vibrational level. The constant K_n is given by

$$K_n = \frac{h}{(2\pi\mu)^{1/2}(C_n)^{1/n}}\left[\frac{n\Gamma(1+1/n)}{\Gamma(1/2+1/n)}\right], \tag{6.72}$$

where h is the Planck constant and μ is the reduced mass of the pair complex.

For polarization forces $n = 4$, and we therefore see that the slope of a plot of $\Delta G^{4/3}$ vs. $[G(v) - D]$ for high-lying vibrational states should yield the value of K_4, and thereby the interaction constant $C_4 = \alpha e^2/2$. This approach was formulated by LeRoy and Bernstein [640–642], and therefore vibrational energy plots of this type are known as LeRoy-Bernstein plots. Figure 6.13 shows such a plot for three potential energy curves of the $ArNi^+$ molecule. These weakly bound species are held together by the induced polarization of the Ar atom. To produce them it was necessary to use a supersonic beam ion source which generated molecules with very low internal energy ($T < 10K$). The vibrational progressions in all three electronic states follow the same limiting pattern, precisely as expected for a charge-induced dipole interaction. The average value from the spectra of all studied ArM^+ ions was found to be $\alpha_{Ar} = 1.49 \pm 0.10$ Å^3. This is close to the literature value of 1.642 ± 0.001 Å^3; the authors of Ref. [639] suggest that the deviation may be due to higher-order polarization effects.

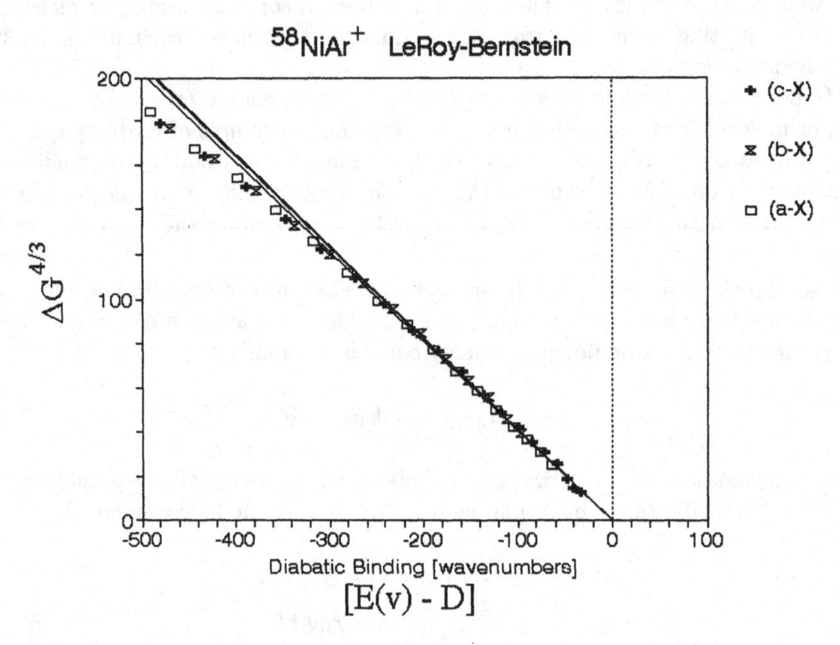

Figure 6.13: A LeRoy-Bernstein plot of the vibrational structure of three quartet electronic states of the $ArNi^+$ ion. For the highest vibrational states of all bands one finds a linear behavior, as expected for a complex held together by a long-range $1/r^4$ force. The limiting slope of the straight lines yields the coefficient K [Eq. (6.72)] and hence the polarizability of argon. (Reprinted from Ref. [639] with the kind permission of Elsevier Science Publishers.)

6.2.2 Interactions between neutral particles

(a) Dispersion forces. In the classical picture one requires the presence of a charge or a dipole moment to set up an attractive force, but quantum-mechanical fluctuations give rise to long-range forces even between neutral particles without permanent moments. As was shown by F. London [643], zero-point motion of the charge distribution produces instantaneous dipoles, and in the second order of perturbation theory one obtains the following attractive potential between particles A and B:

$$V_{AB} = -\frac{1}{R^6}\frac{3e^4\hbar^4}{2m_e^2}\sum_{i,j\neq 0}\frac{f_i^A f_j^B}{(W_0^A - W_i^A)(W_0^B - W_j^B)(W_0^A + W_0^B - W_i^A - W_j^B)}, \tag{6.73}$$

where W_i^A is the energy and f_i^A is the dipole oscillator strength of the i-th state of particle A, and similarly for particle B. This equation is a spherical average and holds for particles which do not have any permanent moments. It can be recast into the following form:

$$V_{AB} = -\frac{3\hbar}{\pi R^6}\int_0^\infty \alpha^A(i\omega)\alpha^B(i\omega)d\omega, \tag{6.74}$$

where $\alpha(i\omega)$ are the dynamical polarizabilities of the two particles, Eq. (2.2), evaluated for an imaginary frequency. (For conciseness, we will limit the discussion to isotropic polarizabilities and interactions between particles in the ground state.). This is a very pretty result, because (i) it shows explicitly that the long-range attraction is connected with polarization and polarizability, and (ii) the factor of $\hbar$ underscores the purely quantum-mechanical origin of the interaction.

The famous $1/R^6$ long-range attraction, the so-called van der Waals potential, is just the first term in the series

$$V_{AB} = -\frac{C_6^{AB}}{R^6} - \frac{C_8^{AB}}{R^8} - \frac{C_{10}^{AB}}{R^{10}} - \dots, \tag{6.75}$$

where the dipole/induced-dipole coefficient C_6^{AB} is that written down above, and the higher-order terms correspond to quadrupole/induced-dipole and higher-order interactions. All of these coefficients involve multipole transition matrix elements and are therefore intimately related to the phenomena of light dispersion by atoms and molecules. Therefore all interactions of this type are commonly labeled *dispersion forces.*

The subject of dispersion forces is very well developed and there exist a multitude of specialized books and reviews dedicated to its various aspects. These include, e.g., the classic book by Margenau and Kestner [644] which also contains a nice historical overview, as well as Refs. [604, 605, 642, 645–650] and others. Advanced research topics involve "ramifications and extensions to more complicated situations involving higher multipolarities, many-body forces, realistic transition moments, the matching at the electronic boundaries with valence forces, and the applications of these intermolecular potentials to concrete situations" [651]. It would be well beyond the aim of the present book to attempt a recapitulation of the mountain of available information, therefore we shall only touch upon certain features of dispersion forces which directly involve polarizabilities, especially static polarizabilities.

(b) Approximate forms of the interaction. The general expression acquires a transparent form if one can assign a characteristic, or average, transition energy $\hbar\bar{\omega}$ to the spectrum of each interacting particle. Then, from Eq. (6.73) and the relation Eq. (4.7) between the static polarizability α and the oscillator strength one obtains

$$C_6^{AB} \approx -\frac{3}{2}\hbar\alpha^A\alpha^B\frac{\bar{\omega}^A\bar{\omega}^B}{\bar{\omega}^A+\bar{\omega}^B}. \tag{6.76}$$

This equation is also due to F. London. The use of a characteristic excitation energy works well, e.g., for alkali atoms with their dominant absorption lines, or for metal clusters and fullerenes displaying giant dipole resonances (an example is discussed in Section 6.2.2(c) below). In other cases, the average energy $\hbar\bar{\omega}$ is often approximated by the ionization potential.

Expression (6.76) can be approximated further if one makes use of a sum rule estimate similar to Eq. (4.14). Then the *Slater-Kirkwood formula* is obtained:

$$C_6^{AB} \approx -\frac{3}{2}\left(\frac{e^2\hbar^2}{m_e}\right)^{1/2} \frac{\alpha^A\alpha^B}{(\alpha^A/N^A)^{1/2}+(\alpha^B/N^B)^{1/2}}, \tag{6.77}$$

where N is the number of valence electrons. For many molecules and atoms, other than the alkalis, the London and Slater-Kirkwood formulas tend to predict interactions energies which are somewhat too small [652]. To cast Eq. (6.76) in yet another form, one can go back to the expression for the ground-state static polarizability in terms of dipole matrix elements:

$$\alpha = \frac{2}{3}e^2 \sum_{k\neq 0} \frac{\left|\langle k|\sum_i \mathbf{r}_i|0\rangle\right|^2}{\hbar\omega_{k0}}. \tag{6.78}$$

If an average transition energy $\hbar\bar{\omega}$ is factored out of the sum, then by the closure relation one obtains

$$\begin{aligned} \alpha &\approx \frac{2}{3}\frac{e^2}{\hbar\bar{\omega}} \sum_{k\neq 0} \left|\langle k|\sum_i \mathbf{r}_i|0\rangle\right|^2 \\ &= \frac{2}{3}\frac{e^2}{\hbar\bar{\omega}} \left[\left\langle 0\left|\left(\sum_i \mathbf{r}_i\right)^2\right|0\right\rangle - \left\langle 0\left|\sum_i \mathbf{r}_i\right|0\right\rangle^2\right] \\ &= \frac{2}{3}\frac{e^2}{\hbar\bar{\omega}} \left\langle\left(\sum_i \mathbf{r}_i\right)^2\right\rangle, \end{aligned} \tag{6.79}$$

where the last term in the second line vanished because we took the origin at the center of symmetry, and in the third line we used the short-hand notation "<>" for the ground-state expectation value. We can write

$$\left\langle\left(\sum_i \mathbf{r}_i\right)^2\right\rangle = \sum_i \left\langle \mathbf{r}_i^2\right\rangle + \sum_{i\neq j} \langle \mathbf{r}_i \cdot \mathbf{r}_j\rangle, \tag{6.80}$$

where the first term is the sum of mean square radii of the electron orbits, and the second is the so-called *correlation term.* In the Hartree approximation, which neglects the Pauli exclusion principle, the latter would vanish due to the symmetry of ground-state wave functions. In reality, however, this term is non-negligible and negative. Let us introduce a parameter β which is a measure of ground-state electron correlations:

$$\left\langle\left(\sum_i \mathbf{r}_i\right)^2\right\rangle \equiv \beta \sum_i \left\langle \mathbf{r}_i^2\right\rangle. \tag{6.81}$$

The mean-square distance is directly related to the molar diamagnetic susceptibility:

$$\chi_{dia} = -\frac{N_0 e^2}{6m_e c^2} \sum_i \left\langle \mathbf{r}_i^2\right\rangle, \tag{6.82}$$

where N_0 is Avogadro's number.

Putting together the London equation Eq. (6.76), the relations Eq. (6.79,6.81) between $\alpha, \hbar\bar{\omega}$, and the mean-square distance, and the above expression for the diamagnetic susceptibility,we can write

$$C_6^{AB} \approx -\frac{6m_e c^2}{N_0} \frac{\alpha^A \alpha^B}{\frac{\alpha^A}{\beta^A \chi^A} + \frac{\alpha^B}{\beta^B \chi^B}}. \tag{6.83}$$

If one assumes that all $\beta = 1$, this becomes the so-called *Kirkwood-Müller formula*[2]. However, as mentioned above, the Pauli principle causes the correlation term in Eq. (6.80) to be negative and therefore β may be significantly less than 1. For example, for N non-interacting electrons in a three-dimensional harmonic oscillator potential one finds [653]

$$\beta = \frac{3}{4} \frac{N}{\sum_{n,l} (2l+1)(2n+l-\frac{1}{2})}, \tag{6.84}$$

where the sum is over the quantum numbers of all occupied levels. For example, for $N = 8$ this yields $\beta = 2/3$ and for $N = 20, \beta = 1/2$. Switching on the electron-electron interaction may decrease the value of β even further. Thus, whereas in the hydrogen molecule β happens to be close to 1, in O_2 it is reduced to approximately 0.3 [652]. For the metal clusters Na_6–Na_{20}, β ranges from 0.5 to 0.16 [653]. As a consequence, the Kirkwood-Müller formula gives only an *upper* bound to the dispersion energy. For comparison, recall that the London and Slater-Kirkwood formulas tend to give a *lower* bound. For a further discussion of these approximations, including generalization to axial molecules, see [652, 654].

(c) Manifestation of long-range forces in collisions and in molecular spectra. Obvious manifestations of dispersion potentials can be found in molecular bonding and in atomic and molecular scattering. Excellent and detailed reviews of both fields abound, and here we shall only mention two examples. One is cluster scattering; the other is long-range molecular states.

Atomic and molecular beam scattering is a powerful tool for investigating interaction forces (see, e.g., [655–657], and references therein). By a proper deconvolution of integral and differential cross sections one can reconstruct the entire shape of the potential. At low impact velocities, collision cross sections are sensitive mostly to the long-range van der Waals tail of the interaction. For elastic scattering in a pure $-C_6/R^6$ potential, one finds the following expression for the *semiclassical* integral cross section:

$$\sigma = 8.083 \left(\frac{C_6}{\hbar v}\right)^{2/5}, \tag{6.85}$$

where v is the collision velocity. From the London equation Eq. (6.76) we see that collisions between strongly polarizable species can give rise to very substantial cross

[2]Combining this approximation with the sum-rule estimate for α, cf. Eq. (6.77), one obtains the so-called Kirkwood relation $\chi \propto \sqrt{\alpha}$. As with the Slater-Kirkwood and Kirkwood-Müller formulas for C_6, this relation represents only a rough estimate.

sections. An example is given in Fig. 6.14 which shows that in collisions involving metallic clusters, which have high polarizabilities, one indeed observes very large cross sections.[3]

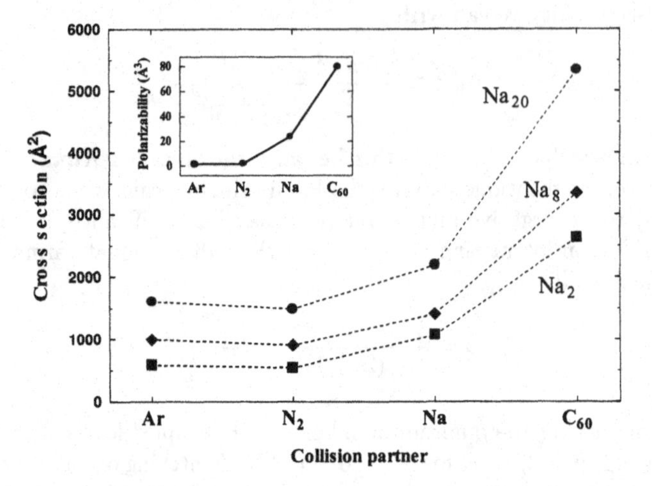

Figure 6.14: Integral center-of-mass cross sections for elastic collisions of Na_2, Na_8, and Na_{20} with Ar, N_2, Na atoms, and C_{60} (after [220]). The cross sections, shown for collision velocity $v \approx 1000$ m/sec, are based on C_6^{AB} parameters extracted from scattering measurements. The insert shows the static electric polarizabilities of the collision partners. Note that the cross sections increase both with the polarizability of the cluster (i.e., with cluster size, as discussed in Section 3.5.1) and with the polarizability of the collision partner, in agreement with Eqs. (6.76) and (6.85). Reflecting the long range of the van der Waals force, these data exceed the hard-sphere geometrical cross sections by factors of $\sim$ 15-20.

Alternatively, if the collision velocity is low enough, even low-polarizability inert atoms can produce giant cross sections. For example, He-He collisions utilizing a slow atomic beam with velocities from 80 m/s down to $\approx$ 40 m/s gave s-wave scattering cross sections which rose from 200 to 1000 Å^2 as the collision energy was reduced [661]. In order to produce such a slow beam, a source was built into a dilution refrigerator at temperatures down to 0.5 K. Good agreement with calculations based on known interatomic potentials was found. While the semiclassical Eq. (6.85) is not directly applicable in this case, it is qualitatively apparent that the weak polarizability of He ($\alpha_{He} = 0.01\alpha_{Na}$, see Section 4.10) has been more than made up for by the low collision velocity.

In passing, we mention that interest in very slow atomic collisions has been raised dramatically by the explosion of work on atoms in traps. Such experimental phenomena as photoassociation spectroscopy [662–664] and atomic Bose-Einstein condensation ([566, 577, 665, 666], and references therein) are sensitive to s-wave scattering

[3]For discussions of Van der Waals energy calculations for clusters see, e g., [193, 658] (interaction between metal clusters), [659] (interaction between a metallic sphere and a rare-gas atom), and [660] (interaction between fullerene molecules and between C_{60} and a substrate), and references therein.

and therefore to the scattering length A. Theoretical methods for approximating this quantity, which were described in Section 6.2.1(c) in the context of polarization potentials, are applicable to the present situation as well.

Matter wave interferometry is another novel technique which can probe long-range interactions. Filling one leg of the atom interferometer described in Section 5.2.5 (see Fig. 5.16) with a gas is analogous to introducing a medium with a complex index of refraction into an optical interferometer. The resulting shift in the beam diffraction pattern is related to the forward scattering amplitude and thereby to the interatomic potential. Measurements on thermal beams of Na [667] and Na_2 [668] showed a qualitative signature of an attractive van der Waals interaction; further detailed studies are required, however, to gain a full quantitative understanding of the data (see also [669]).

Elastic scattering probes van der Waals forces by looking at low-energy translational motion, that is, at the *continuum* part of the spectrum. Alternatively, these forces can be probed on the other side of the threshold by spectroscopic measurements of *bound* vibrational states of long-range molecules. In particular, a large number of studies have considered inert-gas dimers, which are systems bound purely by weak dispersion forces [642], as well as alkali dimers near the dissociation threshold. The main experimental tool is the LeRoy-Bernstein analysis described in Section 6.2.1(f) above. For a potential of the form Eq. (6.70) with $n = 6$, Eq. (6.71) states that the C_6 parameter can be determined from the limiting slope of the plot of $[\Delta G(v)]^{3/2}$ vs. $[G(v) - D]$. Some of a large number of recent publications discussing van der Waals potential curves of dimers of alkalis, noble gases, and their combinations, and providing relevant references are [392, 408, 433, 434, 670–677].

6.2.3 Many-atom, macroscopic, and retarded interactions

(a) Non-additivity of dispersion forces. The discussion so far has focused on long-range potentials between two atoms, two molecules, or two clusters. When three or more particles interact the complexity increases because, in contrast to elementary electric or magnetic forces, the dispersion interaction is not purely additive. Recall that the basic equation Eq. (6.73) was the result of a second-order perturbation calculation. When three particles interact, then to second order the van der Waals potential is still the sum over all pairs, but third-order perturbation theory introduces a correction which mixes all the ingredients [644] :

$$V(\mathbf{R}_{AB}, \mathbf{R}_{BC}, \mathbf{R}_{CA}) = C^{ABC} \frac{1 + 3\cos\beta_{AB}\cos\beta_{BC}\cos\beta_{CA}}{R_{AB}^3 R_{BC}^3 R_{CA}^3}. \tag{6.86}$$

Here β are the angles of the triangle formed by the particles. The full expression for the coefficient C^{ABC} involves a triple sum of products of dipole matrix elements and a set of energy denominators. It can also be written as an integral similar to Eq. (6.74) :

$$C^{ABC} = \frac{3\hbar}{\pi} \int_0^\infty d\omega \alpha^A(i\omega)\alpha^B(i\omega)\alpha^C(i\omega). \tag{6.87}$$

Thus there is an elegant parallel with the two-body expressions. A London-like formula holds if all the particles are isotropic and can be assigned a characteristic dipole frequency:

$$C^{ABC} \approx \frac{3}{2}\hbar\alpha^A\alpha^B\alpha^C \frac{\bar{\omega}^A\bar{\omega}^B\bar{\omega}^C(\bar{\omega}^A+\bar{\omega}^B+\bar{\omega}^C)}{(\bar{\omega}^A+\bar{\omega}^B)(\bar{\omega}^B+\bar{\omega}^C)(\bar{\omega}^C+\bar{\omega}^A)}, \tag{6.88}$$

cf. Eq. (6.76)[4] [644]. For interaction between three identical particles this simplifies to [679]

$$C^{AAA} \approx \frac{3}{4}\alpha^A C_6^{AA}, \tag{6.93}$$

where C_6^{AA} is the two-body interaction coefficient.

Many-body forces, while relatively short-ranged, can make a noticeable contribution in situations when three-body collisions are important (e.g., the third virial coefficient for gases) or in the condensed phase: interactions between macroscopic bodies, surface adsorption, solvent media, etc.

(b) Additive models for interaction of macroscopic objects. Hamaker constant. Despite the preceding statement, it is a common first-order approximation to treat forces between macroscopic objects as arising from simple pairwise summation between their constituent atoms. As pointed out in [650], the limitation of this procedure is not just in neglecting the non-additivity of van der Waals forces, but also in neglecting changes in the energy levels which occur when separate atoms join together to form particles of condensed matter. Indeed, Eq. (6.74) makes it clear that the interaction strength will change if the excitation spectrum is modified. For instance, in metal and carbon microclusters (Section 3.5) electron delocalization and the appearance of giant collective resonances in the dipole spectrum are qualitatively new features resulting from the agglomeration of atoms. This reasoning extends all the way up to macroscopic bodies.

---

[4] Other approximate "combination rules," accurate to better than a few percent for simple atoms, have been formulated (see, e.g., the references and recent calculations of three-dipole interaction coefficients in [433, 678]). One writes

$$C^{ABC} \approx \frac{Q^A Q^B Q^C (Q^A+Q^B+Q^C)}{(Q^A+Q^B)(Q^B+Q^C)(Q^C+Q^A)}, \tag{6.89}$$

with the following proposed forms for the parameters Q:

$$\frac{1}{Q^i} = \frac{1}{C_6^{ij}\alpha^k} + \frac{1}{C_6^{ki}\alpha^j} - \frac{1}{C_6^{kj}\alpha^i}, \tag{6.90}$$

or

$$Q^i = \frac{\alpha^j\alpha^k}{\alpha^i} C_6^{ii}, \tag{6.91}$$

or

$$Q^i = \frac{4}{3}\frac{\alpha^j\alpha^k}{(\alpha^i)^2} C^{iii}, \tag{6.92}$$

with $i, j, k = A, B, C$. These forms have the same basic London-equation character, and permit one to construct a three-body interaction coefficient based on other known one-, two-, and three-particle parameters.

Having said all this, the fact remains that in many cases additivity models predict the correct distance dependence of the interaction as well as the right order of magnitude of its strength. Frequently, the imprecisely known numerical values can be absorbed into an experimentally determined coefficient of order unity.

Thus, consider the problem of a neutral atom or small molecule, A, located a distance, d, above a planar surface made up of atoms, B. If one integrates the van der Waals interaction, $-C_6^{AB}/r^6$, over the entire surface, the following net attractive potential is obtained [605]:

$$V_{atom-wall} = -\frac{\pi C_6^{AB} \rho^B}{6d^3}, \tag{6.94}$$

where ρ^B is the number of atoms per unit volume of the wall. The $1/d^3$ dependence is reasonable from the electrostatics point of view [644]: it is the energy of the attraction of a dipole to its image charge. As we shall see below, more sophisticated treatments of the problem retain this dependence.

To calculate the interaction between an extended sphere of radius R and a planar surface, one can now integrate Eq. (6.94) over the volume of a sphere made up of atoms A. For small separations $d << R$ one finds

$$V_{sphere-wall} = -HR/6d, \tag{6.95}$$

where H is the so-called *Hamaker constant*:

$$H \equiv \pi^2 C_6^{AB} \rho^A \rho^B. \tag{6.96}$$

The value of defining this quantity is two-fold. First, integrated interactions for other macroscopic shapes are also expressible as H times a function of the separation distance (see, e.g., [605, 648] for analytical expressions for such cases as two close spheres, two cylinders, two surfaces, etc.). Second, it turns out that the typical value of the Hamaker constant is of the same order of magnitude for a wide range of condensed phases, solid or liquid. For interactions across vacuum, $H \sim 10^{-19}$ Joules. This similarity, even for very different substances, can be rationalized by the following argument [605]: the intermolecular interaction coefficient is proportional to the constituents' polarizabilities ($C_6^{AB} \sim \alpha^A \alpha^B$), which in turn scale roughly with the molecular volume ($\alpha \sim V$). At the same time, the densities of the media are inversely proportional to V ($\rho \sim 1/V$). Hence

$$H \propto C_6^{AB} \rho^A \rho^B \sim \alpha^A \alpha^B \cdot \rho^A \rho^B \sim V^A V^B \cdot (V^A V^B)^{-1} \approx constant. \tag{6.97}$$

As a result, an approximate knowledge of the experimental geometry and an estimate of the Hamaker constant provide one with a valuable estimate of the interaction force. This is of great utility, for example, in interpreting tip-surface interactions in atomic-force microscopy (see Section 6.2.4 below).

(c) Accurate theories of long-range forces involving macroscopic objects. Atom-surface interactions and dielectric functions. The additive models described above

are convenient, but their accuracy is limited. Consistent extensions of the theory are required for a description of long-range forces involving macroscopic objects, dielectric media, etc. The core insight into approaching this problem is due to Casimir [680], who pointed out that dispersion forces can be viewed as originating from shifts in the energy of electromagnetic field fluctuations due to the introduction of material bodies. The resulting forms of the interaction have been derived and re-derived by a number of authors using a variety of methods, including those of the statistical mechanics of fluctuations, field-theoretic techniques, etc. These approaches and results have been extensively discussed in the literature; we refer the reader to the references already cited in Section 6.2.3(a) as well as to the reviews in [347, 681–685] and other references given therein.

The theories of dispersion forces in macroscopic media are of relevance in a variety of fields, including surface science, physisorption, fluids, colloids, and solvation, as well as atomic and molecular spectroscopy. Leaving the bulk of the subject to the specialized literature, we cite below a few results which directly involve atomic polarizabilities. The formulas apply for relatively short separations; corresponding expressions for so-called retarded interactions will be discussed in Section 6.2.3(e) below.

The attraction of an atom of dynamical polarizability $\alpha(\omega)$ to a dielectric or metallic plane a distance d away is given by the following potential:

$$V_{atom-wall} = -\frac{\hbar}{4\pi d^3}\int_0^\infty d\omega\alpha(i\omega)\frac{\varepsilon(i\omega)-1}{\varepsilon(i\omega)+1}, \tag{6.98}$$

where $\varepsilon(\omega)$ is the dielectric function of the wall (assumed to be a local function). This is the same $1/d^3$ dependence as given by the additive model, Eq. (6.94); the dielectric function effectively plays the role of the polarizability of the surface.

It is instructive [686] to consider this equation for the special case of a nearly-free-electron metal surface whose dielectric function can be approximated by the Drude formula

$$\varepsilon(\omega) = 1 - \frac{\omega_p^2}{\omega^2}, \tag{6.99}$$

where ω_p is the plasma frequency, Eq. (3.88). If this function is substituted back into Eq. (6.98) together with the familiar atomic ground-state dynamic polarizability function

$$\alpha(\omega) = \frac{e^2}{m_e}\sum_{k>0}\frac{f_{0k}}{\omega_{0k}^2-\omega^2} \tag{6.100}$$

(the summation is over all the excited states of the atom) and the replacement $\omega \to i\omega$ is made, then the integral can be carried out and one finds

$$V = -\frac{e^2}{12d^3}\sum_k \frac{\omega_{sp}/\omega_{0k}}{1+\omega_{sp}/\omega_{0k}}|\langle\mathbf{r}\rangle_{0k}|^2. \tag{6.101}$$

Here $\omega_{sp} = \omega_p/\sqrt{2}$ is the so-called planar-surface plasmon frequency; its appearance illustrates that the van der Waals interaction proceeds via virtual excitations of collective surface resonances. The strength of the interaction depends on the relative magnitude of ω_{sp} and the characteristic atomic frequency $\bar{\omega}$ contributing to Eq. (6.100). If $\omega_{sp} \gg \bar{\omega}$, i.e., the surface can respond practically instantaneously to the quantum fluctuations of the atomic dipole moment, then the frequency factor in Eq. (6.101) is approximately unity, and from the closure relation one obtains

$$V \approx -\frac{e^2}{12d^3}\left\langle r^2\right\rangle, \tag{6.102}$$

where “<>” denotes the ground-state expectation value, as in Eq. (6.79).

Conversely, if the reaction of the metallic electrons is relatively sluggish ($\omega_{sp} << \bar{\omega}$) then the coefficient of the $1/d^3$ term in Eq. (6.101) goes to zero and the attraction vanishes. These results demonstrate the richness of possibilities contained in the general expression Eq. (6.98); for real materials one has to take into account the correct dynamical behavior of the specific surface.

Eq. (6.102) indicates that the van der Waals attraction grows with the size of the atomic dipole, i.e., with the orbit radius. A direct way to study this dependence is by exciting atoms into Rydberg states with a high effective quantum number n: the radius of the Rydberg atom scales as n^2, and the attractive force should then scale as n^4. Such an experiment was reported in Ref. [687]. A beam of Cs or Na atoms excited into states with $12 < n < 30$ was sent into a tunnel between two parallel gold-coated mirrors spaced by a gap of width $\approx 2-8\mu$m. The stronger the atom-surface attraction, the less likely is it that an atom, flying near the surface of one of the mirrors, will make it all the way through the tunnel without striking a mirror. By measuring the transmission as a function of the quantum state of the atom, the authors demonstrated that the interaction was indeed much larger than for ground-state atoms (by 3-4 orders of magnitude!), scaled appropriately with n, and confirmed the van der Waals dependence on the distance d. The same group also measured the atom-metal surface interaction spectroscopically by recording the energy shifts of sodium atom nS levels ($n = 10-13$) in a parallel-plate cavity $\approx 1-2.5\mu$m wide [688]. This spectroscopic technique is highly sensitive to the interaction strength, and thus allows one to determine the position of an atom inside a cavity to an accuracy of a few tens of nanometers.

An earlier experiment [689] studied the van der Waals force by sending beams of alkali atoms over the surface of a gold-coated cylinder and measuring the beam deflection. The results were found to be fully consistent with a $1/d^3$ interaction potential, although the measured interaction constants turned out to be noticeably smaller than the theoretically calculated values. Subsequent theoretical papers [690–692] highlighted a number of possible needed corrections (e.g., effects of surface roughness), but no clear resolution of the experimental discrepancy has appeared.

The experiments described so far were devoted to the attraction between an atom and a conducting wall. For interactions with a dielectric wall, a “mechanical” technique was demonstrated [693] which takes advantage of the progress in the field of laser-cooled atoms and atomic mirrors. A cloud of cold Rb atoms from a magneto-optic trap is dropped onto the surface of a glass prism; an evanescent wave propagating over the surface of the prism serves as an "atomic mirror". The potential seen by the atoms is the sum of the reflecting dipole potential of the evanescent wave and the van der Waals attraction of the glass surface. The atoms will bounce back up if the resulting potential barrier is greater than their kinetic energy (Fig. 6.15) This experiment provided a quantitative measurement of the van der Waals potential (accuracy $\approx 30\%$), and also indicated the presence of retardation effects (see below).

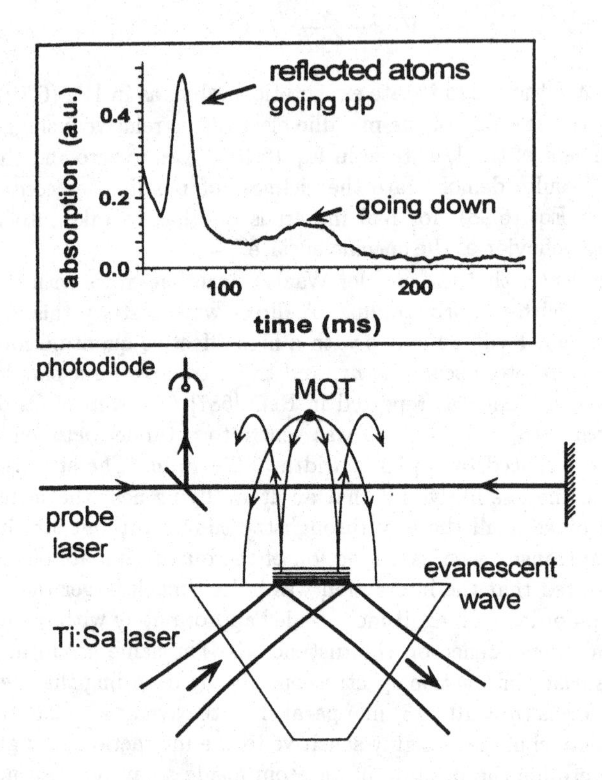

Figure 6.15: Outline of an experimental measurement of the van der Waals interaction of Rb atoms with a glass surface. The atoms dropped from the magneto-optic trap onto an atomic mirror will bounce if the repulsive potential due to the evanescent wave overcomes the van der Waals attraction of the atoms to the prism. (Reprinted with permission from [693].)

If the atoms near a conducting or dielectric surface are prepared not in their ground state but in an excited state, the van der Waals force may become either attractive or repulsive, see [686, 694] and references therein. The latter paper also discusses the cases of an anisotropic polarizability tensor and/or an anisotropic (birefringent) surface.

(d) Interactions of particles in a dielectric medium. Wetting. When two particles A and B are immersed in a medium, their dispersion interaction is of course strongly modified by the surrounding molecules, that is, by the dielectric susceptibility ε_0 of the medium. In many cases, one is interested in the thermodynamics of the solvation

process, in which case the appropriate quantity to consider is the change in free energy resulting from the immersion of the two particles. The result, extensively discussed in the literature cited earlier, is

$$\Delta F = -\frac{6k_BT}{R^6} \sum_{n=0}^{\infty}{}' \frac{\tilde{\alpha}^A(i\omega_n)\tilde{\alpha}^B(i\omega_n)}{\varepsilon_0^2(i\omega_n)}. \tag{6.103}$$

Here the summation is over the discrete frequencies $\omega_n = n(2\pi k_BT/\hbar)$ and the prime over the summation sign denotes that the $n = 0$ term is to be multiplied by 1/2. At very low temperatures the summation over n can be converted into integration over the variable ω, giving an interaction energy of

$$V_{AB} = -\frac{3\hbar}{\pi R^6} \int_0^{\infty} \frac{\tilde{\alpha}^A(i\omega)\tilde{\alpha}^B(i\omega)}{\varepsilon_0^2(i\omega)} d\omega. \tag{6.104}$$

This looks very much like a screened version of the vacuum van der Waals interaction, Eq. (6.74) . However, there is an important difference. The quantities $\tilde{\alpha}$ appearing in the above equations are not the innate polarizabilities of an isolated particle, but rather *effective polarizabilities* which measure the degree to which the introduction of a particle into the medium alters the dielectric function of the solution:

$$\tilde{\alpha}^A(\omega) = \frac{1}{4\pi} \left. \frac{\partial\varepsilon(\omega, n_A)}{\partial n_A} \right|_{n_A \to 0}, \tag{6.105}$$

where n_A is the concentration of particles A.

This equals the bare molecular polarizability only for a very dilute medium, when interactions with the surrounding molecules can be neglected. Furthermore, as pointed out in [605], the definition Eq. (6.105) represents the *excess polarizability* of the particle over that of the solvent, reflecting the fact that if the immersed particles have the same properties as the solvent, they are "invisible" and do not experience a net force.

For a small dielectric sphere of radius a and dielectric constant ε, the effective polarizability in a medium with dielectric constant ϵ_0 is given by [605] [cf. Eq. (3.83)]:

$$\alpha(\omega) = a^3 \cdot \varepsilon_0(\omega) \frac{\varepsilon(\omega) - \varepsilon_0(\omega)}{\varepsilon(\omega) + 2\varepsilon_0(\omega)}, \tag{6.106}$$

hence the interaction energy between two such particles is proportional to terms of the form

$$\Delta F_{AB} \propto -\sum_{n=0}^{\infty} \left\{\varepsilon^A(i\omega_n) - \varepsilon_0(i\omega_n)\right\} \left\{\varepsilon^B(i\omega_n) - \varepsilon_0(i\omega_n)\right\}. \tag{6.107}$$

This leads to a very important conclusion: van der Waals forces between dissimilar molecules in a medium are not necessarily attractive. For example, if the dielectric susceptibility of the solvent is intermediate between those of particles A and B, the interaction will be repulsive. Interaction between identical molecules, however, is always attractive.

Similar to the case of interactions in vacuum, the formalism outlined here can be generalized to include other geometries, such as larger colloidal particles, surfaces, thin films and multilayers, etc., with manifold applications in such areas as solvation and aggregation, wetting, adhesion, etc. (see, e.g., [605, 648, 649, 695–697]). These effects are quite sensitive to various system parameters, such as the dielectric functions of all constituents and their shapes, as well as to other forces present (short-range and electrostatic).

In his review article on the statics and dynamics of wetting (Ref. [698], which also provides additional references on the subject) de Gennes formulates some criteria for the wetting of solids by liquids based on a simple approximation to the van der Waals interaction between the two media. These estimates are based on considering the balance of interfacial solid-liquid, solid-vapor, and liquid-vapor energies. Using the approximation $V_{ij} \approx k\alpha_i\alpha_j$ for the van der Waals coupling energy between two media i and j (where k is roughly independent of i and j), he derives for the liquid-solid contact angle θ_e:

$$\cos\theta_e \approx \frac{2\alpha_S}{\alpha_L} - 1, \tag{6.108}$$

where α_S and α_L are average specific polarizabilities of the solid and the liquid, respectively. Complete wetting corresponds to $\theta_e = 0$, or $\cos\theta_e = 1$, whence a condition for complete wetting is obtained: $\alpha_S > \alpha_L$. In terms of the surface tension γ, this condition is shown to imply that for complete wetting one should choose a liquid with $\gamma \lesssim \gamma_c = \frac{1}{2}k\alpha_S^2$. This explains the experimental observation that for simple molecular liquids (for which van der Waals interactions dominate) γ_c is essentially independent of the nature of the liquid and is primarily a characteristic of the solid alone. While the above estimates are rough, depending on a simple approximation to the dispersion force and neglecting the additional forces present in the system, they are valuable for being expressed purely in terms of the specific polarizabilities and for providing insight into the physics of the liquid-solid interface. Recently, these formulas were used to analyze "nanocapillarity": filling of carbon fullerene nanotubes by molten silver nitrate [699].

(e) Retarded interactions. All of the results discussed so far implicitly assumed instantaneous correlations between charge fluctuations in the interacting objects. This is a valid picture only if the separations R involved are sufficiently short, so that one can neglect the time it takes the electromagnetic action to propagate from one object to the other. The time lag in vacuum is on the order of R/c (c is the speed of light), while the fluctuation time for the dipoles is $\sim 1/\nu_0$, where $\nu_0 = W_0/h$ is the characteristic excitation frequency of the system. Thus the critical distance is given by

$$R \sim \frac{c}{\nu_0} = \lambda_0, \tag{6.109}$$

where λ_0 is the wavelength of light emitted in the corresponding transitions. For atomic transitions the relevant distances may be in the $10^1 - 10^2$ nm range, while in a medium with a high refractive index they may be up to an order of magnitude larger.

At such large separations the dispersion forces will be modified drastically. Long-range interactions in this regime are called *retarded*, or *Casimir*, interactions. They have been the focus of a great deal of research activity and an extraordinary number of monographs, reviews, and articles. In addition to the sources listed in Sections 6.2.2(a) and 6.2.3(c), see, e.g., [685, 700–702], and many references therein. As pointed out in [685], retardation forces are remarkable in that they represent a pronounced relativistic effect even when the particles involved are not moving at relativistic speeds. Furthermore, by virtue of their importance in surface, colloidal, and related phenomena, including biological processes, these forces may provide one of the most dramatic manifestations of quantum electrodynamics in everyday life. As we just stated, the functional form of interaction potentials described in Sections 6.2.2 and 6.2.3(c,d) above is strongly modified when retardation is taken into account. For asymptotically large distances, the interaction between two particles in vacuum becomes

$$V_{AB} \underset{R\to\infty}{\longrightarrow} -\frac{23\hbar c}{4\pi R^7}\alpha_0^A\alpha_0^B, \tag{6.110}$$

where α_0 are the static polarizabilities. This is the limiting form of a result valid for all distances:

$$V_{AB} = -\frac{\hbar}{\pi R^6}\int_0^\infty d\omega\alpha^A(i\omega)\alpha^B(i\omega)e^{-2\frac{\omega R}{c}}P\left(\frac{\omega R}{c}\right), \tag{6.111}$$

where $P(x) = x^4 + 2x^3 + 5x^2 + 6x + 3$ (see, e.g., Ref. [703]). For $(\omega R/c) >> 1$ this leads to Eq. (6.110), while for $(\omega R/c) << 1$ we recover the London interaction, Eq. (6.74).

For two particles submerged in a medium with dielectric function ε, the corresponding limit turns out to be

$$V_{AB} \underset{R\to\infty}{\longrightarrow} -\frac{23\hbar c}{4\pi R^7}\frac{\alpha_0^A\alpha_0^B}{\varepsilon^{3/2}(0)}. \tag{6.112}$$

For the interaction between an atom and a dielectric surface, one finds

$$V_{atom-wall} \underset{R\to\infty}{\longrightarrow} \frac{3\hbar c}{8\pi d^4}\alpha_0\frac{\varepsilon(0)-1}{\varepsilon(0)+1}\phi\left[\varepsilon(0)\right], \tag{6.113}$$

where ϕ is a function such that $\phi(\varepsilon\to 1)\to 23/30$ and $\phi(\varepsilon\to\infty)\to 1$ (see, e.g., [684]). The latter limit corresponds to the case of a perfect conductor, in which case

$$V_{atom-conducting\ wall} \underset{R\to\infty}{\longrightarrow} \frac{3\hbar c\alpha_0}{8\pi d^4}. \tag{6.114}$$

We see that in all cases the distance dependence of the interaction has acquired an extra power in the denominator. This weakening of the dispersion forces at large distances arises because retardation reduces the correlation (i.e., alignment) of the fluctuating dipoles.

Atom-surface Casimir interactions have been studied using the techniques described previously in Section 6.2.3(c): measuring the intensity of an atomic beam passing through a micron-sized parallel plate cavity [704] and colliding laser-cooled atoms with an evanescent-wave atomic mirror [693, 705]. Ref. [703] discusses calculations of two- and three-body atom-atom and atom-wall potentials (van der Waals and retarded) for sodium, making use of the available theoretical and experimental data on oscillator strengths, energy levels, and photoionization cross sections.

In addition to defining the principal interaction between neutral objects at large distances, retardation effects also show up in electron-ion and electron-atom interactions. There is a correction to the polarization potential, Eq. (6.50), whose asymptotic form is [706]

$$V = \frac{11}{4\pi}\frac{\hbar}{m_e c}\frac{\alpha e^2}{r^5}. \tag{6.115}$$

This interaction has been probed by precision spectroscopy of high angular momentum Rydberg states ($n = 10, L \geq 7$) of the helium atoms [707]. The distance at which retardation effects become important in He is about $35a_0$, a region to which the aforementioned Rydberg states are effectively confined.

6.2.4 Long-range forces and the atomic force microscope

The atomic force microscope (AFM) [708] is a high-precision device capable of creating three-dimensional surface images with sub-nanometer resolution. The AFM measures the deflection of a sharp probe tip as it approaches and retracts from the surface. In other words, its operation is defined by long- and short-range tip-surface forces. In the "non-contact" mode of operation, the long-range van der Waals and Casimir interactions are of primary importance. This warrants the inclusion of a few brief comments on the AFM in this chapter.

In just a few years after its invention, the AFM became one of the primary probes in materials science, surface chemistry, colloid science, biological studies, etc. A detailed description of all aspects of AFM theory and applications, together with a tabulation of many experimental results, is given in [709]; see also [710–712], and references therein. In addition, extensive bibliographies and technical notes are published by microscope manufacturers [713, 714]. An outline of various stages of AFM operation is shown in Fig. 6.16. A flexible cantilever, with a sharp tip mounted on its free end, is gradually lowered towards the surface with the help of a piezoelectric translator. The local force experienced by the tip is translated into a bending of the cantilever, which can be measured by monitoring the angle of reflection of a laser beam off the cantilever end. In this way, the "force curve" (force vs. distance) can be mapped out.

The non-contact region at the top of the figure is the one where long-range forces dominate. As the tip-surface separation d is reduced, a point may come when the elastic force of the cantilever ($F_{elastic} \propto z - d$, where z is the undeflected position of the lever) can no longer keep up with the growing van der Waals force ($F_{vdW} \sim 1/d^n$, where the power n depends upon the geometry). At this point, typically 1-10 nm away from the surface, the tip suddenly jumps into contact with the sample ("jump to contact" point). The actual shape of the force curve depends upon the type and

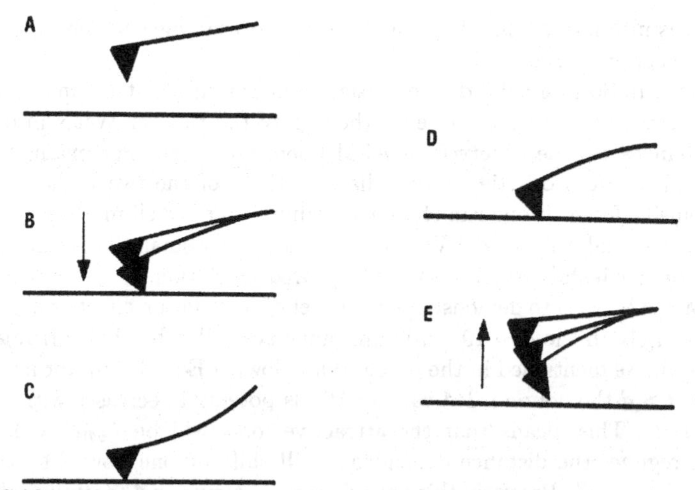

Figure 6.16: A schematic diagram of several stages along the force curve of an atomic force microscope: (A) the approach (non-contact) region, (B) jump to contact, (C) contact, (D) adhesion, and(E) pulloff. (Drawing courtesy of Digital Instruments [713].)

magnitude of all the forces present: the stiffness of the cantilever, whether the surface is charged, and whether the apparatus is immersed in a liquid.

In situations where static charges and electrolytes are absent, the long-range force between the tip and the sample is a pure case of a dispersion interaction between two macroscopic bodies. Within the additive model described in Section 6.2.3(b), the interaction energy has the form

$$V_{tip-sample} = -\frac{CH}{d^n}, \tag{6.116}$$

where H is the Hamaker constant and the constants C and n depend on the geometry. For example, if the setup can be approximated as a spherical-tipped probe close to a planar surface, then $C = R/6$ (R is the sphere radius) and $n = 1$, see Eq. (6.95). With a wide choice of tip and substrate materials available, the AFM provides a powerful means of quantitatively measuring Hamaker constants for a diverse set of systems [715].

For more sophisticated geometries one may need to appeal to detailed calculations. The sources cited in the beginning of this section review a number of examples. The specific case of tip-cluster interactions is analyzed, e.g., in [716] for a thin diamond tip over thin films of C_{60}, C_{70}, and fullerene tubules. This paper provides references to experimental work and discusses the influence of molecular shape on the form of AFM images (see also [159]).

The Hamaker model is also useful in estimating the resolution capabilities of an AFM in non-contact mode. For the ideal situation of an infinitely thin needle a height d above a surface, the calculated full width at half-maximum intensity for the image of a point particle on the surface is $w = 0.8d$ [712]. Thus the resolution in the non-contact mode is ultimately limited by the size of the tip-sample gap. For a treatment of other probe shapes see, e.g., [717].

As described in Section 6.2.3(d), immersion of interacting systems in a liquid can dramatically affect the strength and even the sign of the van der Waals interaction. Analogous changes can be observed in AFM operation [718]. For example, if the medium has dielectric properties intermediate to those of the two bodies, this will lead to a repulsive force. If one can choose a medium that is well-matched to the tip or the sample material, the van der Waals force will be very small. This can eliminate the snap-to-contact instability and lead to improved resolution.

The AFM can be used to demonstrate the onset of retardation effects which should take place precisely in the $d \sim 10 - 100$ nm range accessible by the instrument, see Section 6.2.3(e). As mentioned in the paragraph following Eq. (6.116), for a spherical tip of radius $R > d$ the nonretarded van der Waals potential decreases with distance as $V_{vdW} \propto 1/d$. This means that the attractive force will be $F_{vdW} \propto 1/d^2$. In the retarded regime, the distance dependence will shift by one power, so that one expects $F_{Casimir} \propto 1/d^3$. Precisely this behavior was observed in a carefully calibrated measurement shown in Fig. 6.17 [719]. In connection with this illustration, it is interesting to note that the Casimir force between a sphere and a plate was also recently measured in a precise torsion-pendulum arrangement [720] for separations from 0.6 to 6 micron (i.e., distances a factor of 10^3 greater than those studied with the AFM); agreement with the theory at the level of 5% was obtained.

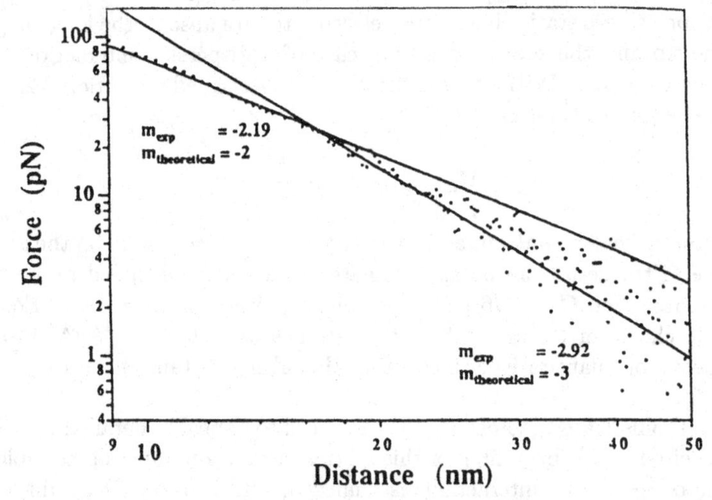

Figure 6.17: Force measurement between a Si_3N_4 AFM tip and a mica sample in air. The force is determined as the cantilever deflection multiplied by the spring constant (0.072 N/m). The lines indicate power-law fits to the data in the range 9-16 nm and in the range above 20 nm. The fall-off of the force is in agreement with that expected for the dispersion interaction of a sphere close to a flat surface. (Reprinted with permission from [719].)

Finally, we would like to mention another scanning-probe microscopy technique which makes direct use of probe and sample polarizabilities. The "scanning interferometric apertureless microscopy" [721] combines an AFM with a laser interferometer. For the tip located over a particular feature on the surface, the tip-feature system is modeled as two polarizable spheres of radius a a distance d from each other. They are illuminated by a laser beam, and the scattered light is probed by combining it with a reference beam in the interferometer. The interferometer output is very sensitive to shifts in the scattering matrix S of the illuminated system. It can be shown that these shifts are related to the dynamical polarizabilities of the AFM tip and of the surface feature:

$$\Delta S \propto \frac{\alpha_{tip}\alpha_{feature}}{(d^2 + a^2)^{3/2}}. \tag{6.117}$$

Thus as the AFM tip is scanned over the surface from one feature to another, the separation d and/or the sample polarizability $\alpha_{feature}$ change, and these changes are picked up by the interferometer. Both the phase and the amplitude variation of the scattered signal can be observed, corresponding to the real and imaginary parts of the polarizabilities. A spatial resolution of 10 Å has been demonstrated.

Chapter 7

Conclusions

In this book we have described polarizabilities, including their tensor nature, their frequency dependence, and their representation as quantum-mechanical operators. Additionally, we have carefully elucidated the relation between the polarizability and other important physical quantities such as the absorption cross section, refractive index, dielectric constant, susceptibility, and van der Waals coupling constants. The polarizabilities for different free particles (atoms, molecules, and clusters) were treated. We reviewed theoretical and experimental techniques for determining electric-dipole polarizabilities of atoms, molecules, and clusters.

From a theoretical standpoint, the increasing speed of modern computers has made possible many calculations that were once deemed too time-consuming. In addition, computers have made it possible to perform reasonable polarizability calculations using large basis sets. Calculations involving Gaussian basis wavefunctions are widespread. The application of statistical methods (such as the local density approximation) and many-body methods (including the diagrammatic approaches) to atoms, molecules, and clusters is increasing.

New experimental techniques have recently been developed for measuring both dc and ac polarizabilities. Atom interferometry has been used to measure the dc polarizability of sodium to high accuracy. A light-force technique has been used to measure the ac polarizability of a refractory element: uranium. Additionally, a universal detector has been used in conjunction with a traditional deflection technique to measure a wide range of aluminum cluster polarizabilities. Much work on polarizabilities has been recently carried out using Dispersive Fourier-Transform Spectroscopy. Simultaneously, traditional techniques still play an important role in polarizability research. Recent application of beam deflection techniques has yielded α_0 results for clusters of semiconducting materials.

We have also reviewed the importance of the polarizability to recent developments in atom trapping and cooling, optical tweezers, nanostructure fabrication, and the behavior of particles interacting via long-range forces.

Much work remains to be done. In atoms, the large value of α_2 for gallium ($\alpha_2/\alpha_0 = 5.67/8.12 = 0.7$) predicted by calculations[25] is a challenge to experiment. Also theory predicts a reduction in α_0 in going from cesium to francium due to relativistic effects [26]. This prediction, which is the reverse of the trend for the rest of the alkali atoms, remains unconfirmed experimentally.

Clusters present an ongoing challenge to both theorists and experimenters. Cluster polarizabilities are an important yardstick for understanding the transition from gas-phase particles to solids. The stubborn discrepancy between measured and calculated alkali-cluster polarizabilities calls for further experimental data and for consistent theoretical models incorporating shell structure, many-body effects, and nonjellium behavior. Semiconductor clusters display fascinating variations of the polarizability with size and with temperature, reflecting a non-trivial interplay between electronic and ionic dynamics. New theoretical ideas are needed for a thorough understanding of the physics involved. Further precise polarizability measurements on these clusters may help settle the debate over the existence and shapes of structural isomers. Techniques for handling large cluster structures such as the higher fullerenes, bucky-onions, and buckytubes are also an active area of research. As mentioned in Section 4.10.5, the large size and high polarizability of C_{60} may lead to deviations from the often- used Clausius-Mossotti relation. This would have important consequences for the procedure of extracting polarizability parameters from the dielectric constant of fullerite crystals.

Photoabsorption spectra of silver clusters [722] revealed that *d*-shell electrons play a significant role in the dipole response of these species. The dynamics of the cluster valence electrons is strongly influenced by the polarization of the ionic cores (see also [723, 724]). This observation underscores the fact that ion polarizabilities, although difficult to measure, are important quantities. They are also prominent in solid state physics, where it has long been recognized that core polarization affects the dielectric susceptibility of crystals and plasma resonances in metals [171, 725–727]. Polarizability measurements on noble-metal clusters and cluster ions would contribute to our understanding of core susceptibility effects. Two contributions to the polarizability are significant for ions: the standard contribution from oscillator strengths and energy levels as indicated in Eq. (2.5) *and* the contribution of the motion of the base ion in the ac electric field. The latter contribution is discussed (for electrons) in Section 3.1.

The measurement of tensor components of the polarizability also would be a useful guide to understanding the physical structure of many large molecules and clusters. Another important issue common to these two fields is the problem of accounting for vibrational and rotational degrees of freedom in a consistent way.

The polarizability is important in so many areas of biology, chemistry, and physics that it has not been possible to adequately review all of the significant topics in this book. No mention has been made of excited-state atom polarizabilities. Two areas where such polarizabilities are important are in Rydberg atoms and in atomic parity nonconservation experiments (see, e.g., such recent publications as [728–730] and references therein). In fact, as an outgrowth of work on parity nonconservation in atoms during the mid-1980's, Wieman and co-workers developed a clever technique for making accurate measurements of excited-state atomic polarizabilities [731]. Subsequently, Hunter and co-workers used a heterodyne method to make neat measurements in alkali-metal atoms that yielded even more accurate results [732, 733].

Level-crossing experiments can also be used to deduce excited-state atomic polarizabilities [734]. Another interesting area not covered is that of pressure-induced shifts. Collisions redistribute the oscillator strength of particles and shift the energy levels — both effects profoundly affect the shape and size of the frequency-dependent polarizability. Interesting effects on the nonlinear susceptibility (polarizability) arising from pressure-induced shifts have been studied by Bloembergen and others [735].

A very broad area that should rightly be given a much deeper treatment is that of Raman polarizabilities. In chemistry, such Raman polarizabilities are important in light scattering and molecular interaction phenomena; hence they play a role in atmospheric scattering, in pollution monitoring and control, and in understanding chemical dynamics. A good starting point for a discussion on Raman scattering is the text by Colthup *et al.* [736] and references therein (see also Ref. [72]).

Optical rotation is a useful diagnostic for the presence and behavior of specific molecules in biological and chemical systems. A good discussion of this topic can be found in one of the many textbooks that review light scattering — see, for example, Refs. [72, 125, 737].

A complete treatment of polarizabilities would include a brief review of the Stark shift of hydrogen and helium atoms. Adequate treatments of these two topics can be found in many books on quantum mechanics (e.g., [40, 275, 738]). Also missing is a summary of polarizability data on molecular hydrogen. Many published values for this simplest molecule can be found in the literature (see, e.g., [382, 506]). Alkali dimers are also of significant interest and their polarizabilities are discussed, for example, in [408, 432, 475].

Diamagnetic effects and their relation to the polarizability are treated by Barron [72]. This is just another instance where molecular polarizabilities manifest themselves in a bulk property. Another example, which was beyond the scope of this book, is the phenomenon of solvation — in particular the reaction of a polarizable fluid to a solvated ion or electron. The issue of solvation in microscopic clusters has attracted a lot of attention [153, 160].

Finally, the topic of nonlinear polarizabilities is fascinating and deserving of a book on its own. The literature on this subject is scattered, and a thorough and consistent treatment awaits the future.

To conclude, we remind the reader that the handprint of polarizabilities surround them in the natural phenomena they see everyday. The world around us utilizes polarizabilities in ways that are inspiring and beneficial. For example, polarizabilities play an important role in the twinkling of stars (light scattering), absorption of radiation by ozone and greenhouse gases, and in the properties of materials (such as why we can see through a window and not through a wall). They define van der Waals forces, which control the appearance of a drop of water on a surface, the behavior of a colloidal solution, and the functioning of the membrane of a living cell. The world is replete with physical effects that rely, at their foundation, on polarizabilities!

Appendix A

Notation and Units

In this book, we give equations and data in cgs units, unless otherwise noted. The electric dipole moment **p** is represented by many other symbols in the literature including $\boldsymbol{\mu}$, **d**, and **D**. Since both the electric field and the energy occur so frequently in this book, the notation for each has to be kept clear. The energy is represented throughout by W (for Work). The electric field is generally represented by an upper case E. Specifically, we have tried to adhere to the notation that a dc electric field is denoted by E, an ac field is given by [see Eqs. (2.41), (2.99), (3.1), (6.1), (6.20), and (6.28)]

$$\mathbf{E}(\mathbf{x},t) = \frac{1}{2}\left[\mathbf{E}_0 e^{-i\omega t} + c.c.\right], \tag{A.1}$$

where

$$\mathbf{E}_0 = \mathbf{E}_0(\mathbf{x}) = \mathcal{E}e^{i\mathbf{k}\cdot\mathbf{x}}. \tag{A.2}$$

Here $\mathbf{E}_0 = \mathbf{E}_0(\mathbf{x})$ is a space-dependent amplitude, whereas $\mathcal{E}$ is simply referred to as the amplitude.

The notation adopted by different authors for the polarizability depends on whether they want to emphasize the tensor rank of each component $\alpha_0, \alpha_1, \alpha_2$ or whether they want to use the physical names associated with objects that belong to these ranks of scalar, vector, and tensor. Hence we have the relations for the scalar part

$$\alpha_0 = \alpha_{sc} = \alpha_s = \bar{\alpha}. \tag{A.3}$$

The quantity $\bar{\alpha}$ is the average polarizability. In atomic physics, this reflects the physical fact that the scalar polarizability is simply the average of the diagonal terms of the polarizability tensor for a given magnetic sublevel and it is the average over all the magnetic sublevels for any diagonal component, that is

$$\alpha_0 = \bar{\alpha} = \frac{1}{3}\sum_{i=x,y,z}\alpha_{ii}(m) = \frac{1}{g_J}\sum_m \alpha_{zz}(m). \tag{A.4}$$

Similarly, the relations for the vector part of the polarizability are

$$\alpha_1 = \alpha_{vec} = \alpha_v, \tag{A.5}$$

and for the tensor polarizability

$$\alpha_2 = \alpha_{tens} = \alpha_t, \tag{A.6}$$

In most cases, polarizability values are quoted in terms of either 'natural' units or Systeme Internationale (SI) units. In terms of the most fundamental quantities, the polarizability has units of volume or length cubed (L^3). The length scale of atoms is angstroms (Å). Hence, in 'natural' units polarizability values are quoted in $\text{Å}^3 = 10^{-24}$ cm^3. However, since the mean radius of an electron in the ground state of hydrogen is one Bohr radius ($a_B = 0.52917725$ Å), then another 'natural' unit for polarizability would be $a_B^3 = 0.14818474\ \text{Å}^3$. Theory papers frequently state they are using atomic units (a.u.). The reader has to exercise some care in determining if the units are Å^3 or a_B^3. In this book atomic units refers to a_B^3.

Some authors prefer to use SI units. The relation between polarizability values in SI units and cgs units can be easily derived. In SI units the polarizability is defined by the definition that relates the induced dipole moment to the inducing field strength [see Eq. (2.1)], i.e.

$$[\alpha] = \frac{[\mathrm{p}]}{[\mathrm{E}]}, \tag{A.7}$$

where [...] means consider the units of the quantity. In SI units the electric dipole has units C·m (charge × length) and the field has units of V/m. Thus α has SI units of $C{\cdot}m^2/V$. To convert polarizability values from cgs to SI units we have

$$\begin{aligned} \alpha(\mathrm{SI}) &= 4\pi\epsilon_0 \times 10^{-6} \times \alpha(\mathrm{cgs}) \\ &= 1.112650056 \times 10^{-16} \alpha(\mathrm{cm}^3). \end{aligned} \tag{A.8}$$

Here the constant ϵ_0 is the electric permitivity of vacuum. The conversion is easy to see since there are 10^{-6} m^3/cm^3 and, from the potential V of a point charge Q at a distance r in SI units

$$V = \frac{Q}{4\pi\epsilon_0 r}, \tag{A.9}$$

we have that there are $4\pi\epsilon_0$ C/V per m. Hence, in terms of atomic units (a.u.) defined as a_B^3, we have

$$1\ \text{a.u.} = 0.14818474\ \text{Å}^3 = 0.16487776 \times 10^{-40} \mathrm{C\ m^2/V}. \tag{A.10}$$

There are also cases in the literature where polarizabilities are reported in terms of $\mathrm{MHz/(V/cm)^2}$, a unit more natural for Stark shift measurements at optical frequencies. The polarizability is derived from the frequency corresponding to the Stark shift of the level in an electric field, i.e.

$$\alpha = -\frac{2W}{h\mathrm{E}^2}, \tag{A.11}$$

where W/h corresponds to the frequency shift of a level placed in a field of amplitude E. The conversion between these units and cgs is $\alpha(\mathrm{cm}^3) = 5.95531 \times 10^{-16} \times \alpha$ $[\mathrm{MHz/(V/cm)^2}]$.

Appendix B

Time-Dependent Perturbation Theory

B.1 Off-Resonant Interaction with Light

B.1.1 Two-level system

Here we use time-dependent perturbation theory to explicitly derive expressions needed for polarizabilities. The particle is treated quantum-mechanically, but the radiation field is treated classically. A detailed derivation of time-dependent perturbation theory using second quantization, where both the particle *and* the radiation field are quantized can be found in the elegant presentation by Sakurai[41]. We are ultimately interested in expressing $\alpha(\omega)$ in terms of dipole matrix elements and unperturbed stationary state energies. Thus, we need to find the first-order correction to the dipole moment of a particle in a time-varying electric field.

The electromagnetic field of the light wave is represented by the vector potential

$$\mathbf{A} = \frac{1}{2}\left[\mathcal{A}e^{i(\mathbf{k}\cdot\mathbf{x}-\omega t)} + c.c.\right]. \tag{B.1}$$

In the Coulomb gauge, where the scalar potential ϕ is zero, the fields can be written

$$\mathbf{E} = -\frac{1}{c}\frac{\partial}{\partial t}\mathbf{A} \tag{B.2}$$

$$\mathbf{B} = \nabla \times \mathbf{A}, \tag{B.3}$$

where c is the speed of light, $\mathbf{E}$ is the electric field, and $\mathbf{B}$ is the magnetic field. The perturbing Hamiltonian that represents the interaction between the particle and the light is

$$\hat{H}_I(t) = -\frac{e}{2mc}\sum_j\left[\hat{\boldsymbol{p}}_j\cdot\mathbf{A} + \mathbf{A}\cdot\hat{\boldsymbol{p}}_j - \frac{e}{c}\mathbf{A}\cdot\mathbf{A}\right], \tag{B.4}$$

where $\hat{\boldsymbol{p}}_j$ corresponds to the linear momentum operator of the jth molecular electron. An equivalent way of representing the interaction Hamiltonian is

$$\hat{H}_I(t) = \hat{V}e^{-i\omega t} + h.c., \tag{B.5}$$

where $h.c.$ denotes the Hermitian conjugate of the preceding terms. Here the quantity $\hat{V}$ is

$$\hat{V} = -\frac{1}{2}\hat{\mathbf{p}} \cdot \mathbf{E}_0, \tag{B.6}$$

where the electric field is given by

$$\mathbf{E} = \frac{1}{2}\left(\mathbf{E}_0 e^{-i\omega t} + c.c.\right). \tag{B.7}$$

The quantity $\hat{\mathbf{p}}$ represents the electric dipole moment operator, which can be expressed as

$$\hat{\mathbf{p}} = \sum_n q_n \hat{\boldsymbol{r}}_n, \tag{B.8}$$

where q_n is the charge on the nth particle (electron or ion) in the molecule and $\hat{\boldsymbol{r}}_n$ is the position operator of the nth particle. We note that the two approaches, using the two different Hamiltonian forms, are equivalent (see [39] or [72]).

Let ϕ_k, $k = 1, 2, 3...$ represent the energy eigenfunctions of the unperturbed Hamiltonian $\hat{H}_0$, that is ϕ_k satisfies the time-independent Schrödinger equation for the unperturbed Hamiltonian

$$\hat{H}_0\phi_k = W_k\phi_k. \tag{B.9}$$

Here W_k is the eigenenergy corresponding to the unperturbed eigenfunction ϕ_k.

For simplicity, assume the molecule is initially in the quantum state k and the light couples this state to state l. The new molecular wavefunction can be written

$$\psi(t) = a_k(t)\phi_k e^{-iW_k t/\hbar} + a_l(t)\phi_l e^{-iW_l t/\hbar}, \tag{B.10}$$

where the coefficients a_k, a_l will be chosen so that the new wavefunction satisfies the Schrödinger equation

$$i\hbar\frac{\partial}{\partial t}\psi = (\hat{H}_0 + \hat{H}_I)\psi. \tag{B.11}$$

To solve for the time dependence of the amplitudes a_l in this interaction picture, substitute the wavefunction Eq. (B.10) into the Schrödinger equation Eq. (B.11) and find

$$i\hbar\dot{a}_l = V_{lk}\, e^{i(\omega_{lk}-\omega)t} + V_{lk}^{\dagger}\, e^{i(\omega_{lk}+\omega)t}. \tag{B.12}$$

The angular transition frequencies are given by the usual Bohr formula

$$\hbar\omega_{lk} = W_l - W_k. \tag{B.13}$$

Suppose that at time t_0 the molecule was entirely in the state ϕ_k, so that $|a_k(t_0)| = 1$ and $|a_l(t_0)| = 0$. Then the first-order solution to Eq. (B.12) is

$$a_l(t) = -a_k(t_0)\left\{V_{lk}\left[\frac{e^{i(\omega_{lk}-\omega)t} - e^{i(\omega_{lk}-\omega)t_0}}{\hbar(\omega_{lk}-\omega)}\right] + V_{lk}^{\dagger}\left[\frac{e^{i(\omega_{lk}+\omega)t} - e^{i(\omega_{lk}+\omega)t_0}}{\hbar(\omega_{lk}+\omega)}\right]\right\}. \tag{B.14}$$

We are interested in the matrix element of the electric-dipole operator $\hat{\mathbf{p}}$ while the system is experiencing the interaction represented by the perturbing Hamiltonian.

Thus we will derive a result for a general operator $\hat{G}$ that is independent of time $\hat{G} = \hat{G}(t)$, and we shall discuss the specific case of the dipole operator at the very end. The matrix element of an operator $\hat{G}$ for a system described by the perturbed wavefunction $\psi(t)$ [Eq (B.10)] is given by

$$\int \psi^* \hat{G} \psi \; d^3\boldsymbol{x} = |a_k|^2 G_{kk} + a_k^* a_l G_{kl} e^{i\omega_{kl} t} + a_k a_l^* G_{lk} e^{i\omega_{lk} t} + |a_l|^2 G_{ll}, \tag{B.15}$$

where

$$G_{kl} = \int \phi_k^* \hat{G} \phi_l \; d^3\boldsymbol{x}, \tag{B.16}$$

is time-independent and $d^3\boldsymbol{x}$ is the volume element. Note that we are only interested in calculating the first-order corrections to the matrix elements of an operator due to the perturbing Hamiltonian. The first term in Eq. (B.15) is the lowest, zeroth-order term and it is therefore independent of the perturbing Hamiltonian (which only occurs in the coefficients a_l). The last term in Eq. (B.15) is second-order in the perturbing Hamiltonian. Hence only the middle two terms contribute and we can substitute $|a_k(t_0)| = |a_k(t)| = 1$ and from Eq. (B.14) for $a_l(t)$. Making these substitutions and retaining only those terms harmonic in (ωt) yields the result

$$\begin{aligned} \int \psi^* \hat{G} \psi \; d^3\boldsymbol{x} \Big|_{1st-order} &= -\left[e^{-i\omega t} \left\{ \frac{G_{kl} V_{lk}}{\hbar(\omega_{lk} - \omega)} + \frac{G_{lk} V_{kl}}{\hbar(\omega_{lk} + \omega)} \right\} \right. \\ &\left. + \, e^{i\omega t} \left\{ \frac{G_{kl} V_{kl}^*}{\hbar(\omega_{lk} + \omega)} + \frac{G_{lk} V_{lk}^*}{\hbar(\omega_{lk} - \omega)} \right\} \right]. \end{aligned} \tag{B.17}$$

B.1.2 Multi-level system

We now consider a multilevel system, where the perturbing Hamiltonian couples the quantum level k to many quantum levels l. In this case, the wavefunction is described by a sum over many quantum numbers and their corresponding wavefunctions and time-dependent coefficients, so that Eq (B.10) becomes

$$\psi_k(t) = \sum_l a_{kl}(t) \phi_l e^{-iW_l t/\hbar}, \tag{B.18}$$

Under these more general considerations we can find the new coefficients in exactly the same manner as before

$$a_{kl}(t) = -a_{kk}(t_0) \left\{ V_{lk} \left[\frac{e^{i(\omega_{lk}-\omega)t} - e^{i(\omega_{lk}-\omega)t_0}}{\hbar(\omega_{lk} - \omega)} \right] + V_{lk}^\dagger \left[\frac{e^{i(\omega_{lk}+\omega)t} - e^{i(\omega_{lk}+\omega)t_0}}{\hbar(\omega_{lk} + \omega)} \right] \right\}. \tag{B.19}$$

Similarly an expression for the first-order correction to the matrix element of an operator can be found. Generalizing the matrix element expression to now include *off-diagonal* matrix elements, we find

$$\begin{aligned} \int \psi_m^* \hat{G} \psi_k \; d^3\boldsymbol{x} \Big|_{1st-order} &= -e^{i\omega_{mk} t} \sum_l \left\{ e^{-i\omega t} \left[\frac{G_{ml} V_{lk}}{\hbar(\omega_{lk} - \omega)} + \frac{G_{lk} V_{ml}}{\hbar(\omega_{lm} + \omega)} \right] \right. \\ &\left. + \; e^{i\omega t} \left[\frac{G_{ml} V_{kl}^*}{\hbar(\omega_{lk} + \omega)} + \frac{G_{lk} V_{ml}^*}{\hbar(\omega_{lm} - \omega)} \right] \right\}. \end{aligned} \tag{B.20}$$

The previous case, where we only considered diagonal terms, is the special case where $m = k$.

B.2 Near-Resonant Interaction with Light

So far, the results in this appendix have ignored the case of resonance. Here the frequency of the light field is close to the difference in Bohr frequencies between two levels, e.g. when $\omega \simeq \omega_{lk}$ in Eq. (B.19) or in Eq. (B.20). In this case, a denominator gets very small, making the resulting quantity unphysically large. Obviously the amplitude $a_{lk}(t)$ must remain finite. The amplitude of an excited state decays, in the absence of a radiation field, according to $a_{lk} \approx e^{-\Gamma_l t/2\hbar}$, where Γ_l represents a decay rate of level l due to spontaneous emission or due to collisions. Although the original Schrödinger equation does not contain a Hamiltonian that can account for such a decay, such a term is accommodated by adding an anti-Hermitian part to the Hamiltonian. Recall that an anti-Hermitian operator obeys the rule

$$H_{ij}^* = -H_{ij}. \tag{B.21}$$

For example, Eq. (B.12) would be modified to

$$i\hbar\dot{a}_l = V_{lk}\, e^{i(\omega_{lk}-\omega)t} + V_{lk}^\dagger\, e^{i(\omega_{lk}+\omega)t} - i\Gamma_l/2a_l. \tag{B.22}$$

In this case, the dominant contribution to the perturbed wavefunction would come from the intermediate quantum state involved in the near-resonance. The effect on matrix elements is to cause the substitution for the energy of the intermediate state (excited state)

$$W_l \rightarrow W_l - i\Gamma_l/2\hbar. \tag{B.23}$$

The new terms arising due to the $-i\Gamma_l/2\hbar$ substitution correspond physically to absorption of photons or light from the radiation field. The molecule, left in an excited state, can relax via collisions or due to spontaneous emission.

Appendix C

Derivation of the Kramers-Krönig Relations

A neutral particle located in an electric field will respond by setting up an induced dipole moment. The dipole moment p at time t will only depend on the electric field E applied at *earlier times*, i.e.

$$\mathrm{p}(t) = -\int_0^\infty d\tau \alpha(\tau) \mathrm{E}(t-\tau). \tag{C.1}$$

Since p and E are real, this implies that the response function $\alpha(\tau)$ must also be real. In the weak-field limit the electronic response exhibits *linearity* and *causality*. Time-dependent phenomena are conveniently expressed using Fourier methods where time-dependent functions are expressed as sums of amplitudes of monochromatic waves. For example, the part of $\mathrm{E}(t)$ oscillating at frequency ω is expressed as

$$\tfrac{1}{2}\left[\mathrm{E}(\omega)e^{-i\omega t} + \mathrm{E}^*(\omega)e^{i\omega t}\right].$$

Expressing the Fourier components of the dipole moment p in a similar way and using Eq. (C.1), we can conclude

$$\mathrm{p}(\omega) = \alpha(\omega)\mathrm{E}(\omega), \tag{C.2}$$

where the frequency-dependent polarizability is given by

$$\alpha(\omega) = \int_0^\infty \alpha(t) e^{i\omega t} dt. \tag{C.3}$$

Since $\alpha(t)$ is real, it is clear from Eq. (C.3) that $\alpha(\omega)$ is complex and we can write

$$\alpha(\omega) = \alpha'(\omega) + i\alpha''(\omega). \tag{C.4}$$

However, for mathematical convenience we will assume that the frequency is also complex $\omega = \omega' + i\omega''$. From Eq. (C.3) it is clear that

$$\alpha^*(\omega) = \alpha(-\omega^*). \tag{C.5}$$

The integral given in Eq. (C.3) disappears as $\omega'' \to \infty$ due to the rapid exponential decay of $e^{-\omega'' t}$. The integral also approaches zero as $\omega' \to \pm\infty$ due to the very rapid oscillations of the term $e^{i\omega' t}$. These rapid oscillations will cause the integral to average to zero. Thus if we restrict our attention to the upper half of the complex frequency plane, then

$$\lim_{\omega\to\infty} \alpha(\omega) = 0. \tag{C.6}$$

From the definition Eq. (C.3), we can conclude that $\alpha(\omega)$ is finite everywhere in the upper half of the complex frequency plane. This is also reflected in the expressions for $\alpha(\omega)$ given in Section (2.5), where poles only occur in the lower-half of the frequency plane at $\omega = \pm\omega_{lk} - i\Gamma_l/2$. The location of these poles in the complex frequency plane is indicated in Fig. C.1(a).

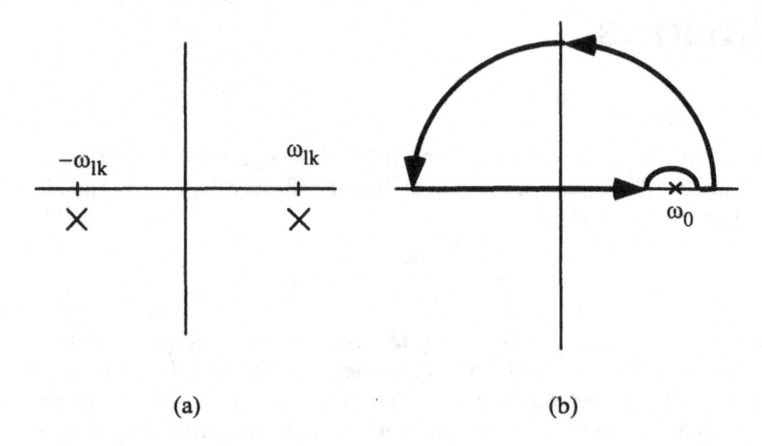

Figure C.1: In (a), the poles of the polarizability $\alpha(\omega)$ are shown. In (b), there is a sketch of the integration contour for performing the integral in Eq. (C.7).

For any *real* frequency ω we have

$$\oint \frac{\alpha(\omega)}{\omega - \omega_0} d\omega = 0. \tag{C.7}$$

The path of integration is sketched in Fig. C.1(b), where we have avoided the pole at ω_0 by taking a small semicircular detour around it.

The large semicircle contributes nothing to the integral since the polarizability approaches zero as the radius of the semicircle approaches infinity, i.e. $\alpha(\omega) \to 0$ as $\omega \to \infty$. For the small semicircle of radius ρ, as $\rho \to 0$ we can replace $\alpha(\omega)$ inside the integrand with $\alpha(\omega_0)$ and remove it from the integral. From this procedure we obtain

$$\int_{-\infty}^{\omega_0-\rho} \frac{\alpha(\omega)}{\omega - \omega_0} d\omega + \alpha(\omega_0) \int_{\omega_0-\rho}^{\omega_0+\rho} \frac{1}{\omega - \omega_0} d\omega + \int_{\omega_0+\rho}^{\infty} \frac{\alpha(\omega)}{\omega - \omega_0} d\omega = 0. \tag{C.8}$$

From the Cauchy residue theorem, it can be shown that the integral around the small semicircle is $-i\pi$ and then we can write

$$\mathcal{P} \int \frac{\alpha(\omega)}{\omega - \omega_0} d\omega = i\pi\alpha(\omega_0), \tag{C.9}$$

where the principal value of an integral is defined as

$$\mathcal{P}\int \frac{\alpha(\omega)}{\omega-\omega_0}d\omega = \lim_{\rho\to 0}\left[\int_{-\infty}^{\omega_0-\rho}\frac{\alpha(\omega)}{\omega-\omega_0}d\omega + \int_{\omega_0+\rho}^{\infty}\frac{\alpha(\omega)}{\omega-\omega_0}d\omega\right]. \tag{C.10}$$

Note that the path of integration is along the real axis. Using the expression Eq. (C.9), the real and imaginary parts of the polarizability are related by

$$\alpha'(\omega_0) = \frac{1}{\pi}\mathcal{P}\int_{-\infty}^{+\infty}\frac{\alpha''(\omega)}{\omega-\omega_0}d\omega, \tag{C.11}$$

$$\alpha''(\omega_0) = -\frac{1}{\pi}\mathcal{P}\int_{-\infty}^{+\infty}\frac{\alpha'(\omega)}{\omega-\omega_0}d\omega. \tag{C.12}$$

The formulas Eq. (C.11) and Eq. (C.12) are the Kramers-Krönig relations. The physical implications of these results are discussed in Section 2.5.

To see why the Raman polarizabilities do not satisfy the Kramers-Krönig relations, consider the response of a particle that has an electric field applied to it that is monochromatic at frequency ω. The resulting dipole moment was given in Eq. (2.57) of Section 2.4 and is reproduced here:

$$(\mathrm{p}_i)_{mk}(t) = \frac{1}{2}(\alpha_{ij})_{mk}\mathcal{E}_j\, e^{-i(\omega_{km}+\omega)t} + c.c. \tag{C.13}$$

Here the induced dipole oscillates at a frequency $\omega_{km}+\omega$ in response to a driving electric field at frequency ω. Causality still holds but *linearity* does not. The dipole is not oscillating at the driving field but gets shifted in frequency. Hence there is no linear relationship between the applied field and the induced dipole, like the general form given in Eq. (C.1).

Bibliography

[1] K.D.Bonin and M.A.Kadar-Kallen, Int. J. Mod. Phys. B **24**, 3313 (1994).

[2] T.M.Miller, in *CRC Handbook of Chemistry and Physics*, 77th ed., edited by D.R.Lide (CRC Press, Boca Raton, 1996), pp. 10–199 to 10–214.

[3] P. Lambin, A.A.Lucas, and J.P.Vigneron, Phys. Rev. B **46**, 1794 (1992).

[4] G.G.Stokes, Phil. Trans. 463 (1852).

[5] F.-P. Roux, Ann. de Chimie et de Physique, Ser. 3 **61**, 285 (1861).

[6] P. Kundt, Ann. Phys. und Chemie **142**, 163 (1871).

[7] P. Kundt, Ann. Phys. und Chemie **143**, 149 (1871).

[8] P. Kundt, Ann. Phys. und Chemie **143**, 259 (1871).

[9] J.C.Maxwell, Cambridge Mathematical Tripos Examination(1869), see the discussion in W. Wilson, *A Hundred Years of Physics* (Duckworth, London, 1950), p.98.

[10] Sellmeier, see the discussion in W. Wilson, *A Hundred Years of Physics* (Duckworth, London, 1950), p.98.

[11] W. Voigt, Ann. Phys. **iv**, 197 (1901).

[12] A. L. Surdo, Atti Reale Accademia Lincei **22**, 664 (1913).

[13] J. Stark, Sitz. Akad. Wiss. Berlin **47**, 932 (1913).

[14] P.S.Epstein, Phys. ZS **xvii**, 148 (1916).

[15] P.S.Epstein, Ann. Phys. **1**, 489 (1916).

[16] K. Schwarzschild, Berlin Sitz. 548 (April 1916).

[17] H.A.Kramers, Zeit. f. Phys. **3**, 199 (1920).

[18] E. Hochheim, Ver. Deutsch. Phys. Ges. **40**, 446 (1908).

[19] C. Cuthbertson and M. Cuthbertson, Proc. Roy. Soc. London **84A**, 13 (1911).

[20] J. Koch, Nov. Act. Soc. Ups. **2**, 61 (1909).

[21] H. Scheffers and J. Stark, Phys. Z. **35**, 625 (1934).

[22] M.A.Kadar-Kallen and K.D.Bonin, Phys. Rev. Lett. **72**, 828 (1994).

[23] R.Schäfer, J.Woenckhaus, J.A.Becker, and F.Hensel, Z. Naturforsch. **50a**, 445 (1995).

[24] D. Goebel, U. Hohm, and G. Maroulis, Phys. Rev. A **54**, 1973 (1996).

[25] T.M.Miller and B. Bederson, Adv. At. Mol. Phys. **13**, 1 (1977).

[26] T.M.Miller and B. Bederson, Adv. At. Mol. Phys. **25**, 37 (1989).

[27] A.D.Buckingham, in *Physical Chemistry, Series 1*, edited by D. Ramsay (Butterworths, London, 1972), pp. 3–73 to 3–117.

[28] A.D.Buckingham, in *Intermolecular Forces*, edited by J.O.Hirschfelder (Wiley, New York, 1967), pp. 107–142.

[29] M.P.Bogaard and B.J.Orr, in *Physical Chemistry, Series 2*, edited by A. Buckingham (Butterworths, London, 1975), pp. 2–149 to 2–194.

[30] R.R.Teachout and R.T.Pack, At. Data **3**, 195 (1971).

[31] P. W. Atkins, *Molecular Quantum Mechanics*, 2 ed. (Oxford University Press, Oxford, 1983).

[32] C. Böttcher, O. van Belle, P. Bordewijk, and A. Rip, *Theory of Electric Polarization*, 2 ed. (Elsevier, Amsterdam, 1973), Vol. 1.

[33] C. Böttcher and P. Bordewijk, *Theory of Electric Polarization*, 2 ed. (Elsevier, Amsterdam, 1978), Vol. 2.

[34] J. Bertin and J. Loeb, *Experimental and Theoretical Aspects of Induced Polarization* (Gebrüder Borntraeger, Berlin, 1976), Vol. 1 & 2.

[35] S.H.Ward, in *Methods of Experimental Physics: Geophysics*, edited by C. Sammis and T. Henyey (Academic Press, Orlando, 1987), Vol. 24 - B, pp. 265–376.

[36] M. Weissbluth, *Atoms and Molecules* (Academic Press, New York, 1978).

[37] W. Happer and B.S.Mathur, Phys. Rev. **163**, 12 (1967).

[38] G. Placzek, in *Handbuch der Radiologie*, edited by A. V. V. Erich Marx, Leipzig (UCRL-Trans-526(L), National Technical Information Service, U.S. Atomic Energy Commission, Springfield, VA 22151, 1934), pp. 2–209 to 2–374.

[39] V.B.Berestetskii, E.M.Lifshitz, and L.P.Pitaevskii, *Quantum Electrodynamics*, 2nd ed. (Pergamon Press, Oxford, 1980), pp. 221–246.

[40] A.S.Davydov, *Quantum Mechanics*, 2nd ed. (Pergamon Press, Oxford, 1976).

[41] J.J.Sakurai, *Advanced Quantum Mechanics* (Addison-Wesley, Reading, 1967), pp. 53–57.

[42] O. Klein, Z. Physik **41**, 407 (1927).

[43] G. Placzek, in *Handbuch der Radiologie*, edited by A. V. V. Erich Marx, Leipzig (UCRL-Trans-526(L), National Technical Information Service, U.S. Atomic Energy Commission, Springfield, VA 22151, 1934), p. 29.

[44] R. Krönig, J. Opt. Soc. Am. **12**, 547 (1926).

[45] H.A.Kramers, Att. Congr. dei Fisici Como **12**, 545 (1927).

[46] H.A.Kramers, *Collected Scientific Papers* (North Holland, Amsterdam, 1956), Vol. 12, p. 333.

[47] W. Demtröder, *Laser Spectroscopy* (Springer-Verlag, New York, 1981).

[48] P.N.Butcher and D. Cotter, *Elements of Nonlinear Optics* (Cambridge University Press, Cambridge, 1990), pp. 314–315.

[49] D.J.Caldwell and H.Eyring, *Theory of Optical Activity* (Wiley, New York, 1971).

[50] P. Roman, *Advanced Quantum Theory* (Addison-Wesley, Reading, 1965).

[51] H.M.Nussenzveig, *Causality and Dispersion Relations* (Academic, New York, 1972).

[52] N. Bloembergen, *Nonlinear Optics* (Benjamin Addison Wesley, Reading, 1965), pp. 45–46.

[53] M. Faraday, Phil. Trans. **128**, 1 (1837/1838).

[54] M. Faraday, Phil. Trans. **128**, 79 (1837/1838).

[55] M. Faraday, Phil. Trans. 265 (1837/1838).

[56] J.D.Jackson, *Classical Electrodynamics* (Wiley, New York, 1975).

[57] F.F.Cap, *Handbook on Plasma Instabilities* (Academic Press, New York, 1976), Vol. 1-3, see pp.59 of Vol. 1 for a discussion of the nonrelativistic pondermotive force; Volume 3 includes a discussion of the relativistic pondermotive force. It is called the Miller force in the reference cited here.

[58] J. Schmiedmayer, P. Riehs, J.A.Harvey, and N.W.Hill, Phys. Rev. Lett. **66**, 1015 (1991).

[59] F.J.Federspiel *et al.*, Phys. Rev. Lett. **67**, 1511 (1991).

[60] V.A.Petrun'kin, Sov. J. Part. Nuc. **12**, 278 (1981).

[61] J.L.Friar, in *Proceedings of the Workshop on Electron-Nucleus Scattering*, edited by A. Fanbrocini *et al.* (World Scientific, Singapore, 1989).

[62] V. Bernard, N. Kaiser, and U.-G. Meissner, Phys. Rev. Lett. **67**, 1515 (1991).

[63] V. Bernard, N. Kaiser, and U.-G. Meissner, Nucl. Phys. B **373**, 346 (1992).

[64] W. Broniowski, M.K.Banerjee, and T.D.Cohen, Phys. Lett. B **283**, 22 (1992).

[65] O.L.Zhizhimov and I.B.Khriplovich, Sov. Phys. JETP **55**, 4 (1982).

[66] B.W.Shore, *Theory of Coherent Atomic Excitation* (Wiley, New York, 1990), p. 1322 to 1332.

[67] D.M.Brink and G.R.Satchler, *Angular Momentum* (Oxford University Press, Oxford, 1994).

[68] J.R.P.Angel and P.G.H.Sandars, Proc. Roy. Soc. A. **305**, 125 (1968).

[69] S. Bhagavantam and D. Suryanarayana, Acta Cryst. **2**, 21 (1949).

[70] H.A.Jahn, Acta Cryst. **2**, 30 (1949).

[71] F. Diederich and R.L.Whetten, Acc. Chem. Res. **25**, 119 (1992).

[72] L.D.Barron, *Molecular Light Scattering and Optical Activity* (Cambridge Univ. Press, Cambridge, 1982), pp. 94–95.

[73] K.J.Miller, J. Am. Chem. Soc. **112**, 8543 (1990).

[74] M. Breiger, Chem. Phys. **89**, 275 (1984).

[75] D.M.Bishop, Rev. Mod. Phys. **62**, 343 (1990).

[76] M. Eckert and G. Zundel, J. Phys. Chem. **91**, 5170 (1987).

[77] M. Eckert and G. Zundel, J. Phys. Chem. **92**, 7016 (1988).

[78] W.Q.Cai *et al.*, Phys.Rev. A **36**, 4722 (1987).

[79] Z. Lu and D.P.Shelton, J. Chem. Phys. **87**, 1967 (1987).

[80] D.B.Whitehouse and A.D.Buckingham, Chem. Phys. Lett. **207**, 332 (1993).

[81] N.J.Bridge and A.D.Buckingham, Proc. Roy. Soc. A **295**, 334 (1966).

[82] A.D.Buckingham, Disc. Faraday Soc. **40**, 171 (1965).

[83] P. Debye, *Polar Molecules* (Dover, New York, 1929).

[84] G.Arfken, *Mathematical Methods for Physicists* (Academic Press, New York, 1995).

[85] G. Herzberg, *The Spectra and Srtucture of Simple Free Radicals* (Cornell University Press, Ithaca, 1971).

[86] G.W.King, *Spectroscopy and Molecular Structure* (Holt, Rinehart, and Winston, New York, 1964).

[87] M. Mizushima, *The Theory of Rotating Diatomic Molecules* (Wiley, New York, 1975).

[88] J. V. Vleck, *The Theory of Electric and Magnetic Susceptibilities* (Oxford University Press, Oxford, 1932).

[89] G. Herzberg, *Molecular Spectra and Molecular Structure, II. Infrared and Raman Spectra of Polyatomic Molecules* (Van Nostrand Reinhold, New York, 1945).

[90] S. Golden and J. E.B.Wilson, J. Chem. Phys. **16**, 669 (1948).

[91] C.H.Townes and A.L.Schawlow, *Microwave Spectroscopy* (McGraw-Hill, New York, 1955).

[92] B.S.Ray, Z. Physik **78**, 74 (1932).

[93] P.C.Cross, R.M.Hanier, and G.W.King, J. Chem. Phys. **12**, 210 (1944).

[94] W.F.Murphy, J. Raman Spectrosc. **11**, 339 (1981).

[95] G.W.King, R.M.Hanier, and P.C.Cross, J. Chem. Phys. **11**, 27 (1943).

[96] H.C.Allen and P.C.Cross, *Molecular VibRotors* (Wiley, New York, 1962).

[97] J.M.Fernandez-Sanchez and W.F.Murphy, J. Mol. Spectrosc. **156**, 444 (1992).

[98] M.T.Zhao, B.P.Singh, and P.N.Prasad, J. Chem. Phys. **89**, 5535 (1988).

[99] B.P.Singh and P.N.Prasad, J. Opt. Soc. Am. B **5**, 453 (1988).

[100] H.Thienpont *et al.*, Phys. Rev. Lett. **65**, 2141 (1990).

[101] B. Champagne, D.H.Mosley, and J.-M.Andre, J. Chem. Phys. **100**, 2034 (1994).

[102] H.A.Pohl, Int. J. Quantum Chem., Quant. Biol. Symp. **71**, 411 (1980).

[103] W.G.J.Hol, P. Duijnen, and H.J.C.Berendsen, Nature **273**, 443 (1978).

[104] K.R.Shoemaker *et al.*, Nature **326**, 563 (1987).

[105] H. DeVoe, J. Chem. Phys. **41**, 393 (1964).

[106] H. DeVoe, J. Chem. Phys. **43**, 3199 (1965).

[107] W. Rhodes and M. Chase, Rev. Mod. Phys. **39**, 348 (1967).

[108] C.L.Cech, W. Hug, and J. I. Tinoco, Biopolymers **15**, 131 (1976).

[109] A.J.Sadlej, Collection Czech. Chem. Commun. **53 Part I**, 1995 (1988).

[110] A.J.Sadlej, Theor. Chim. Acta **79**, 123 (1991).

[111] A.I.Vogel, J. Chem. Soc. **1833**, (1948).

[112] K.G.Denbigh, Trans. Faraday Soc. **36**, 936 (1940).

[113] C.G.LeFevre and R.J.W.LeFevre, Rev. Pure Appl. Chem. **5**, 261 (1955).

[114] J. Applequist, J.R.Carl, and K.-K. Fung, J. Amer. Chem. Soc. **94**, 2952 (1972).

[115] K.J.Miller and J.A.Savchik, J. Am. Chem. Soc. **101**, 7206 (1979).

[116] K.J.Miller, J. Am. Chem. Soc. **112**, 8533 (1990).

[117] Y.K.Kang and M.S.Jhon, Theoret. Chim. Acta **61**, 41 (1982).

[118] *Organic Materials for Nonlinear Optics* (Academic, New York, 1987), Vol. I-II.

[119] P. Prasad and D.J.Williams, *Introduction to Nonlinear Effects in Molecules and Polymers* (Wiley, New York, 1991).

[120] L.Salem, *The Molecular Orbital Theory of Conjugated Systems* (Benjamin, New York, 1966).

[121] G.J.Visser, G.J.Heeres, J.Wolters, and A. Vos, Acta Crystallogr Sect. B **24**, 1671 (1968).

[122] F. van Blohius *et al.*, Synth. Metals **15**, 353 (1986).

[123] G. Barbarella, M. Zambianchi, A. Bongini, and L. Antolini, Adv. Mater. **4**, 282 (1992).

[124] B.J.Caldwell and H. Eyring, *The Theory of Optical Activity* (Wiley Interscience, New York, 1971).

[125] S. F. Mason, *Molecular Optical Activity and the Chiral Discriminations* (Cambridge University Press, Cambridge, 1982).

[126] R.A.Goldbeck and D.S.Kliger, Spectrum **8**, 13 (1995).

[127] H. Ito and Y.J.I'Haya, Chem. Phys. Lett. **142**, 25 (1987).

[128] D.B.Shapiro *et al.*, Appl. Opt. **33**, 5733 (1994).

[129] D.B.Shapiro, P.G.Hull, J.E.Hearst, and A.J.Hunt, J. Chem. Phys. **101**, 4214 (1994).

[130] C.L.Cech and J. I. Tinoco, Biopolymers **16**, 43 (1977).

[131] H. Ito and Y.J.I'Haya, J. Chem. Phys. **77**, 6270 (1982).

[132] J.C.Sutherland, K.P.Griffin, P.C.Keck, and P.Z.Takacs, Proc. Nat. Acad. Sci. US **78**, 4801 (1981).

[133] M.J.Bethe, Biopolymers **24**, 519 (1986).

[134] S. Goto, Biopolymers **23**, 2211 (1984).

[135] L. Chinsky *et al.*, Biopolymers **23**, 1931 (1984).

[136] M.-F.Hacques and C. Marion, Biopolymers **25**, 2281 (1986).

[137] H. Basch *et al.*, Chem. Phys. Lett. **163**, 514 (1989).

[138] M.G.Papadpoulos and J. Waite, J. Mol. Structure **170**, 189 (1988).

[139] C.J.F.Böttcher, *Theory of Electric Polarization*, 1st ed. (Elsevier, Amsterdam, 1952), p. 289.

[140] M.-F.Hacques and C. Marion, Biochem. Biophys. Res. Comm. **166**, 1140 (1990).

[141] C. Marion, J. Biomol. Struct. Dynam. **2**, 303 (1984).

[142] C. Marion *et al.*, Biophys. Chem. **22**, 53 (1985).

[143] C. Marion, B. Perot, B. Roux, and J.C.Bernengo, Makromol. Chem. **185**, 1665 (1984).

[144] S. Diekmann *et al.*, Biophys. Chem. **15**, 157 (1982).

[145] D. Porschke, Biopolymers **28**, 1383 (1989).

[146] S. Diekmann and D. Porschke, Biophys. Chem. **26**, 207 (1987).

[147] J.A.A.J.Perenboom, P.Wyder, and F.Meier, Phys. Rep. **78**, 171 (1981).

[148] U.Kreibig and L.Genzel, Surf. Sci. **156**, 678 (1985).

[149] W.P.Halperin, Rev. Mod. Phys. **58**, 533 (1986).

[150] E.W.Becker, Z. Phys. D **3**, 101 (1986).

[151] M.Kappes and S.Leutwyler, in *Atomic and Molecular Beam Methods*, edited by G.Scoles (Oxford, New York, 1988), Vol. 1.

[152] W. Heer, Rev. Mod. Phys. **65**, 611 (1993).

[153] *Clusters of Atoms and Molecules*, edited by H.Haberland (Springer, Berlin, 1994).

[154] T.P.Martin, Phys. Rep. **273**, 199 (1996).

[155] S.Sugano, *Microcluster Physics* (Springer, Berlin, 1991).

[156] P.Joyes, *Les Agregats Inorganiques Elementares* (Les editions de Physique, Les Ulis, 1990).

[157] M.Moskowitz, Annu. Rev. Phys. Chem. **42**, 465 (1991).

[158] U.Kreibig and M.Vollmer, *Optical Properties of Metal Clusters* (Springer, Berlin, 1995).

[159] M.S.Dresselhaus, G.Dresselhaus, and P.C.Eklund, *Science of Fullerenes and Carbon Nanotubes* (Academic, San Diego, 1996).

[160] A. Castleman, Jr. and K. Bowen, Jr., J. Phys. Chem. **100**, 12911 (1996).

[161] *Atomic and Molecular Clusters*, edited by E.R.Bernstein (Elsevier, Amsterdam, 1990).

[162] Phase Transitions **24-26**, (1990).

[163] *Advances in Metal and Semiconductor Clusters*, edited by M.A.Duncan (JAI Press, Greenwich, 1993-95), Vol. 1-3.

[164] *Proceedings of International Symposia on Small Particles and Inorganic Clusters (ISSPIC)*, edited by K.H.Bennemann and J.Koutecký (Surf. Sci., **156**, 1985).

[165] *Proceedings of International Symposia on Small Particles and Inorganic Clusters (ISSPIC)*, edited by C.Chapon, M.F.Gillet, and C.R.Henry (Z. Phys. D, **12**, 1989).

[166] *Proceedings of International Symposia on Small Particles and Inorganic Clusters (ISSPIC)*, edited by O.Echt and E.Recknagel (Z. Phys. D, **19-20**, 1991).

[167] *Proceedings of International Symposia on Small Particles and Inorganic Clusters (ISSPIC)*, edited by R.S.Berry, J.Burdett, and A.W.Castleman (Z. Phys. D, **26**, 1993).

[168] *Proceedings of International Symposia on Small Particles and Inorganic Clusters (ISSPIC)*, edited by Y. Nishina and S. Sugano (Surf. Rev. Lett., **3**, 1996).

[169] *Physics and Chemistry of Finite Systems: From Clusters to Crystals*, edited by P.Jena, S.N.Khanna, and B.K.Rao (Kluwer Academic, Dordrecht, 1992).

[170] N.W.Ashcroft and N.D.Mermin, *Solid State Physics* (Holt, Rinehart and Winston, New York, 1976).

[171] W.A.Harrison, *Electronic Structure and the Properties of Solids: The Physics of the Chemical Bond* (Dover, New York, 1989).

[172] W. Heer, W.D.Knight, M.Y.Chou, and M.L.Cohen, in *Solid State Physics*, edited by H.Ehrenreich and D.Turnbull (Academic, New York, 1987), Vol. 40.

[173] S.Bjørnholm, Comments At. Mol. Phys. **31**, 137 (1995).

[174] V.V.Kresin and W.D.Knight, Z. Phys. Chem. **203**, (1998).

[175] W.D.Knight, in *The Chemical Physics of Atomic and Molecular Clusters*, edited by G.Scoles (North-Holland, Amsterdam, 1990), p. 413.

[176] K.Clemenger, Phys. Rev. B **32**, 1359 (1985), this ellipsoidal model is based on a similar approach to nuclear structure, see Ref. [739].

[177] W.D.Knight *et al.*, Phys. Rev. Lett. **52**, 2141 (1984).

[178] C.Bréchignac and Ph.Cahuzac, Comments At. Mol. Phys. **31**, (1995), special issue, *Nuclear Aspects of Simple Metal Clusters.*

[179] T.P.Martin *et al.*, Chem. Phys. Lett. **186**, 53 (1991).

[180] C. Bréchignac *et al.*, J. Chem. Phys. **93**, 7449 (1990).

[181] M. Seidl, K.-H. Meiwes-Broer, and M. Brack, J. Chem. Phys. **95**, 1295 (1991).

[182] C.L.Pettiette *et al.*, J. Chem. Phys. **88**, 5377 (1988).

[183] J.G.Eaton *et al.*, in *Nuclear Physics Concepts in the Study of Atomic Cluster Physics*, Vol. 404 of *Lecture Notes in Physics*, edited by R.Schmidt, H.O.Lutz, and R.Dreizler (Springer, Berlin, 1992).

[184] U. Näher *et al.*, Phys. Rep. **285**, 245 (1997).

[185] V.V.Kresin, Phys. Rep. **220**, 1 (1992).

[186] V.O.Nesterenko, Sov. J. Part. Nucl. **23**, 726 (1992).

[187] M.Brack, Rev. Mod. Phys. **65**, 677 (1993).

[188] V. Bonačić-Koutecký, P. Fantucci, and J. Koutecký, Phys. Rev. B **37**, 4369 (1988).

[189] L.D.Landau, E.M.Lifshitz, and L.P.Pitaevskii, *Electrodynamics of Continuous Media*, 2nd ed. (Pergamon, Oxford, 1984).

[190] J.A.Osborn, Phys. Rev. **67**, 351 (1945).

[191] E.C.Stoner, Philos. Mag. **36**, 803 (1945).

[192] L.P.Gor'kov and G.M.Eliashberg, Sov. Phys. JETP **21**, 940 (1965).

[193] A.A.Lushnikov, V.V.Maksimenko, and A.J.Simonov, in *Electromagnetic Surface Modes*, edited by A. Boardman (Wiley, New York, 1982).

[194] R.Dupree and M.A.Smithard, J. Phys. C **5**, 408 (1972).

[195] F.Meier and P.Wyder, Phys. Lett. A **39**, 51 (1972).

[196] A.A.Lushnikov and A.J.Simonov, Phys.Lett. A **41**, 45 (1973).

[197] M.J.Rice, W.R.Schneider, and S.Strässler, Phys. Rev. B **8**, 474 (1973).

[198] K. Clemenger, Ph.D. thesis, University of California, Berkeley, 1985.

[199] W.D.Knight, K.Clemenger, W. Heer, and W.A.Saunders, Phys. Rev. B **31**, 2539 (1985).

[200] D.Pines, *Elementary Excitations in Solids* (Benjamin, Reading, 1963).

[201] *Theory of the Inhomogeneous Electron Gas*, edited by S.Lundqvist and N.H.March (Plenum, New York, 1983).

[202] S.Bjørnholm *et al.*, Phys. Rev. Lett. **65**, 1627 (1990).

[203] J.Pedersen *et al.*, Nature **353**, 733 (1991).

[204] K.Selby *et al.*, Phys. Rev. B **43**, 4565 (1991).

[205] J. Serri *et al.*, J. Chem. Phys. **74**, 5116 (1981).

[206] A.Maiti and L.M.Falicov, Phys. Rev. A **45**, 6918 (1992).

[207] N.D.Lang, in *Theory of the Inhomogeneous Electron Gas*, edited by S.Lundqvist and N.H.March (Plenum, New York, 1983). See Ref. [201].

[208] C.Guet and W.R.Johnson, Phys. Rev. B **45**, 11283 (1992).

[209] D.R.Snider and R.S.Sorbello, Phys. Rev. B **28**, 5702 (1983).

[210] W. Heer, P.Milani, and A.Châtelain, Phys. Rev. Lett. **63**, 2834 (1989).

[211] P.Milani, I. Moullet, and W. Heer, Phys. Rev. A **42**, 5150 (1990).

[212] C.R.C.Wang *et al.*, Z. Phys. D **19**, 13 (1991).

[213] H.Fallgren and T.P.Martin, Chem. Phys. Lett. **168**, 233 (1990).

[214] A.Kawabata and R.Kubo, J. Phys. Soc. Japan **21**, 1765 (1966).

[215] G.Mie, Ann. der Phys. **25**, 377 (1908).

[216] S.Fedrigo, W.Harbich, J. Belyaev, and J.Buttet, Chem. Phys. Lett. **211**, 166 (1993).

[217] K.J.Taylor *et al.*, J. Chem. Phys. **93**, 7515 (1990).

[218] V.V.Kresin, A.Scheidemann, and W.D.Knight, in *Electron Collisions with Molecules, Clusters, and Surfaces*, edited by H.Ehrhardt and L.A.Morgan (Plenum, New York, 1994).

[219] V.V.Kresin, A.Scheidemann, and W.D.Knight, Phys. Rev. A **49**, 2696 (1994).

[220] A.A.Scheidemann, V.V.Kresin, and W.D.Knight, Phys. Rev. A **49**, R4293 (1994).

[221] M.Sundaram, S.A.Chalmers, P.F.Hopkins, and A.C.Gossard, Science **254**, 1326 (1991).

[222] T.Chakraborty, Comments Cond. Mat. Phys. **16**, 35 (1992).

[223] Physics Today **46**, (June 1993), special issue: Optics of Nanostructures, edited by D.S. Chemla.

[224] A.D.Yoffe, Adv. Phys. **42**, 173 (1993).

[225] *Microcrystalline and Nanocrystalline Semiconductors*, edited by R.W.Collins *et al.* (Materials Research Society, Pittsburgh, 1995).

[226] N.F.Johnson, J. Phys.: Condens. Matter **7**, 965 (1995).

[227] M.F.Jarrold, Science **252**, 1085 (1991).

[228] M.L.Mandich, W.D.Rents, and V.D.Bondybey, in *Atomic and Molecular Clusters*, edited by E.R.Bernstein (Elsevier, Amsterdam, 1990). See Ref. [161].

[229] K.Raghavachari, Phase Trans. **24-26**, 61 (1990).

[230] D.A.Jelski, T.F.George, and J.M.Vienneau, in *Clusters of Atoms and Molecules*, edited by H.Haberland (Springer, Berlin, 1994). See Ref. [153].

[231] E.C.Honea *et al.*, Nature **366**, 42 (1993).

[232] M.R.Pederson *et al.*, Phys. Rev. B **54**, 2863 (1996).

[233] J.M.Alford, R.T.Laaksonen, and R.E.Smalley, J. Chem. Phys. **94**, 2618 (1991).

[234] O.Cheshnovsky *et al.*, Chem. Phys. Lett. **138**, 119 (1987).

[235] L.-W.Wang and A.Zunger, Phys. Rev. Lett. **74**, 1039 (1994).

[236] R.Schäfer, S.Schlecht, J.Woenckhaus, and J.A.Becker, Phys. Rev. Lett. **76**, 471 (1996).

[237] M.F.Jarrold, J. Phys. Chem. **99**, 11 (1995).

[238] T.T.Rantala, M.I.Stockman, D.A.Jelski, and T.F.George, J. Chem. Phys. **93**, 7427 (1990).

[239] S.Adachi, *GaAs and Related Materials* (World Scientific, Singapore, 1994).

[240] M.A.Al-Laham and K.Raghavachari, Chem. Phys. Lett. **187**, 13 (1991).

[241] L.Lou, P.Nordlander, and R.E.Smalley, J. Chem. Phys. **97**, 1858 (1992).

[242] R.Schäfer and J.A.Becker, Phys. Rev. B **54**, 10296 (1996).

[243] A.Rubio *et al.*, Phys. Rev. Lett. **77**, 5442 (1996).

[244] S.Schlecht, R.Schäfer, J.Woenckhaus, and J.A.Becker, Chem. Phys. Lett. **246**, 315 (1995).

[245] J.C.Phillips, *Bonds and Bands in Semiconductors* (Academic, New York, 1973).

[246] C.Jin, K.J.Taylor, J.Conceicao, and R.E.Smalley, Chem. Phys. Lett. **175**, 17 (1990).

[247] L.Wang *et al.*, Chem. Phys. Lett. **172**, 335 (1990).

[248] S.C.O'Brien *et al.*, J. Chem. Phys. **84**, 4074 (1986).

[249] H.W.Kroto *et al.*, Nature **318**, 162 (1985).

[250] W.Krätschmer, L.D.Lamb, K.Fostiropoulos, and D.R.Huffman, Nature **347**, 354 (1990).

[251] *Fullerene Fundamentals*, edited by H.Ehrenreich and F.Spaepen (Academic, New York, 1994), Vol. 48.

[252] B.Shanker and J.Applequist, J. Phys. Chem. **98**, 6486 (1994).

[253] P.A.Gravil *et al.*, Surf. Sci. **329**, 199 (1995).

[254] R.J.Knize, Optics Commun. **106**, 95 (1994).

[255] G.A.Gallup, Chem. Phys. Lett. **187**, 187 (1991).

[256] A.Bulgac and N.Ju, Phys. Rev. B **46**, 4297 (1992).

[257] Y.Wang, G.F.Bertsch, and D.Tománek, Z. Phys. D **25**, 181 (1993).

[258] I.V.Hertel *et al.*, Phys. Rev. Lett. **68**, 784 (1992).

[259] D. Tománek, Comments At. Mol. Phys. **31**, 337 (1995).

[260] E.Westin, A.Rosén, G. T. Velde, and E.J.Baerends, J. Phys. B: At. Mol. Opt. Phys. **29**, 5087 (1996).

[261] L.-S.Wang, J. Conceicao, C. Jin, and R.E.Smalley, Chem. Phys. Lett. **182**, 5 (1991).

[262] A.A.Vostrikov, D. Y. Dubov, and A.A.Agarkov, Tech. Phys. Lett. **21**, 517 (1995).

[263] C.D.Finch, R.A.Popple, P. Nordlander, and F.B.Dunning, Chem. Phys. Lett. **244**, 345 (1995).

[264] J.M.Weber, M.-W.Ruf, and H.Hotop, Z. Phys. D **37**, 351 (1996).

[265] K.L.Han, H. Lin, E.B.Gallogy, and W.M.Jackson, Chem. Phys. Lett. **235**, 211 (1995).

[266] P.C.Ecklund *et al.*, Thin Solid Films **257**, 211 (1995).

[267] J.Baker *et al.*, Chem. Phys. Lett. **184**, 182 (1991).

[268] F.Willaime and L.M.Falicov, J. Chem. Phys. **98**, 6369 (1993).

[269] T.P.Martin, U.Näher, H.Schaber, and U.Zimmermann, Phys. Rev. Lett. **70**, 3079 (1993).

[270] K. Hansen, H. Hohmann, R. Müller, and E.E.B.Campbell, J. Chem. Phys. **105**, 6088 (1996).

[271] D.Ugarte, Nature **359**, 707 (1992).

[272] T.P.Martin *et al.*, J. Chem. Phys. **99**, 4210 (1993).

[273] J.Cioslowsky and K.Raghavachari, J. Chem. Phys. **98**, 8734 (1993).

[274] A.Rubio, J.A.Alonso, J.M.Lopez, and M.J.Stott, Phys. Rev. B **49**, 17397 (1994).

[275] H.A.Bethe and E.E.Salpeter, *Quantum Mechanics of One-and Two-Electron Atoms* (Springer, Berlin, 1957).

[276] L.C.Allen, Phys. Rev. **118**, 167 (1960).

[277] A.Dalgarno, Adv. Phys. **11**, 281 (1962).

[278] G.D.Mahan and K.R.Subbaswamy, *Local Density Theory of Polarizability* (Plenum, New York, 1990).

[279] A. Dalgarno and A.E.Kingston, Proc. Phys. Soc. London **73**, 455 (1959).

[280] M.A.Kadar-Kallen and K.D.Bonin, Phys. Rev. A **47**, 944 (1993).

[281] H.A.Bethe and R.Jackiw, *Intermediate Quantum Mechanics*, 3rd ed. (Benjamin, Menlo Park, 1986).

[282] M.Marinescu, H.R.Sadeghpour, and A.Dalgarno, Phys. Rev. A **49**, 982 (1994), and references therein.

[283] P.Ring and P.Schuck, *The Nuclear Many-Body Problem* (Springer, New York, 1980).

[284] O.Bohigas, A.M.Lane, and J.Martorell, Phys. Reports **51**, 267 (1979).

[285] E.Lipparini and S.Stringari, Phys. Reports **175**, 103 (1989).

[286] G.F.Bertsch and R.A.Broglia, *Oscillations in Finite Quantum Systems* (Cambridge, New York, 1994).

[287] E.Lipparini and S.Stringari, Z. Phys. D **18**, 193 (1991).

[288] N. Van Giai and E.Lipparini, Z. Phys. D **27**, 193 (1993).

[289] A. Kumar and W.J.Meath, Can. J. Chem. **63**, 1616 (1985).

[290] O. Goscinski, Int. J. Quantum Chem. **2**, 761 (1968).

[291] U. Hohm, Chem. Phys. Lett. **183**, 304 (1991).

[292] U. Hohm, Chem. Phys. Lett. **211**, 498 (1993).

[293] M. Sampoli *et al.*, Phys. Rev. A **42**, 6910 (1992).

[294] F. Barocchi, F. Hensel, and M. Sampoli, Chem. Phys. Lett. **232**, 445 (1995).

[295] D. Goebel *et al.*, J. Phys. Chem. **98**, 13123 (1994).

[296] D. Goebel and U. Hohm, J. Phys. Chem. **100**, 7710 (1996).

[297] R.E.Christofferson, *Basic Principles and Techniques of Molecular Quantum Mechanics* (Springer, New York, 1989).

[298] F.Herman and S.Skillman, *Atomic Structure Calculations* (Prentice-Hall, Englewood Cliffs, 1963).

[299] E.U.Condon and H.Odabaşi, *Atomic Structure* (Cambridge, New York, 1980).

[300] G.G.Hall, Proc. R. Soc. (London) **A205**, 541 (1951).

[301] C.C.J.Roothan, Rev. Mod. Phys. **23**, 69 (1951).

[302] C.C.J.Roothan, Rev. Mod. Phys. **32**, 179 (1960).

[303] E.Clementi, IBM J. Res. Dev. **9**, 2 (1965).

[304] E.Clementi and C.Roetti, Atom. Data Nuclear Data Tables **14**, 177 (1974).

[305] S.F.Boys, Proc. R. Soc. (London) **A200**, 542 (1950).

[306] *Gaussian Basis Sets for Molecular Calculations*, Vol. 16 of *Physical Sciences Data*, edited by S. Huzinaga (Elsevier, Amsterdam, 1984).

[307] R. Poirier, R. Kari, and I.G.Csizmadia, *Handbook of Gaussian Basis Sets: A Compendium for Ab-Initio Molecular Orbital Calculations*, Vol. 24 of *Physical Sciences Data* (Elsevier, Amsterdam, 1985).

[308] A.Szabo and N.S.Ostlund, *Modern Quantum Chemistry: Introduction to Advanced Electronic Structure Theory* (McGraw-Hill, New York, 1989), reprinted by Dover, New York, 1996.

[309] E.R.Davidson and D.Feller, Chem. Rev. **86**, 681 (1986).

[310] W.J.Hehre, L.Radom, P.V.R.Scheleyer, and J.A.Pople, *Ab Initio Molecular Orbital Theory* (Wiley, New York, 1986).

[311] L.Szasz, *Pseudopotential Theory of Atoms and Molecules* (Wiley, New York, 1985).

[312] P.Fulde, *Electron Correlations in Molecules and Solids*, 3rd ed. (Springer, Berlin, 1995).

[313] A.Dalgarno, Proc. Roy. Soc. **A251**, 282 (1959).

[314] S.Kaneko, J. Phys. Soc. Japan **14**, 1600 (1959).

[315] A.Dalgarno, in *Intermolecular Forces*, edited by J.O.Hirschfelder, Advances in Chemical Physics **XII** (Wiley, New York, 1967), p. 143.

[316] S.Kaneko and S.Arai, J. Phys. Soc. Japan **26**, 170 (1969).

[317] R.M.Sternheimer, Phys. Rev. **96**, 951 (1954).

[318] *Methods of Electronic Structure Theory*, edited by H. Schaefer, III (Plenum, New York, 1977).

[319] I.Prigogine, S.Rice, and K.P.Lawley, (eds.), *Ab Initio Methods in Quantum Chemistry-II*, Advances in Chemical Physics **LXIX** (Wiley, Chichester, 1987).

[320] M.J.Frisch *et al.*, Gaussian 94, Gaussian Inc., Pittsburgh, PA; http://www.gaussian.com.

[321] J. Stewart *et al.*, MOPAC93, Stewart Computational Chemistry, Colorado Springs, CO; http://iti2.net/jstewart/.

[322] M.W.Schmidt *et al.*, J. Comput. Chem. **14**, 1347 (1993), http://www.msg.ameslab.gov/gamess/gamess.html.

[323] V.Bonačić-Koutecký, P.Fantucci, and J.Koutecký, Chem. Rev. **91**, 1035 (1991).

[324] V.Bonačić-Koutecký *et al.*, Comments At. Mol.Phys. **31**, 233 (1995).

[325] H.D.Cohen and C.C.J.Roothaan, J. Chem. Phys. **43**, 534 (1965).

[326] J.S.Craw, A.Hinchliffe, and J.J.Perez, in *Self- Consistent Field, Theory and Applications*, edited by R.Carb and M.Klobukowski (Elsevier, Amsterdam, 1990), p. 866.

[327] A.D.McLean and M.Yoshimine, J. Chem. Phys. **46**, 3682 (1967).

[328] M.J.S.Dewar and J.J.P.Stewart, Chem. Phys. Lett. **111**, 416 (1984).

[329] *Theory of the Inhomogeneous Electron Gas*, edited by S.Lundqvist and N.H.March (Plenum, New York, 1983).

[330] R.G.Parr and Y. Weitao, *Density Functional Theory of Atoms and Molecules* (Oxford, New York, 1994).

[331] R.M.Dreizler and E.K.U.Gross, *Density Functional Theory* (Springer, Berlin, 1990).

[332] L.D.Landau and E.M.Lifshitz, *Quantum Mechanics*, 3rd ed. (Pergamon, Oxford, 1977).

[333] S.Flügge, *Practical Quantum Mechanics* (Springer, Berlin, 1974).

[334] J.C.Slater, *Solid-State and Molecular Theory* (Wiley, New York, 1975).

[335] W.Hanke, N.Mekini, and H.Weiler, in *Electronic Structure, Dynamics, and Quantum Structural Properties of Condensed Matter*, edited by J.T.Devreese and P. V. Camp (Plenum, New York, 1985), p. 113.

[336] M.Brewczyk and M.Gajda, Phys. Rev. A **40**, 3475 (1989).

[337] M.Madjet, C.Guet, and W.R.Johnson, Phys. Rev. A **51**, 1327 (1995).

[338] W.Ekardt, J.M.Pacheco, and Z.Penzar, Comments At. Mol.Phys. **31**, 291 (1995).

[339] C.Møller and M.S.Plesset, Phys. Rev. **46**, 618 (1934).

[340] A.L.Fetter and J.D.Walecka, *Quantum Theory of Many-Particle Systems* (McGraw-Hill, New York, 1971).

[341] J.-M.André, C.Barbier, V.Bodart, and J.Delhalle, in *Nonlinear Optical Properties of Organic Molecules and Crystals*, edited by D.S.Chemla and J.Zyss (Academic, Orlando, 1987), Vol. 2, p. 137.

[342] H.P.Kelly, in *Correlation Effects in Atoms and Molecules*, edited by R.Lefebvre and C.Moser, Adv. Chem. Phys. **XIV** (Wiley, London, 1969), p. 129.

[343] M.Amusia, *Atomic Photoeffect* (Plenum, New York, 1990).

[344] V.M.Galitskii and A.B.Migdal, Sov. Phys. -JETP **7**, 96 (1958).

[345] A.B.Migdal, *Theory of Finite Fermi Systems and Applications to Atomic Nuclei* (Wiley, New York, 1967).

[346] D.Pines, *The Many-Body Problem* (Benjamin, New York, 1961).

[347] A.A.Abrikosov, L.P.Gorkov, and I.E.Dzyaloshinskii, *Methods of Quantum Field Theory in Statistical Physics* (Prentice Hall, Englewood Cliffs, 1963), reprinted by Dover, New York, 1975.

[348] R.D.Mattuck, *A Guide to Feynman Diagrams in the Many-Body Problem*, 2nd ed. (McGraw-Hill, New York, 1976), reprinted by Dover, New York, 1992.

[349] G.D.Mahan, *Many-Particle Physics*, 2nd ed. (Plenum, New York, 1990).

[350] H.Warston, I.Lindgren, and S.Salomonson, Phys. Rev. A **55**, 2757 (1997).

[351] D.Pines, *Elementary Excitations in Solids* (Benjamin, New York, 1963).

[352] D.J.Rowe, *Nuclear Collective Motion* (Methuen, London, 1970).

[353] M.J.Stott and E.Zaremba, Phys. Rev. A **21**, 12 (1980).

[354] A.Zangwill and P.Soven, Phys. Rev. A **21**, 1561 (1980).

[355] C.Guet, Comments At. Mol.Phys. **31**, 305 (1995).

[356] J.Oddershede, in *Ab Initio Methods in Quantum Chemistry - II*, edited by K.P.Lawley, Adv. Chem. Phys. **LXIX** (Wiley, Chichester, 1987), p. 201.

[357] A. Steiger, Berichte **54**, 1381 (1921).

[358] C. Smyth, Phil. Mag. **50**, 361 (1925).

[359] Y.K.Kang and M.S.Jhon, Theor. Chim. Acta **56**, 191 (1980).

[360] H.E.Watson and K.L.Ramaswamy, Proc. Roy. Soc., Ser. A **156**, 144 (1936).

[361] H.A.Stuart and S.v.Schieszl, Ann. Phys. **2**, 321 (1948).

[362] J. Applequist, J. Phys. Chem. **97**, 6016 (1993).

[363] A.J.Stone, *Theory of Intermolecular Forces* (Oxford University, Oxford, 1996).

[364] M.L.Olson and K.R.Sundberg, J. Chem. Phys. **69**, 5400 (1978).

[365] B.Shanker and J. Applequist, J. Phys. Chem. **98**, 6486 (1994).

[366] B.Friedrich and D.Herschbach, Z. Phys. D **36**, 221 (1996).

[367] K.T.Chung, Phys. Rev. **166**, 1 (1968).

[368] P.H.S.Martin, W.H.Henneker, and V. McKoy, J. Chem. Phys. **62**, 69 (1975).

[369] E.A.Reinsch, J. Chem. Phys. **83**, 5784 (1985).

[370] Z.W.Liu and H.P.Kelly, Theo. Chim. Acta **80**, 307 (1991).

[371] Y.B.Malykhanov, Opt. Spectrosc. **52**, 105 (1982).

[372] Y.Y.Dmitriev, Y.B.Malykhanov, and B. Rus, Opt. Spectrosc. **57**, 334 (1984).

[373] D.M.Bishop and J. Pippin, J. Chem. Phys. **97**, 3374 (1992).
[374] R.M.Glover and F. Weingold, J. Chem. Phys. **65**, 4913 (1976).
[375] A.C.Newell and R.C.Baird, J. Appl. Phys. **36**, 3751 (1965).
[376] P.W.Langhoff and M. Karplus, J. Opt. Soc. Am. **59**, 863 (1969).
[377] C. Cuthbertson and M. Cuthbertson, Proc. R. Soc. London Ser. **A 135**, 40 (1932).
[378] J.E.Rice, P.R.Taylor, T.J.Lee, and J. Almlöf, J. Chem. Phys. **94**, 4972 (1991).
[379] C. Huiszoon and W.J.Briels, Chem. Phys. Lett. **203**, 49 (1993).
[380] M. Caffarel, M. Rérat, and C. Pouchan, Phys. Rev. A **47**, 3704 (1993).
[381] D.M.Bishop and S. M.Cybulski, Chem. Phys. Lett. **211**, 255 (1993).
[382] D.M.Bishop and J. Pipin, Int. J. Quantum Chem. **45**, 349 (1993).
[383] D.E.Woon and T.H.Dunning, J. Chem. Phys. **100**, 2975 (1994).
[384] M.K.Harbola, Chem. Phys. Lett. **217**, 461 (1994).
[385] A.K.Bhatia and R.J.Drachman, J. Phys. B **27**, 1299 (1994).
[386] M.J.Jamieson, G.W.F.Drake, and A. Dalgarno, Phys. Rev. A **51**, 3358 (1995).
[387] M.-K. Chen, J. Phys. B **28**, 1349 (1995).
[388] M.-K. Chen and K.T.Chung, Phys. Rev. A **53**, 1439 (1996).
[389] W.R.Johnson and K.T.Cheng, Phys. Rev. A **53**, 1375 (1996).
[390] M.K.Harbola and A. Banerjee, Phys. Rev. A **54**, 283 (1996).
[391] L.R.Pendrill, J. Phys. B **29**, 3581 (1996).
[392] Z.-C. Yan, J.F.Babb, A. Dalgarno, and G.W.F.Drake, Phys. Rev. A **54**, 2824 (1996).
[393] D. Gugan and G.W.Michel, Mol. Phys. **39**, 783 (1980).
[394] D. Gugan, Metrologia **28**, 405 (1991).
[395] K. Grohmann and H. Luther, *Temperature — Its Measurement and Control in Science and Industry* (AIP, New York, 1992), Vol. 6, p. 21.
[396] A.D.Buckingham and P.G.Hibbard, Symp. Faraday Soc. **2**, 41 (1968).
[397] R.H.Orcutt and R.H.Cole, J. Chem. Phys. **46**, 697 (1967).
[398] R. Luyckx, P. Coulon, and H.N.W.Lekkerkerker, Chem. Phys. Lett. **48**, 187 (1977).
[399] A.J.Thakkar, J. Chem. Phys. **75**, 4496 (1981).
[400] A.D.Buckingham, communication of A. D. Buckingham (as quoted in Ref.[399]).
[401] F. Visser, P.E.S.Wormer, and P. Stam, J. Chem. Phys. **79**, 4973 (1983).
[402] J.F.Thornbury and A. Hibbert, J. Phys. B **20**, 6447 (1987).
[403] F. Khan and G.S.Khandelwal, Phys. Rev. A **38**, 6159 (1988).
[404] D.M.Bishop and J. Pippin, J. Chem. Phys. **91**, 15 (1989).
[405] D.M.Bishop and M. Rérat, J. Chem. Phys. **91**, 5489 (1989).
[406] R.J.Tweed and J. Langlois, J. Phys. B **26**, 1779 (1991).
[407] N. E. B. Kassimi *et al.*, J. Mol. Struct. (Theochem) **254**, 177 (1992).
[408] D.Spelsberg, T.Lorenz, and W.Meyer, J. Chem. Phys. **99**, 7845 (1993).

[409] W.L.Wiese, M.W.Smith, and B.M.Miles, *Atomic Transition Probabilities, II, National Stand. Ref. Data Ser., Nat. Bur. Stand. 22* (Dept. of Commerce, Washington, D.C., 1969).

[410] G.V.Marr and D.M.Creek, Proc.R. Soc. London Ser. A **304**, 233 (1967).

[411] C.R.Ekstrom *et al.*, Phys. Rev. A **51**, 3883 (1995).

[412] M. Cohen and G.W.F.Drake, Proc. Phys. Soc. London **92**, 23 (1967).

[413] W. Muller, J. Flesch, and W. Meyer, J. Chem. Phys. bf **80**, 3297 (1984).

[414] W.D.Hall and J.C.Zorn, Phys. Rev. A **10**, 1141 (1974).

[415] R.W.Molof, H.L.Schwartz, T.M.Miller, and B.Bederson, Phys. Rev. A **10**, 1131 (1974).

[416] M. Yoshimine and R.P.Hurst, Phys. Rev. **135**, 612 (1964).

[417] P.W.Langhoff and R.P.Hurst, Phys. Rev. **139**, 1415 (1965).

[418] R.M.Sternheimer, Phys. Rev. **183**, 112 (1969).

[419] P.K.Mukherjee, R.K.Moitra, and A. Mukherji, Int. J. Quantum Chem. **5**, 637 (1971).

[420] S.A.Adelman and A. Szabo, J. Chem. Phys. **58**, 687 (1973).

[421] K.T.Tang, J.M.Norbeck, and P.R.Certain, J. Chem. Phys. bf **64**, 3063 (1976).

[422] E.A.Reinsch and W. Meyer, Phys. Rev. A **14**, 915 (1976).

[423] E.A.Reinsch and W. Meyer, Mol. Phys. **31**, 855 (1976).

[424] N.L.Manakov and V.D.Ovsiannikov, J. Phys. B **10**, 569 (1977).

[425] F. Maeder and W. Kutzelnigg, Chem. Phys. **42**, 95 (1979).

[426] P. Fuenteabla, J. Phys. B **15**, L555 (1982).

[427] G.S.Soloveva, P.F.Gruzdev, and A.I.Sherstyuk, Opt. Spectrosc. (USSR) **57**, 473 (1984).

[428] B. Kundu, D. Ray, and P.K.Mukherjee, Phys. Rev. A **34**, 62 (1986).

[429] L. Windholz and M. Musso, Phys. Rev. A **39**, 2472 (1989).

[430] A.J.Sadlej and M. Urban, J. Mol. Struct. (Theochem) **234**, 147 (1991).

[431] P. Fuenteabla and O. Reyes, J. Phys. B **26**, 2245 (1993).

[432] J.Guan, M.E.Casida, A.M.Köster, and D.R.Salahub, Phys. Rev. B **52**, 2184 (1995).

[433] S.H.Patil and K.T.Tang, Phys. Rev. A **106**, 2298 (1997).

[434] M.Mérawa and M.Rérat, J. Chem. Phys. **106**, 3658 (1997).

[435] Y. Luo, O. Vahtras, H. Agren, and P. Jorgensen, Chem. Phys. Lett. **205**, 555 (1993).

[436] H.J.A.Jensen, P. Jorgensen, T. Helgaker, and J. Olsen, Chem. Phys. Lett. **162**, 355 (1989).

[437] S.R.Langhoff, J. C.W.Bauschlicher, and D.P.Chong, J. Chem. Phys. **78**, 5287 (1983).

[438] C.J.Jameson and P.W.Fowler, J. Chem. Phys. **85**, 3432 (1986).

[439] G. Maroulis and A.J.Thakkar, J. Chem. Phys. **88**, 7623 (1988).

[440] W. Rijks and P.E.S.Wormer, J. Chem. Phys. **88**, 5704 (1988).

[441] M.A.Morrison and P.J.Hay, J. Phys. B **10**, L647 (1977).

[442] G.R.Alms, A.K.Burnham, and W.H.Flygare, J. Chem. Phys. **63**, 3321 (1975).

[443] G.D.Zeiss and W.J.Meath, Mol. Phys. **33**, 1155 (1977).

[444] J. Kendrick, J. Phys. B **11**, L601 (1978).

[445] M. Morrison and P. Hay, J. Chem. Phys. **70**, 4034 (1979).

[446] E.N.Svendsen and J. Oddershede, J. Chem. Phys. **71**, 3000 (1979).

[447] R.D.Amos, Mol. Phys. **39**, 1 (1980).

[448] J. Oddershede and E.N.Svendsen, Chem. Phys. **64**, 359 (1982).

[449] C.E.Dykstra, J. Chem. Phys. **82**, 4120 (1985).

[450] G. Maroulis and D.M.Bishop, Mol. Phys. **58**, 273 (1986).

[451] I. Cernusak, G.H.F.Diercksen, and A.J.Sadlej, Chem. Phys. **108**, 45 (1986).

[452] W.J.Meath and A. Kumar, Int. J. Quantum Chem. Symp. **24**, 501 (1990).

[453] D. Spelsberg and W. Meyer, J. Chem. Phys. **101**, 1282 (1994).

[454] C. Jamorski, M.E.Casida, and D.R.Salahub, J. Chem. Phys. **104**, 5134 (1996).

[455] L. Serra *et al.*, J.Phys.:Condens. Matter **1**, 10391 (1989).

[456] D.E.Beck, Phys. Rev. B **30**, 6935 (1984).

[457] A.Rubio, L.C.Balbás, Ll.Serra, and M.Barranco, Phys. Rev. B **42**, 10950 (1990).

[458] M.K.Harbola, Solid State Commun. **98**, 629 (1996).

[459] V.Kresin, Phys. Rev. B **42**, 3247 (1990).

[460] C.Yannouleas, E.Vigezzi, and R.A.Broglia, Phys. Rev. B **47**, 9849 (1993).

[461] F.Alasia *et al.*, Phys. Rev. B **52**, 8488 (1995).

[462] P.-G.Reinhard, M.Brack, and O.Genzken, Phys. Rev. A **41**, 5568 (1990).

[463] W.Ekardt, Phys. Rev. B **31**, 6360 (1985).

[464] J.M.Pacheco and W.Ekardt, Ann. Physik (Leipzig) **1**, 254 (1992).

[465] P.Stampfli and K.H.Bennemann, Phys. Rev. A **39**, 1007 (1989).

[466] P.Sheng, M.Y.Chou, and M.L.Cohen, Phys. Rev. B **34**, 732 (1986).

[467] A. Rubio, L. Balbás, and J. Alonso, Z. Phys. D **19**, 93 (1991).

[468] I.Moullet, J.L.Martins, F.Reuse, and J.Buttet, Phys. Rev. Lett. **65**, 476 (1990).

[469] I.Moullet, J.L.Martins, F.Reuse, and J.Buttet, Phys. Rev. B **42**, 11598 (1990).

[470] A.Rubio *et al.*, Phys. Rev. Lett. **77**, 247 (1996).

[471] J.M.Pacheco and J.L.Martins, J. Chem. Phys. **106**, 6039 (1997).

[472] C.Yannouleas, F.Catara, and N. V. Giai, Phys. Rev. B **51**, 4569 (1995).

[473] M.Membrado, A.F.Pacheco, and J. Sanudo, Phys. Rev. B **41**, 5643 (1990).

[474] A.Bulgac and C.Lewenkopf, Europhys. Lett. **31**, 519 (1995).

[475] V.Tarnovsky, M.Bunimovicz, L.Vušcović, and B.Bederson, J. Chem. Phys. **98**, 3894 (1993).

[476] P. Norman, Y. Luo, and H. Agren, J. Chem. Phys. **106**, 8788 (1997).

[477] M.R.Pederson and A.W.Quong, Phys. Rev. B **46**, 13584 (1992).

[478] T.Rabenau *et al.*, Z. Phys. B **90**, 69 (1993).

[479] Z. Kafafi *et al.*, Chem. Phys. Lett. **188**, 492 (1992).

[480] A. Hebard, R. Fleming, and A. Karton, Appl. Phys. Lett. **59**, 2109 (1991).

[481] P.L.Hansen, P.J.Fallon, and W.Krätschmer, Chem. Phys. Lett. **181**, 367 (1991).

[482] G.F.Bertsch, A.Bulgac, D.Tománek, and Y.Wang, Phys. Rev. Lett. **67**, 2690 (1991).

[483] Ph.Lambin, A.A.Lucas, and J.-P.Vigneron, Phys. Rev. B **46**, 1794 (1992).

[484] A. Bulgac and N. Ju, Phys. Rev. B **46**, 4297 (1992).

[485] P.W.Fowler, P. Lazzeretti, and R. Zanasi, Chem. Phys. Lett. **165**, 79 (1990).

[486] M. Matsuzawa and D.A.Dixon, J. Phys. Chem. **96**, 6872 (1992).

[487] H. Weiss, R. Ahlrichs, and N. Häser, J. Chem. Phys. **99**, 1262 (1993).

[488] R.L.Hettich, R.N.Compton, and R.H.Ritchie, Phys. Rev. Lett. **67**, 1242 (1990).

[489] G.B.Talapatra *et al.*, J. Phys. Chem. **96**, 5206 (1992).

[490] A.A.Quong and M.R.Pederson, Phys. Rev. B **46**, 12906 (1992).

[491] F.Alasia *et al.*, J. Phys. B: At. Mol. Opt. Phys. **27**, L643 (1994).

[492] M.J.Puska and R.M.Nieminen, Phys. Rev. A **47**, 1181 (1993).

[493] B.Vasvári, Z. Phys. B **100**, 223 (1996).

[494] G.Wendin and B.Wästberg, Phys. Rev. B **48**, 14764 (1993).

[495] Z.Shuai and J.L.Bredas, Phys. Rev. B **46**, 16135 (1992).

[496] A.N.M.Barnes, D.J.Turner, and L.E.Sutton, Trans. Faraday Soc. **67**, 2902 (1971).

[497] H. Sutter and R.H.Cole, J. Chem. Phys. **52**, 132 (1970).

[498] T.K.Bose, J.S.Sochanski, and R.H.Cole, J. Chem. Phys. **57**, 3592 (1972), see also Ref. [397].

[499] C. Meyer and G. Morrison, J. Phys. Chem. **95**, 3860 (1991).

[500] A.R.H.Goodwin and G. Morrison, J. Phys. Chem. **96**, 5521 (1992).

[501] A.D.Buckingham and J.A.Pople, Trans. Faraday Soc. **51**, 1029 (1955).

[502] *Collision- and Interaction-induced Spectroscopy*, Vol. 452 of *NATO ASI*, edited by G.C.Tabisz and M.N.Neuman (Kluwer, Dordrecht, 1995).

[503] V.A.Zamkov, in *Molecular Liquids: New Perspectives in Physics and Chemistry*, Vol. 379 of *NATO ASI*, edited by J. J. C. Teixeira-Dias (Kluwer, Dordrecht, 1992).

[504] H.B.Chae, J.W.Schmidt, and M.R.Moldover, J. Phys. Chem. **94**, 8840 (1990).

[505] U. Hohm and K. Kerl, Mol. Phys. **58**, 541 (1986).

[506] U. Hohm and K. Kerl, Mol. Phys. **69**, 803 (1990).

[507] U. Hohm and K. Kerl, Mol. Phys. **69**, 819 (1990).

[508] R.A.Alpher and D.R.White, Phys.Fluids **2**, 153 (1959).

[509] U. Hohm and U. Trümper, J. Chem. Soc. Faraday Trans. **91**, 1277 (1995).

[510] H.G.Sutter, *Dielectric and Related Molecular Processes* (The Chemical Society, London, 1972), Vol. 1, p. 64.

[511] H.W.Kroto, A.W.Allaf, and S.P.Balm, Chem. Rev. **91**, 1213 (1991).

[512] N.J.Bridge and A.D.Buckingham, Proc. R. Soc. London **295**, 334 (1966).

[513] M.P.Bogaard, A.D.Buckingham, R.K.Pierens, and A.H.White, J. Chem. Soc. Faraday Trans. I **74**, 3008 (1978).

[514] V.A.Maltsev, O.A.Nerushev, S.A.Novopashin, and B.A.Selivanov, Chem. Phys. Lett. **212**, 480 (1993).

[515] A.D.Buckingham and J.A.Pople, Proc. Phys. Soc. A **68**, 905 (1955).

[516] A.D.Buckingham and B.J.Orr, Proc. Roy. Soc. A **305**, 259 (1968).

[517] A.D.Buckingham, Proc. Roy. Soc. A **267**, 271 (1962).

[518] A.D.Buckingham and D.A.Dows, Discussions Faraday Soc. **35**, 48 (1963).

[519] J.M.Brown, A.D.Buckingham, and D.A.Ramsay, Can. J. Phys. **49**, 914 (1971).

[520] J.H.Williams, Chem. Phys. Lett. **147**, 585 (1988).

[521] H. Scheffers and J. Stark, Phys. Z. **37**, 217 (1936).

[522] H. Scheffers, Phys. Z. **41**, 399 (1940).

[523] R. Kremens *et al.*, J. Chem. Phys. **81**, 1676 (1984).

[524] B. Bederson, J. Eisinger, K. Rubin, and A. Salop, Rev. Sci. Instrum. **31**, 852 (1960).

[525] A. Salop, E. Pollack, and B. Bederson, Phys. Rev. **124**, 1431 (1961).

[526] J. Levine, R.J.Celotta, and B. Bederson, Phys. Rev. **171**, 31 (1968).

[527] T. Guella *et al.*, Phys. Rev. A **29**, 2977 (1984).

[528] L. Nelissen, J. Reuss, and A. Dymanus, Physica **42**, 619 (1969).

[529] N.F.Ramsey, *Molecular Beams* (Clarendon, Oxford, 1956), Chap. 1.

[530] H. Gould, Phys. Rev. A. **14**, 922 (1976).

[531] H. Gould, E. Lipworth, and M.C.Weisskopf, Phys. Rev. **188**, 24 (1969).

[532] T.C.English and K.B.MacAdam, Phys. Rev. Lett. **24**, 555 (1970).

[533] K.B.MacAdam and N.F.Ramsey, Phys. Rev. A **6**, 898 (1972).

[534] P.K.Tien, R. Ulrich, , and R.J.Martin, Appl. Phys. Lett. **14**, 291 (1969).

[535] R. Ulrich and R. Torge, Appl. Opt. **12**, 2901 (1973).

[536] K. Tanaka, Appl. Phys. Lett. **34**, 672 (1979).

[537] P.P.Herrmann, Appl. Opt. **19**, 3261 (1980).

[538] T.-N.Ding and E. Garmire, Appl. Opt. **22**, 3177 (1983).

[539] J.R.Birch, Mikrochim. Acta **III**, 105 (1987).

[540] T.J.Parker, Contemp. Phys. **31**, 335 (1990).

[541] J.R.Birch and J.Yarwood, in *Spectrosopy and Relaxation of Molecular Liquids*, edited by D. Steele and J. Yarwood (Elsevier, Amsterdam, 1991), p. 174.

[542] D. Goebel and U. Hohm, Phys. Rev. A **52**, 3691 (1995).

[543] D. Goebel, U. Hohm, and K. Kerl, J. Mol. Struct. **349**, 253 (1995).

[544] D. Goebel and U. Hohm, J. Phys. D: Appl. Phys. **29**, 3132 (1996).

[545] U. H. G. Maroulis, C. Makris and D. Goebel, J. Phys. Chem. A **101**, 953 (1997).

[546] D. Goebel and U. Hohm, J. Phys. Chem. A **101**, 953 (1997).

[547] K. Kerl and H. Häusler, Infrared Phys. **24**, 297 (1984).

[548] U. Hohm and K. Kerl, Meas. Sci. Technol. **1**, 329 (1990).

[549] U. Hohm and U. Trümper, J. Raman Spectrosc. **26**, 1059 (1995).

[550] K. Kerl, U. Hohm, and H. Varchmin, Ber. Bunsenges. Phys. Chem. **96**, 728 (1992).

[551] U. Hohm, Mol. Phys. **78**, 929 (1993).

[552] U. Hohm and U. Trümper, Chem. Phys. **189**, 443 (1994).

[553] W. Heer and P. Milani, Rev. Sci. Instrum. **62**, 670 (1991).

[554] M.A.Kadar-Kallen, Ph.D. thesis, Princeton University, 1992.

[555] B.J.Chang, R. Alferness, and E.N.Leith, Appl. Optics **14**, 1592 (1975).

[556] M. Weissbluth, *Photon-Atom Interactions* (Academic Press, New York, 1989).

[557] J.J.Sakurai, *Advanced Quantum Mechanics* (Addison-Wesley, Reading, 1967), pp. 59–64.

[558] H. de Hulst, *Light Scattering by Small Particles* (Wiley, New York, 1957).

[559] D. Marcuse, *Principles of Quantum Electronics* (Academic Press, New York, 1980).

[560] M. Born and E. Wolf, *Principles of Optics* (Pergamon Press, Oxford, 1980).

[561] R. Loudon, *The Quantum Theory of Light* (Clarendon Press, Oxford, 1984).

[562] C.F.Bohren and D.R.Huffman, *Absorption and Scattering of Light by Small Particles* (Wiley, New York, 1983).

[563] T. Sleator *et al.*, Phys. Rev. Lett. **68**, 1996 (1992).

[564] C. Hoyer, S. Monajembashi, and K.O.Greulich, Sci. Prog. **79**, 233 (1996).

[565] S. Chu and C. Wieman, eds., JOSA B **6**, 2020 (1989).

[566] M.H.Anderson *et al.*, Science **269**, 198 (1995).

[567] S. Chu, J.E.Bjorkholm, A. Ashkin, and A. Cable, Phys. Rev. Lett. **57**, 314 (1986).

[568] C.J.Foot, Contemp. Phys. **32**, 369 (1991).

[569] in *Proceedings of the 1991 Enrico Fermi Summer School on 'Laser Manipulation of Atoms and Ions, Varenna, Italy'*, edited by E. Arimondo and W.D.Phillips (North-Holland, Amsterdam, 1993).

[570] H. Metcalf and P. van der Straten, Phys. Rep. **244**, 203 (1994).

[571] O. Frisch, Zeit. f. Phys. **86**, 42 (1933).

[572] J. Prodan *et al.*, Phys. Rev. Lett. **54**, 992 (1985).

[573] W.D.Phillips, J.V.Prodan, and H.J.Metcalf, JOSA B **2**, 1751 (1985).

[574] J. Dalibard and C. Cohen-Tannoudji, JOSA B **6**, 2023 (1989).

[575] Personal communication, Thad Walker, May, 1997.

[576] W. Petrich, M.H.Anderson, J.R.Escher, and E.A.Cornell, Phys. Rev. Lett. **74**, 3352 (1995).

[577] K. Davis *et al.*, Phys. Rev. Lett. **75**, 3969 (1995).

[578] A. Ashkin and J.P.Gordon, Opt. Lett. **8**, 511 (1983).

[579] D.E.Pritchard *et al.*, Phys. Rev. Lett. **57**, 310 (1986).

[580] E.L.Raab *et al.*, Phys. Rev. Lett. **59**, 2631 (1987).

[581] A.M.Steane, M. Chowdhury, , and C.J.Foot, JOSA B **9**, 2142 (1992).

[582] T. Walker, Laser Phys. **4**, 965 (1994).

[583] T. Takekoshi, J.R.Yeh, and R.J.Knize, Opt. Commun. **114**, 421 (1995).

[584] T. Takekoshi, J.R.Yeh, and R.J.Knize, Opt. Lett. **21**, 77 (1996).

[585] A. Ashkin, Phys. Rev. Lett. **24**, 156 (1970).

[586] A. Ashkin, J.M.Dziedzic, J.E.Bjorkholm, and S. Chu, Opt. Lett. **11**, 288 (1986).

[587] S.B.Smith, Y. Cui, and C. Bustamante, Science **271**, 795 (1996).

[588] S.B.Smith, Y. Cui, and C. Bustamante, Biophys. J. **72**, 1335 (1997).

[589] K. Helmerson, R. Kishore, W.D.Phillips, and H.H.Weetall, Clin. Chem. **43**, 379 (1997).

[590] S.C.Kuo and M.P.Sheetz, Science **260**, 232 (1993).

[591] S.M.Block, in *Noninvasive Techniques in Cell Biology* (Wiley-Liss, New York, 1990).

[592] R.S.Afzal and E.B.Treacy, Rev. Sci. Instrumen. **63**, 2157 (1992).

[593] T. Kasuya and M. Tsukakoshi, in *Laser Science and Technology*, edited by V. Letokhov, C. Shank, and H. Walther (Harwood Academic Publishers, Chur, 1989), Vol. 1.

[594] L.P.Ghislain, N.A.Swirtz, and W.W.Webb, Rev. Sci. Instrumen. **65**, 2762 (1994).

[595] G. Weber and K.O.Greulich, Int. Rev. Cytol. **133**, 1 (1992).

[596] K.O.Greulich and G. Weber, Exp. Tech. Phys. **40**, 1 (1994).

[597] T. Kuroiwa *et al.*, Protoplasma **194**, 275 (1996).

[598] A. Fert and P. Bruno, *Ultrathin Magnetic Structures II* (Springer, Berlin, 1994), p. 82.

[599] R.J.Celotta, R. Gupta, R.E.Scholten, and J.J.McClelland, J. Appl. Phys. **79**, 6079 (1996).

[600] J.J.McClelland, JOSA B **12**, 1761 (1995).

[601] K.K.Berggren *et al.*, Science **269**, 1255 (1995).

[602] E.W.McDaniel, *Atomic Collisions: Electron and Photon Projectiles* (Wiley, New York, 1989).

[603] E.A.Mason and E.W.McDaniel, *Transport Properties of Ions in Gases* (Wiley, New York, 1988).

[604] *Intermolecular Interactions: From Diatomics to Polymers*, edited by B.Pullman (Wiley, Chichester, 1978).

[605] J.N.Israelachvili, *Intermolecular and Surface Forces*, 2nd ed. (Academic, London, 1991).

[606] P.Langevin, Ann. Chim. Phys. **5**, 245 (1905).

[607] E.W.McDaniel, *Collision Phenomena in Ionized Gases* (Wiley, New York, 1964).

[608] E.Vogt and G.H.Wannier, Phys. Rev. **95**, 1190 (1954).

[609] A.Henglein, in *Physical Chemistry - An Advanced Treatise*, edited by W.Jost (Academic, New York, 1975), Vol. VIB.

[610] L.D.Landau and E.M.Lifshitz, *Mechanics*, 3rd ed. (Pergamon, Oxford, 1976).

[611] C.E.Klots, Chem. Phys. Lett. **38**, 61 (1976).

[612] D.Klar *et al.*, Z. Phys. D **31**, 235 (1994).

[613] T.F.O'Malley, Phys.Rev. **137**, A1668 (1965).

[614] E.P.Wigner, Phys. Rev. **73**, 1002 (1948).

[615] H.Hotop and W.C.Lineberger, J. Chem. Phys. **58**, 2379 (1973).

[616] T.F.O'Malley, Phys.Rev. **130**, 1020 (1963).

[617] I.T.Iakubov and A.G.Khrapak, Rep. Prog.Phys. **45**, 697 (1982).

[618] I.T.Iakubov and A.G.Khrapak, Phys. Rev. A **51**, 5043 (1995).

[619] V.M.Atrazhev and I.T.Iakubov, J. Chem. Phys. **103**, 9030 (1995).

[620] G.F.Gribakin and V.V.Flambaum, Phys.Rev. A **48**, 546 (1993).

[621] L.Rosenberg, Phys.Rev. A **55**, 2857 (1997).

[622] I.I.Fabrikant, Comments At. Mol. Phys. **32**, 267 (1996).

[623] in *Linking the Gaseous and Condensed Phases of Matter: The Behavior of Slow Electrons*, edited by L.G.Christophorou, E.Illenberger, and W.F.Schmidt (Plenum, New York, 1994).

[624] N.F.Mott and H.S.W.Massey, *Theory of Atomic Collisions*, 3rd ed. (Oxford, New York, 1965).

[625] *Electron-Molecule Interactions and Their Applications*, edited by L.G.Christophorou (Academic, Orlando, 1984).

[626] G.F.Drukarev, *Collisions of Electrons with Atoms and Molecules* (Plenum, New York, 1987).

[627] P.G.Burke and C.J.Joachain, *Theory of Electron-Atom Collisions* (Plenum, New York, 1995).

[628] A.Chutjian, A.Garskadden, and J.M.Wadehra, Phys. Rep. **264**, 393 (1996).

[629] G.Csanak, D.C.Cartwright, S.K.Srivastava, and S.Trajmar, in *Electron-Molecule Interactions and Their Applications*, edited by L.G.Christophorou (Academic, Orlando, 1984).

[630] C.Joachain, *Quantum Collision Theory* (North-Holland, Amsterdam, 1975).

[631] R.E.Johnson, *Introduction to Atomic and Molecular Collisions* (Plenum, New York, 1982).

[632] E.W.McDaniel *et al.*, *Ion-Molecule Reactions* (Wiley, New York, 1970).

[633] in *Gas Phase Ion Chemistry*, edited by M.T.Bowers (Academic, New York, 1979).

[634] E.W.McDaniel, J.B.A.Mitchell, and M.E.Rudd, *Atomic Collisions: Heavy Particle Projectiles* (Wiley, New York, 1993).

[635] T.F.Gallagher, *Rydberg Atoms* (Cambridge University Press, Cambridge, 1994).

[636] J.E.Mayer and M.G.Mayer, Phys.Rev. **43**, 605 (1933).

[637] V.M.Nabutovskii and D.A.Romanov, Sov. J. Low Temp. Phys. **11**, 277 (1985).

[638] M.Rosenblit and J.Jortner, Phys. Rev. B **52**, 17461 (1995).

[639] R.L.Asher *et al.*, Chem. Phys. Lett. **234**, 113 (1995).

[640] R.J.LeRoy and R.B.Bernstein, J. Chem. Phys. **52**, 3869 (1970).

[641] R.J.LeRoy, J. Chem. Phys. **73**, 6003 (1980).

[642] G.C.Maitland, M.Rigby, E.B.Smith, and W.A.Wakeham, *Intermolecular Forces, Their Origin and Determination* (Oxford University Press, Oxford, 1981).

[643] F.London, Z. Phys. Chem. **B11**, 222 (1930).

[644] H.Margenau and N.R.Kestner, *Theory of Intermolecular Forces* (Pergamon, Oxford, 1969).

[645] A.Dalgarno, Adv. At. Mol. Phys. **2**, 1 (1966).

[646] *Intermolecular Forces*, edited by J.O.Hirschfelder, Advances in Chemical Physics **XII** (Wiley, New York, 1967).

[647] D.Langbein, *Theory of van der Waals Attraction* (Springer, Berlin, 1974).

[648] J.Mahanty and B.W.Ninham, *Dispersion Forces* (Academic, London, 1976).

[649] P.Hobza and R.Zahradník, *Intermolecular Complexes, The Role of van der Waals Systems in Physical Chemistry and in the Biodisciplines* (Elsevier, Amsterdam, 1988).

[650] Yu.S.Barash, *Van der Waals Forces* (Nauka, Moscow, 1988).

[651] E.A.Power, in *Intermolecular Forces*, edited by J.O.Hirschfelder, Advances in Chemical Physics **XII** (Wiley, New York, 1967).

[652] L.Salem, Mol. Phys. **3**, 441 (1960).

[653] V.V.Kresin and E.Lipparini, Phys. Rev. B **46**, 9812 (1992).

[654] J.Goodisman, *Diatomic Interaction Potential Theory* (Academic, New York, 1973), Vol. 2.

[655] J.P.Toennies, in *Physical Chemistry - An Advanced Treatise*, edited by W.Jost (Academic, New York, 1974), Vol. VIA, Chap. Intermolecular Forces.

[656] in *Atom-Molecule Collision Theory: A Guide for the Experimentalist*, edited by R.B.Bernstein (Plenum, New York, 1979).

[657] *Atomic and Molecular Beam Methods*, edited by G.Scoles (Oxford, New York, 1988), Vol. 1.

[658] Ll.Serra and F.Garcias, Phys. Rev. B **53**, 7006 (1996).

[659] C.Girard, S.Maghezzi, and F.Hache, J. Chem.Phys. **91**, 5509 (1989).

[660] Ch.Girard, Ph.Lambin, A.Dereux, and A.A.Lucas, Phys. Rev. B **49**, 11425 (1994).

[661] J.C.Mester *et al.*, Phys. Rev. Lett. **71**, 1343 (1993).

[662] J.Weiner, Adv. At. Mol. Opt. Phys. **35**, 45 (1995).

[663] P.D.Lett, P.S.Julienne, and W.D.Phillips, Ann. Rev. Phys. Chem. **46**, 423 (1995).

[664] F.Martin *et al.*, Phys. Rev. A **55**, 3458 (1997).

[665] G.P.Collins, Phys. Today 18 (1996).

[666] C.C.Bradley, C.A.Sackett, and R.G.Hulet, Phys. Rev. Lett. **78**, 985 (1997).

[667] J.Schmiedmayer *et al.*, Phys. Rev. Lett. **74**, 1043 (1995).

[668] M.S.Chapman *et al.*, Phys. Rev. Lett. **74**, 4783 (1995).

[669] R.C.Forrey, L.You, V.Kharchenko, and A.Dalgarno, Phys. Rev. A **55**, R3311 (1997).

[670] K.T.Tang and J.P.Toennies, J. Chem. Phys. **80**, 3726 (1984).

[671] J.M.Standard and P.R.Certain, J. Chem. Phys. **83**, 3002 (1985).

[672] D.Cvetko *et al.*, J. Chem. Phys. **100**, 2052 (1994).

[673] U.Kleinekathöfer *et al.*, Chem. Phys. Lett. **249**, 257 (1996).

[674] W.Weickenmeier *et al.*, J. Chem. Phys. **82**, 5354 (1985).

[675] W.T.Zemke and W.C.Stwalley, J. Chem. Phys. **97**, 2053 (1993).

[676] W.T.Zemke and W.C.Stwalley, J. Chem. Phys. **100**, 2661 (1994).

[677] W.T.Zemke, C.-C.Tsai, and W.C.Stwalley, J. Chem. Phys. **101**, 10382 (1994).

[678] M.Marinescu and A.F.Starace, Phys. Rev. A **55**, 2067 (1997).

[679] T.Kihara, Adv. Chem. Phys. **1**, 267 (1958).

[680] H.B.G.Casimir, Proc. K. Ned. Akad. Wet. **60**, 793 (1948).

[681] I.E.Dzyaloshinski, E.M.Lifshitz, and L.P.Pitaevskii, Advan. Phys. **10**, 165 (1961).

[682] A.D.McLachlan, Disc. Faraday Soc. **40**, 239 (1966).

[683] Yu.S.Barash and V.L.Ginzburg, in *The Dielectric Function of Condensed Systems*, edited by L.V.Keldysh, D.A.Kirzhnitz, and A.A.Maradudin (North-Holland, Amsterdam, 1989).

[684] E.M.Lifshitz and L.P.Pitaevskii, *Statistical Physics, Part 2* (Pergamon, Oxford, 1980).

[685] L.Spruch, Science **272**, 1452 (1996).

[686] A.Bambini and E.J.Robinson, Phys. Rev. A **45**, 4661 (1992).

[687] A.Anderson *et al.*, Phys. Rev. A **37**, 3594 (1988).

[688] V.Sandoghdar, C.I.Sukenik, E.A.Hinds, and S.Haroche, Phys. Rev. Lett. **68**, 3432 (1992).

[689] A.Shih and V.A.Parsegian, Phys. Rev. A **12**, 835 (1975).

[690] M.J.Mehl and W.L.Schaich, Phys. Rev. A **21**, 1177 (1980).

[691] M.J.Mehl and W.L.Schaich, Surf. Sci. **99**, 553 (1980).

[692] A.M.Marvin and F.Toigo, Phys. Rev. A **25**, 803 (1982).

[693] A.Landragin *et al.*, Phys. Rev. Lett. **77**, 1464 (1996).

[694] M.Fichet, F.Schuller, D.Bloch, and M.Ducloy, Phys. Rev. A **51**, 1553 (1995).

[695] F.M.Fowkes, *Contact Angle, Wettability and Adhesion*, No. 43 in *Advances in Chemistry Series* (American Chemical Society, Washington,DC, 1964).

[696] *Wetting, Spreading, and Adhesion*, edited by J.F.Padday (Academic, New York, 1978).

[697] P.-G. de Gennes and J. Badoz, *Fragile Objects* (Springer, New York, 1996).

[698] P. de Gennes, Rev. Mod. Phys. **57**, 827 (1985).

[699] D.Ugarte, A.Chàtelain, and W. A. de Heer, Science **274**, 1897 (1996).

[700] L.Spruch, Phys. Today 37 (1986).

[701] L.Spruch, J.F.Babb, and F.Zhou, Phys. Rev. A **49**, 2476 (1994).

[702] E.Elizalde and A.Romeo, Am. J. Phys. **59**, 711 (1991).

[703] P.Kharchenko, J.F.Babb, and A.Dalgarno, Phys. Rev. A **55**, 3566 (1997).

[704] C.I.Sukenik *et al.*, Phys. Rev. Lett. **70**, 560 (1993).

[705] M.Kasevich *et al.*, in *Atomic Physics 12*, edited by J.C.Zorn and R.R.Lewis (American Institute of Physics, New York, 1991).

[706] E.J.Kelsey and L.Spruch, Phys. Rev. A **18**, 15 (1978).

[707] S.R.Lundeen, in *Atomic Physics 12*, edited by J.C.Zorn and R.R.Lewis (American Institute of Physics, New York, 1991).

[708] G.Binnig, C.F.Quate, and C.Gerber, Phys. Rev. Lett. **56**, 930 (1986).

[709] D.Sarid, *Scanning Force Microscopy, with Applications to Electric, Magnetic, and Atomic Forces* (Oxford University Press, New York, 1994).

[710] *STM and SFM in Biology*, edited by O.Marti and M.Amrein (Academic, San Diego, 1993).

[711] D.L.Patrick and J. T.P.Beebe, in *Microscopic and Spectroscopic Imaging of the Chemical State*, edited by M.D.Morris (Dekker, New York, 1993).

[712] C.Bustamante and D.Keller, Phys. Today **48**, 32 (1995).

[713] Digital Instruments, Santa Barbara, CA, http://www.di.com.

[714] Park Scientific Instruments, Sunnyvale, CA, http://www.park.com.

[715] H.D.Ackler, R.H.French, and Y.-M.Chiang, J. Coll. Int. Sci. **179**, 460 (1996).

[716] A.Dereux *et al.*, J. Chem. Phys. **101**, 10973 (1994).

[717] U.Hartmann, Phys. Rev. B **43**, 2404 (1991).

[718] J.L.Hutter and J.Bechhoefer, J. Appl. Phys. **73**, 4123 (1993).

[719] J.L.Hutter and J.Bechhoefer, J. Vac. Sci. Technol. B **12**, 2251 (1994).

[720] S.K.Lamoreaux, Phys. Rev. Lett. **78**, 5 (1997).

[721] F.Zenhausern, Y.Martin, and H.K.Wickramasinghe, Science **269**, 1083 (1995).

[722] J. Tiggesbäumker, L. Köller, K.H.Meiwer-Broer, and A. Liebsch, Phys. Rev. A **48**, R1749 (1993).

[723] Ll.Serra and A. Rubio, Phys. Rev. Lett. **78**, 1428 (1997).

[724] V.V.Kresin, Phys. Rev. B **51**, 1844 (1995).

[725] K. Sturm, Adv. Phys. **31**, 1 (1982).

[726] S.R.Nagel and J. T.A.Whitten, Phys. Rev. B **11**, 1623 (1975).

[727] Y. Onodera, Prog. Theor. Phys. **49**, 37 (1973).

[728] I.B.Khriplovich, *Parity Nonconservation in Atomic Phenomena* (Gordon and Breach, Philadelphia, 1991).

[729] D. DeMille, D. Budker, and E.D.Commins, Phys. Rev. A **50**, 4657 (1994).

[730] C.S.Wood *et al.*, Science **275**, 1759 (1997).

[731] C.E.Tanner and C.Wieman, Phys. Rev. A **38**, 162 (1988).

[732] L. Hunter *et al.*, Opt. Commun. **94**, 210 (1993).

[733] K.E.Miller, J. D. Krause, and L. Hunter, Phys. Rev. A **48**, 5128 (1994).

[734] A. Khadjavi, A. Lurio, and W. Happer, Phys. Rev. **167**, 128 (1968).

[735] Y. Prior, A.R.Bogdan, M. Dagenais, and N. Bloembergen, Phys. Rev. Lett. **46**, 111 (1981).

[736] N.B.Colthup, L.H.Daly, and S.E.Wiberley, *Intorduction to Infrared and Raman Spectroscopy* (Academic Press, San Diego, 1990).

[737] D.S.Kliger, J.W.Lewis, and C.E.Randall, *Polarized Light in Optics and Spectroscopy* (Academic Press, San Diego, 1990).

[738] C. Cohen-Tannoudji, B. Diu, and F. Laloë, *Quantum Mechanics* (Wiley Interscience, New York, 1978).

[739] S.G.Nilsson, Det. Kong. Dan. Vidensk, Selsk. Mat.Fys. Medd. **29**, (1955).

Index